TO MY DEAR FRIEND HOWARD

DAVIS, CA.   MAY 2008

THOMAS

Global Development of Organic Agriculture: Challenges and Prospects

Global Development of Organic Agriculture: Challenges and Prospects

Edited by
Niels Halberg
Hugo Fjelsted Alrøe
Marie Trydeman Knudsen
Erik Steen Kristensen

Danish Research Centre for Organic Food and Farming
Foulum
P.O. BOX 50
DK-8830 Tjele
Demark

CABI Publishing

CABI Publishing is a division of CAB International

CABI Publishing
CAB International
Wallingford
Oxfordshire OX10 8DE
UK

CABI Publishing
875 Massachusetts Avenue
7th Floor
Cambridge, MA 02139
USA

Tel: +44 (0)1491 832111
Fax: +44 (0)1491 833508
E-mail: cabi@cabi.org
Website: www.cabi-publishing.org

Tel: +1 617 395 4056
Fax: +1 617 354 6875
E-mail: cabi-nao@cabi.org

©CAB International 2006. All rights reserved. No part of this publication may be reproduced in any form or by any means, electronically, mechanically, by photocopying, recording or otherwise, without the prior permission of the copyright owners.

A catalogue record for this book is available from the British Library, London, UK.

A catalogue record for this book is available from the Library of Congress, Washington, DC.

ISBN   1 84593 078 9
       978 1 84593 078 3

Printed and bound in the UK from copy supplied by the editors by Cromwell Press, Trowbridge.

Contents

Contributors		ix
Preface		xi

1. Global trends in agriculture and food systems — 1
Marie Trydeman Knudsen, Niels Halberg, Jørgen E. Olesen, John Byrne, Venkatesh Iyer and Noah Toly
- Introduction — 3
- World agriculture – trends and impacts — 3
- Global trends in organic agriculture — 26
- Conclusions — 41

2. Globalization and sustainable development: a political ecology strategy to realize ecological justice — 49
John Byrne, Leigh Glover and Hugo F. Alrøe
- Introduction — 50
- Organic farming and the challenge of sustainability — 51
- Political ecology as one approach to globalization and sustainable development — 53
- Commons as the basis of ecological justice — 55
- Overcoming commodification — 63
- Putting ecological justice into practice: guidelines for policy — 68
- Conclusions — 71

3. Organic agriculture and ecological justice: ethics and practice — 75
Hugo F. Alrøe, John Byrne and Leigh Glover
- Introduction — 76
- Sustainability, globalization and organic agriculture — 79
- The ethics and justice of ecological justice — 84
- Challenges for organic agriculture: commodification, externalities and distant trade — 90
- Putting ecological justice into organic practice — 97
- Conclusions — 108

4. Ecological economics and organic farming — 113
Paul Rye Kledal, Chris Kjeldsen, Karen Refsgaard and Peter Söderbaum
- Introduction — 114

	Ecological economics as a trans-disciplinary approach	114
	Political economics and the conception of time and scale	119
	Farming, production time, nature's time and scale	121
	Organic farming: a response to ecological damage caused by growth in scale and shortening of production time	124
	The ecological economic perspective and organic farming	126
	Frameworks for decision-making	127
	Conclusions	130

5. Organic farming in a world of free trade — 135
Christian Friis Bach

	Introduction	136
	The agricultural agenda	136
	Trade and the environment	139
	Conclusions	149

6. Certified and non-certified organic farming in the developing world — 153
Nicholas Parrott, Jørgen E. Olesen and Henning Høgh-Jensen

	Introduction	154
	Certified organic farming in the developing world	155
	Non-certified organic farming	159
	Unlearning from the Northern experience: reconceptualizing the benefits of organic farming from a Southern perspective	164
	Livelihoods	170
	Questions to guide future studies in organic farming	173
	Conclusions	176

7. Possibilities for closing the urban–rural nutrient cycles — 181
Karen Refsgaard, Petter D. Jenssen and Jakob Magid

	Introduction	182
	Recycling nutrients in society – an ecological economics perspective	184
	Basic economic, institutional and social aspects of waste handling	186
	Quantities of nutrients and organic resources – from household to agricultural systems	188
	Ecological handling systems for organic waste and wastewater	191
	The cost of the handling system	196
	Moral and cultural aspects related to recycling urban waste	197
	Health aspects related to recycling urban waste	198
	Recycling nutrients from urban waste – global examples	200
	Conclusions	208

8.	**Soil fertility depletion in sub-Saharan Africa: what is the role of organic agriculture?**	**215**
	John Pender and Ole Mertz	
	Introduction	216
	Causes of soil fertility depletion in SSA	218
	Approaches to restore soil fertility and improve productivity	220
	Conclusions	233
9.	**Sustainable veterinary medical practices in organic farming: a global perspective**	**241**
	Mette Vaarst, Stephen Roderick, Denis K. Byarugaba, Sofie Kobayashi, Chris Rubaire-Akiiki and Hubert J. Karreman	
	Introduction	243
	The potential for organic livestock farming	244
	Disease management in organic livestock production	247
	The use and risks of antimicrobial drugs	248
	Vector-borne diseases and organic livestock farming	254
	Disease control issues associated with land use and land tenure	261
	Developing organic strategies to enhance animal health and livestock production	264
	Moving from an 'organic approach' to 'organic animal production'	268
	Conclusions and recommendations	272
10.	**The impact of organic farming on food security in a regional and global perspective**	**277**
	Niels Halberg, Timothy B. Sulser, Henning Høgh-Jensen, Mark W. Rosegrant and Marie Trydeman Knudsen	
	Introduction	278
	Overview of existing food supply and security projections	281
	Significant factors determining the effect of OF on food supply and food security	287
	Modelling consequences of large scale conversion to OF for food security	301
	Discussion	313
	Conclusions	316
11.	**Towards a global research programme for organic food and farming**	**323**
	Henrik Egelyng and Henning Høgh-Jensen	
	Introduction	324
	Examining the basis for a global organic programme: institutional analysis	325

	The role of – selected – international organizations in promoting organic farming	329
	Research as a support tool for developing organic farming and food systems	331
	Towards a development-oriented research programme for organic agriculture	336
	Conclusions	339
12.	**Synthesis: prospects for organic agriculture in a global context**	**343**
	Niels Halberg, Hugo F. Alrøe and Erik Steen Kristensen	
	Introduction	344
	Three perspectives on the challenges and prospects of organic agriculture	345
	How may certified organic farming meet the challenges of the increased globalization of organic food chains?	347
	What solutions do certified and non-certified organic agriculture offer to sustainability problems in the global food system?	354
	Conclusions	363
Index		**369**

Contributors

Alrøe, H.F., *Danish Research Centre for Organic Food and Farming, Research Centre Foulum, P.O. Box 50, Blichers Allé 20, DK-8830 Tjele, Denmark.*
Bach, C.F., *Vejgaard, Svinemosevej 7, DK-3670 Veksø, Denmark.*
Byarugaba, D.K., *Makerere University, P.O. Box 7062, Kampala, Uganda.*
Byrne, J., *Center for Energy and Environmental Policy (CEEP), University of Delaware, Newark, Delaware 19716, USA.*
Dissing, I., *Fasanvej 11, DK-9460 Brovst, Denmark.*
Dissing, Å., *Fasanvej 11, DK-9460 Brovst, Denmark.*
Egelyng, H., *Danish Institute for International Studies, Department of Development Research: Poverty, Aid and Politics, Strandgade 56, DK-1401 Copenhagen, Denmark.*
Glover, L., *Center for Energy & Environmental Policy, University of Delaware, Newark, Delaware 19716, USA.*
Halberg, N., *Danish Institute of Agricultural Sciences, Department of Agroecology, Research Centre Foulum, P.O. Box 50, Blichers Allé 20, DK-8830 Tjele, Denmark.*
Hauser, M., *University of Natural Resources and Applied Life Sciences, Department of Sustainable Agricultural Systems, Gregor Mendel Straße 33, A-1180 Wien, Austria.*
Høgh-Jensen, H., *The Royal Veterinary and Agricultural University, Department of Agricultural Sciences, Højbakkegård Alle 30, DK-2630 Taastrup, Denmark.*
Iyer, V., *Center for Energy and Environmental Policy (CEEP), University of Delaware, Newark, Delaware 19716, USA.*
Jenssen, P.D., *Norwegian Universtity of Life Sciences, P.O. Box 5003, N-1432 Ås, Norway.*
Karreman, H.J., *Pen Dutch Cow Care, 1272 Mt Pleasant Rd, Quarryville, Pennsylvania, PA 17566, USA.*
Kjeldsen, C., *Danish Institute of Agricultural Sciences, Department of Agroecology, Research Centre Foulum, Blichers Allé 20, Postbox 50, DK-8830 Tjele, Denmark.*
Kledal, P.R., *Danish Research Institute of Food Economics, Rolighedsvej 25, DK-1958 Frederiksberg C, Denmark.*
Knudsen, M.T., *Danish Research Centre for Organic Food and Farming, Research Centre Foulum, P.O. Box 50, Blichers Allé 20, DK-8830 Tjele, Denmark.*

Kobayashi, S., *The Royal Veterinary and Agricultural University, Department of Agricultural Sciences, Højbakkegård Alle 30, DK-2630 Taastrup, Denmark.*

Kristensen, E.S., *Danish Research Centre for Organic Food and Farming, Research Centre Foulum, P.O. Box 50, Blichers Allé 20, DK-8830 Tjele, Denmark.*

Magid, J., *The Royal Veterinary and Agricultural University, Department of Agricultural Sciences, Thorvaldsensvej 40, 3., DK-1871 Frederiksberg C, Denmark.*

Mertz, O., *University of Copenhagen, Institute of Geography, Øster Voldgade 10, DK-1350 Copenhagen K, Denmark.*

Olesen, J.E., *Danish Institute of Agricultural Sciences, Department of Agroecology, Research Centre Foulum, P.O. Box 50, Blichers Allé 20, DK-8830 Tjele, Denmark.*

Parrott, N., *c/o Dept of Rural Sociology, University of Wageningen, Hollandseweg 1, 6706 JW, Wageningen, The Netherlands.*

Pender, J., *International Food Policy Research Institute, Environment and Production Technology Division, 2033 K Street, NW, Washington, DC 20006-1002, USA.*

Pengue, W., *Universidad de Buenos Aires, GEPAMA - FADU, Pabellón III, CIudad Universitaria, 4 PISO - Oficinas 420 AyB, 1428, Ciudad de Buenos Aires, ARGENTINA.*

Refsgaard, K., *Norwegian Agricultural Economics Research Institute, P.O. Box 8024 Dep., N-0030 Oslo, Norway.*

Roderick, S., *Organic Studies Centre, Duchy College, Rosewarne, Camborne, Cornwall, TR14 0AB, UK.*

Rosegrant, M.W., *International Food Policy Research Institute, Environment and Production Technology Division, 2033 K Street NW, Washington, DC 20006-1002, USA.*

Rubaire-Akiiki, C., *Makerere University, P.O. Box 7062, Kampala, Uganda.*

Söderbaum, P., *Mälardalen University, Högskoleplan 2, Gåsmyrevreten, Västerås, Sweden.*

Sulser, T.B., *International Food Policy Research Institute, Environment and Production Technology Division, 2033 K Street NW, Washington, DC 20006-1002, USA*

Toly, N., *Center for Energy and Environmental Policy (CEEP), University of Delaware, Newark, Delaware 19716, USA.*

Vaarst, M., *Danish Institute of Agricultural Sciences, Animal Health, Welfare and Nutrition, Research Centre Foulum, P.O. Box 50, Blichers Allé 20, DK-8830 Tjele, Denmark.*

English language editing of chapters by non-native English speaking authors by www.textualhealing.nl: info@textualhealing.nl

Preface

Modern agriculture and food systems, including organic agriculture, are undergoing a technological and structural modernization and are faced with a growing globalization. Organic agriculture (OA) can be seen as pioneering efforts to create sustainable development based on other principles than mainstream agriculture. There are however large differences between the challenges connected to, on the one hand, modern farming and consumption in high-income countries and, on the other hand, smallholder farmers and resource-poor consumers in low-income countries.

The point of departure is the increasing globalization and the production and trade of food and fodder and how this influences the role of OA. The main aim of this book is to provide an overview of the potential role and challenges of organic agriculture in this global perspective, as seen from different perspectives such as sustainability, food security and fair trade.

Initially, the book provides an overall status of global trends in agriculture followed by discussions of sustainability, globalisation and the relative new concepts of 'ecological justice' and 'political ecology'. Different views on economy and trade are furthermore discussed with a focus on ecological economics. Then, the status and possibilities of organic agriculture in developing countries are discussed, including problems of nutrient cycles and soil depletion plus issues on veterinary medicine. Furthermore, organic farming is related to the world food supply. The possibilities of knowledge exchange in organic agriculture are also evaluated and it is assessed how a large scale conversion to OA would impact on food security. Finally, prospects and challenges of organic farming in a globalized world are discussed in a synthesis chapter. Readers who seek first an overview and summary across the different chapters are recommended to start by reading the synthesis.

The book springs from a so-called 'knowledge synthesis' initiated by The Danish Research Centre for Organic Food and Farming (DARCOF[1]) in January 2004 to provide an overview of the potential role of organic agriculture in a global perspective.

---

[1] DARCOF (Danish Research Centre for Organic Food and Farming) is a so-called 'centre without walls' where the actual research of organic food and farming is performed in interdisciplinary collaboration between the participating research groups. The aim of DARCOF is to elucidate the ideas and problems faced in organic farming through the promotion of high quality research. Knowledge synthesis is one of the central activities of DARCOF to ensure that the research focuses on the most relevant challenges and the projects are undertaken in the most appropriate manner.

In short, a knowledge synthesis analyses, discusses and synthesizes the existing knowledge on a subject not yet clarified and often disputed in relation to the main points of view. This work takes place in a group of experts from different fields that represent the different points of view on the subject. It is therefore important to include experts with different backgrounds and different perceptions of the subject. The work was initiated by five key questions:
To which extent and under which circumstances:

1. Can organic production contribute to global food security? How?
2. Can organic production in developing countries contribute to a sustainable development? How?
3. Can organic certification protect natural resources, improve work conditions, etc.? How?
4. Can a fair global trade with organic products be realized? How?
5. Can organic research in high-income countries benefit organic agriculture in low-income countries? How?

An international workshop, 'Organic farming in a global perspective – globalisation, sustainable development and ecological justice', was held in April 2004 in Copenhagen to provide inputs to the knowledge synthesis, discuss the issue and clarify ambiguous concepts. Experts from USA, Sweden, Austria, The Netherlands and Denmark were invited to give presentations on the international workshop. On the basis of presentations, discussions and group work at the workshop the outline of this book was laid down and the Danish and international experts started preparing the chapters.

The knowledge synthesis on organic agriculture in a global perspective was performed by a group of Danish experts from a wide range of relevant fields in cooperation with international experts (see list of contributors). A website was established to communicate background material and working papers, and facilitate critical comments from other participants. Thus all chapters have been improved from reviews made by other experts, whom we wish to thank here.

DARCOF wishes to thank all contributors to the book; their efforts are most gratefully acknowledged.

Head of DARCOF,
Erik Steen Kristensen   August 2005

1
Global trends in agriculture and food systems

Marie Trydeman Knudsen,* Niels Halberg, Jørgen E. Olesen, John Byrne, Venkatesh Iyer and Noah Toly

Introduction 3
World agriculture – trends and impacts 3
   Trends in agricultural production 5
   Environmental impacts 10
   Socio-economic impacts 16
Global trends in organic agriculture 26
   Status in global distribution of organic farming 26
   Global developments and challenges of organic farming 28
Conclusions 41

Summary

Increasing globalization affects agricultural production and trade and has consequences for the sustainability of both conventional and organic agriculture.

During the last decades, agricultural production and yields have been increasing along with global fertilizer and pesticide consumption. This development has been especially pronounced in the industrialized countries and some developing countries such as China, where cereal yields have increased a remarkable twofold and 4.5-fold respectively since 1961. In those countries, food security has increased, a greater variety of food has been offered and diets have changed towards a greater share of meat and dairy products. However, this development has led to a growing disparity among agricultural systems and population, where especially developing countries in Africa have seen very few improvements in food security and production. The vast majority of rural households in developing countries lack the ecological resources or financial means to shift into intensive modern agricultural practices as well as being integrated into the global markets. At the same time, agricultural development

---

* Corresponding author: Danish Research Centre for Organic Food and Farming (DARCOF), P.O. Box 50, Blichers Allé 20, DK-8830 Tjele, Denmark. E-mail: MarieT.Knudsen@agrsci.dk

has contributed to environmental problems such as global warming, reductions in biodiversity and soil degradation. Furthermore, pollution of surface and groundwater with nitrates and pesticides remains a problem of most industrialized countries and will presumably become a growing problem of developing countries. Nitrate pollution is now serious in parts of China and India. The growing global trade with agricultural products and the access to pesticides and fertilizers have changed agricultural systems. Easier transportation and communication has enabled farms to buy their inputs and sell their products further away and in larger quantities and given rise to regions with specialized livestock production and virtual monocultures of e.g. Roundup Ready soybeans in Argentina. Since 1996, the Argentinean area devoted to soybeans has increased remarkably from 6 to 14 million ha, covering approximately 50% of the land devoted to major crops in 2003. Since 1997, Brazilian Amazon has seen a deforestation of more than 17,000 km^2 each year with medium or large-scale cattle rangers presumably being the key driving force.

Organic farming offers a potentially more sustainable production but has likewise been affected by globalization. Organic farming is practiced in approximately 100 countries of the world and the area is increasing. European countries have the highest percentage of land under organic management, but vast areas under organic management exist in e.g. Australia and Argentina. Europe and North America represents the major markets for certified organic products, accounting for roughly 97% of global revenues. The international trade with organic products has two major strands: i) trade between European and other Western countries (USA, Australia, New Zealand); and ii) South–North trade, involving production sites, most importantly in Latin America, which ship to major Northern organic markets. The recent development holds the risk of pushing organic farming towards the conventional farming model, with specialization and enlargement of farms, increasing capital intensification and marketing becoming export-oriented rather than local. Furthermore, as the organic products are being processed and packaged to a higher degree and transported long-distance, the environmental effects need to be addressed. Organic farming might offer good prospects for marginalized smallholders to improve their production without relying on external capital and inputs, either in the form of uncertified production for local consumption or certified export to Northern markets. However, in order to create a sustainable trade with organic products focus should be given to issues like trade and economics (Chapters 4 and 5), certification obstacles, and ecological justice and fair trade (Chapters 2 and 3). Furthermore, the implications of certified and non-certified organic farming in developing countries need to be addressed (Chapters 6 and 9) including issues on soil fertility (Chapter 8) and nutrient cycles (Chapter 7) and the contribution to food security (Chapter 10).

Introduction

Increasing globalization has been one of the major trends in the latest decades, as a consequence of the dominating technological and social development. Globalization is here understood as 'the erosion of the barriers of time and space that constrain human activity across the earth and the increasing social awareness of these changes' (Byrne and Glover, 2002). The increasing globalization has consequences for the way that we produce and trade agricultural products and thereby also environmental consequences for the climate, biodiversity, and land resources among other things. Globalization has implications for conventional agriculture but contains also specific opportunities and problems for organic farming – related to e.g. trade with organic certified products from developing countries. The idea of 'Sustainable development' has been another key concept in the latest decades and can be seen as reaction to the dominating development. Sustainability is a concept that can have different meanings (Jacobs, 1995; Rigby and Cáceres, 2001). The definitions of sustainability include both the interpretation related to 'functional integrity', where man is seen as an integrated part of nature (Thompson, 1996) and the 'resource sufficiency', which addresses the rate of resource consumption linked to production. In the following, recent trends in agriculture in relation to globalization and sustainability will be presented. Focus will be given to issues that are relevant for the discussion of the role and conditions for further development of organic farming in a global context.

The overall aim with this chapter is to:

- Show global trends in agriculture and food systems related to globalization and their environmental and socio-economic impacts.
- Show global trends in organic farming related to globalization – and indicate potentials and challenges in global organic agriculture related to environmental and socio-economic issues.

World agriculture – trends and impacts

Agriculture and food systems have changed very much over the last 50 years. Agricultural development has seen a rapid advance of agricultural technology in industrialized countries with the green revolution in the 1960s being counteracted by an increasing public awareness of environmental protection and sustainable development that evolved in the 1980s (FAO, 2000). In the 1990s an increasing globalization occurred that has continued into the 21st century. The current wave of globalization was made possible by technological breakthroughs in transportation and communication technologies (notably the Internet, mobile telephone technology and just-in-time systems) and affordable fuel in tandem with various efforts to liberalize international trade and investment flows (FAO, 2003). Increases in long-distance food trade, global concentration in food processing and retail industries and diet change are signs of the globalization of the food system (von Braun, 2003).

In the following, major trends in agricultural production and food systems in relation to globalization will be shown along with environmental and socio-economic impacts. The conceptual model in Figure 1.1 shows the structure in this section and illustrates possible connections in the development of the global agricultural and food systems. The figure is not intended to cover all aspects on global food systems sustainability, but to illustrate possible problematic situations.

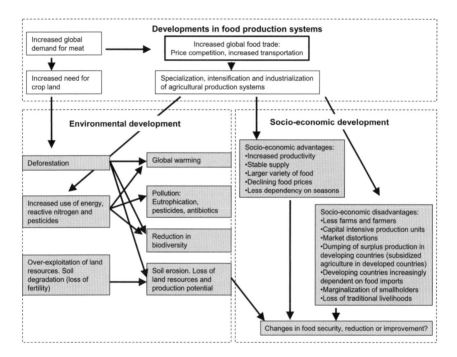

Figure 1.1. Illustration of possible problematic aspects in global food systems sustainability, environmentally and socio-economically. The arrows indicate possible effects.

Trends in agricultural production

Agricultural production has increased greatly over the last decades and in most continents the food production has been able to surpass the population growth. According to FAO (2000), the increase in production is attributable to the following factors, among others:

- the spread in developed countries of the modern agricultural revolution (involving large-scale mechanization, biological selection, use of chemicals, specialization);
- a modern agricultural revolution in some developing countries that is not dependent on heavy motorized mechanization but instead involves the use of chemicals and the selection of varieties;
- the expansion of irrigated surfaces, from about 80 million ha in 1950 to about 270 million ha in 2000;
- the expansion of arable land and land under permanent crops, from some 1330 million ha in 1950 to 1500 million ha in 2000,
- the development of mixed farming systems using high levels of available biomass (combining crop, arboriculture, livestock and, sometimes, fish farming) in the most densely populated areas that lack new land for clearing or irrigation.

The average yields of a milking cow and crop yields per ha and per worker have been increasing over the last 50 years (FAO, 2000). In the past four decades, increasing yields accounted for about 70% of the increase in crop production, compared to expanding the land area or increasing the cropping frequency (often through irrigation). However, yield increases have been most profound in industrial countries and e.g. China, whereas the yield increases in e.g. developing countries in Africa have been very limited (Figure 1.2).

The considerable advances in agriculture cannot hide the fact that most of the world's farmers use inefficient manual tools and their plants and domestic animals have benefited very little from selection. The progress in agricultural production hides a growing disparity among agricultural systems and populations. The gap between the most productive and least productive farming systems has increased 20 fold in the last 50 years (FAO, 2000). The agricultural revolution with all its attributes and especially its motorized mechanization has not extended far beyond the developed countries, with the exception of small portions of Latin America, North Africa, South Africa and Asia, where it has only been adopted by large national or foreign farms that have the necessary capital (FAO, 2000).

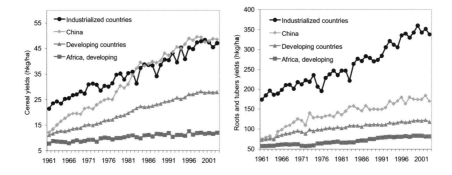

Figure 1.2. Yields of cereals plus roots and tubers in industrialized and developing countries plus in China and Africa, developing from 1961-2003 (hkg/ha) (FAOSTAT data, 2005).

Agricultural intensification

The agricultural revolution and globalization have had an enormous impact on agriculture and food systems in the developed countries. Developments in industry, biotechnology, transport and communication have affected agriculture in different ways.

Industrial developments have provided the means for motorization and large-scale mechanization, mineral fertilization, treatment of pests and diseases (pesticides, veterinary drugs etc.) and the conservation and processing of vegetable and animal products in developed countries. Developments in biotechnology supplied through selection, high yielding plant varieties and animal breeds have been adapted to the new means of production (FAO, 2000). The latest biotechnological developments are the genetically modified crops, grown primarily in USA, Canada and Argentina (see case from Argentina in Box 1.1).

The increase in fertilizer use and the use of improved varieties, through selection, have been among the important factors for the increased food production. A third of the increase in world cereal production in the 1970s and 1980s has been attributed to increased fertilizer use (FAO, 2003). World fertilizer consumption grew rapidly in the 1960s, 1970s and 1980s (Figure 1.3). The fertilizer usages in Europe have slowed down since the 1980s mainly due to reduced government support for agriculture and increased concern over the environmental impact. Fertilizer use in Asia, especially China, has been increasing (FAO, 2003; Figure 1.3), but the level of fertilizer use varies enormously between regions. North America, Western Europe and South–East Asia accounted for four-fifths of world fertilizer use in 1997–99 (FAO, 2003).

The highest rates are applied in East Asia, especially in China, followed by the industrial countries. At the other end of the scale, farmers in sub-Saharan Africa apply much less (FAO, 2003; Figure 1.3). The average fertilizer consumption is predicted to increase in developing countries (FAO, 2003). However, the average figure masks that for many (especially small) farmers the purchase of manufactured fertilizers and pesticides is and will continue to be constrained by their high costs relative to output prices and risks or simply by unavailability (FAO, 2003).

The global usages of pesticides have increased considerably during the second part of the 20th century (Figure 1.4). Some of the problems with diseases and insects have increased with the increased use of nitrogen fertilizers due to a higher susceptibility of the crop to attack at higher nitrogen input (Olesen *et al.*, 2003). Some countries in Western Europe have seen a reduction in pesticide consumption in recent years, primarily due to policies that promote or enforce management strategies with reduced pesticide use (Stoate *et al.*, 2001). Future pesticide consumption is likely to grow more rapidly in developing countries than in developed ones (FAO, 2003). The treatment of pests and diseases, in both plants and livestock, has become more important to safeguard investments in farm output.

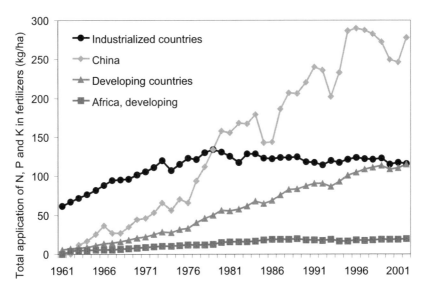

Figure 1.3. Total fertilizer application of N, P and K in industrialized and developing countries plus China and Africa, developing from 1961 to 2002 (kg/ha) (FAOSTAT data, 2005).

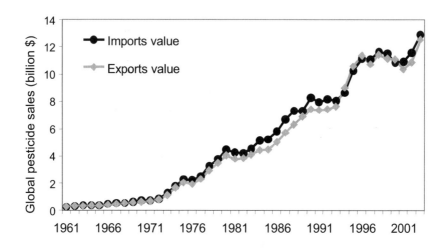

Figure 1.4. Imports and exports value of global pesticide sales from 1961 to 2003 (FAOSTAT data, 2005).

The more expensively bred and fed animal and the larger and more concentrated the animal production, the higher the risks. A great part of the antibiotics produced today are used as treatments against infectious diseases or as growth promoters in animal production, especially for pigs and poultry. Mellon et al. (2001) estimated that 70% of all antibiotics used in the USA are used for non-therapeutic livestock use. JETACAR (1999) found that approximately one-third of the antibiotics imported to Australia is for humans and two-thirds for animals. Denmark became the first country with a significant livestock industry to curtail the use of antibiotic growth-promoters in pig and poultry production in 1998. Approximately 70% of the antimicrobials used in Denmark are for therapeutic veterinary use (Heuer and Larsen, 2003).

Agricultural systems have changed with the introduction of mineral fertilizer, pesticides etc. With the use of mineral fertilizer cash crop production no longer relies on soil fertility building or use of manure. Furthermore with the introduction of mechanization, agriculture has also been freed from the need to produce forage for draught animals. Consequently, agricultural holdings suited for mechanized crop production have been able to abandon fodder and livestock production and specialize in cash crop production while other agricultural holdings have specialized in livestock production, often without sufficient land for manure application (FAO, 2000). Furthermore, the use of agricultural chemicals and GMO crops has partly released agricultural holdings from former crop rotation systems used to control weeds, insects and diseases. As a result cropping systems have been simplified and further specialized, culminating in monocropping or quasi-monocropping.

There has been a trend towards a narrower genetic base used for plant and animal production. Of the 270,000 known species of higher plants only three species (wheat, rice and maize) provide half of the world's plant derived energy intake (FAO, 1997; Cromwell *et al.*, 1999). At a national and regional level, only a few varieties are used over large-scale areas and the same trend can be seen in livestock genetic resources (CBD, 2001). The latest development in this aspect is the rapid spread of GMO crops, where a few (pesticide resistant) varieties of e.g. maize and soybean now cover large areas of land (see case from Argentina in Box 1.1).

Developments in transportation and communication have opened up the farms and agricultural regions and enabled them to procure their fertilizer, feed and other inputs from further away and in larger quantities. It also allowed for the sale of their products in increased amounts and wider areas. An increased globalization has freed agricultural holdings even more from comprehensive localized self-supply and made them able to focus on the most profitable product (or simplified combination of products). Virtual monocultures of soybean, maize, wheat, cotton, vineyards, vegetables, fruit and flowers and specialized productions of pig and poultry have thus spread over entire regions giving rise to new specialized regional agricultural systems (FAO, 2000).

Dietary changes

Just as world average calorie intakes have increased, so have also people's diets changed. Patterns of food consumption are becoming more similar throughout the world, incorporating higher-quality and more expensive foods such as meat and dairy products.

This diet change is partly due to simple preferences by populations. Partly, too, it is due to increased international trade in foods, to the global spread of fast food chains, and to exposure to North American and European dietary habits. Convenience also plays a part, for example the portability and ease of preparation of ready-made bread or pizza, versus root vegetables. Changes in diet closely follow rises in incomes and occur almost irrespective of geography, history, culture or religion (FAO, 2003).

These changes in diet have had an impact on the global demand for agricultural products and will continue to do so. Meat consumption in developing countries, for example, has risen from only 10 kg per person per year in 1964–66 to 26 in 1997–99. It is projected to rise still further, to 37 kg per person per year in 2030. Milk and dairy products have also seen a rapid growth, from 28 kg per person per year in 1964–66 to 45 kg in 1997–99, and with the expected consumption of 66 kg by 2030 in developing countries. The intake of calories derived from sugar and vegetable oils is furthermore expected to increase. However, average human consumption of cereals, pulses, roots and tubers is expected to level off (FAO, 2003).

Environmental impacts

Human activities and in particular the provision of foods for the growing world population put increasing demands on the natural resources of the earth. These effects are seen in several ways (Figure 1.1). In some areas of the world, agricultural land use increases at the expense of forests and other natural terrestrial ecosystems. In other parts of the world there is an overexploitation of the land resources leading to soil degradation and loss of soil fertility. However, the major way used to satisfy the need for food is through intensification of the agricultural production, primarily through the use of fertilizers and pesticides (see 'Trends in agricultural production'). All of these pathways have their own effects on the environment.

Four major indicators of environmental sustainability (EEA, 2005) are considered here as illustrated in Figure 1.1:

- Loss of land resources by soil erosion and soil degradation. Eroded soils are often lost for productive agricultural use for a very long time, whereas soils that are degraded through loss of soil organic matter, soil compaction, nutrient mining or salinization may be restored through proper agricultural management techniques. Loss of land resources has secondary negative effects on biodiversity and global warming.
- Loss of biodiversity involves a reduction in the number of living species on the earth and thus a loss of genetic resources (CBD, 2001) and a loss of ecosystem services in both natural and managed ecosystems (Costanza et al., 1997). Both effects have negative long-term consequences for the interaction between the human population and the environment. Biodiversity is reduced by a number of agricultural activities, such as deforestation, reduction of field margins and hedgerows, drainage of wetlands, genetic uniformity in crop land, pesticides etc. (FAO, 2003).
- Global warming is a consequence of increasing emissions of greenhouse gases (primarily CO_2, CH_4, N_2O and CFCs) to the atmosphere. The global emissions of CO_2 in 1996 (23,900 million t) were nearly four times the 1950 total (UNEP, 1999). The use of fossil fuels is the primary cause of these emissions. However, agricultural production contributes about 39% of the methane and 60% of the nitrous oxide emissions released in OECD countries (OECD, 2000 cf. OECD, 2001b). Methane emissions from agriculture are mainly produced from ruminant animals and the handling of manure, while the main source of nitrous oxide emissions is nitrogen fertilizers (OECD, 2001b). In addition, CO_2 from deforestation, soil degradation and soil erosion also have major contributions to the global greenhouse gas emissions. Furthermore, the use of fertilizer is associated with high energy requirements for their production resulting in CO_2 emissions (Dalgaard et al., 2000).

- The use of fertilizer in high amounts per ha and the large amounts of manure concentrated in specific geographical areas has increased the emission of ammonia and nitrate, which creates eutrophication and acidification in sensitive aquatic and terrestrial environments and pollution of ground and surface water (EEA, 2003; see more below). With increasing load of phosphorus in agricultural soils in particular with intensive livestock farming, there is also a risk of phosphorus losses to sensitive aquatic environments (Novotny, 2005).

Most of the environmental problems have increased considerably in recent decades. These problems are usually externalized, being greater for the society as a whole than for the farms on which they operate, and direct incentives for the farmers to correct them are therefore largely lacking (Stoate et al., 2001). Impacts on biodiversity and global warming are trans-boundary or global in their nature, and efforts to deal with these therefore require international collaboration.

In the following, the effects on the above-mentioned environmental indicators caused by 1) agricultural land use and by agricultural intensification through 2) the global nitrogen cycle and 3) pesticides will be discussed.

Agricultural land use

The world's land area comprises 130.7 million km^2. However, less than half of this land area is suitable for agriculture, including grazing (Kindall and Pimentel, 1994). Nearly all of the world's productive land is already exploited. Thus, only a small increase in agricultural area has been seen over the past 40 years. Most of the unexploited land is too steep, too wet, too dry or too cold for agriculture. For arable crops, soils also limit land use, because many soils are unsuitable for tillage or depleted in nutrients.

Expansion of the cropland has to come at the expense of forest and grassland, which also have essential uses. The net gain in agricultural area comes from adding land through deforestation and loss of land from land degradation and reforestation. It has been estimated that 70–80% of deforestation is associated with agricultural uses (Kindall and Pimentel, 1994). There are several environmental problems associated with deforestation, of which loss of biodiversity and CO_2 emissions are the major ones. It has thus been estimated that CO_2 emissions from land use changes amount to 20% of the emissions associated with fossil energy use (Houghton et al., 2001).

Degradation of existing agricultural land involves loss of productive land. According to some analysts, land degradation is a major threat to food security and it is getting worse (Pimentel et al., 1995; UNEP, 1999; Bremen et al., 2001). Others believe that the seriousness of the situation has been overestimated at the global and local level (Crosson, 1997; Scherr, 1999; Lindert, 2000; Mazzucato and Niemeijer, 2001). Brown (1984) estimated that about 10 million ha of agricultural land was lost by soil erosion every year, corresponding to 0.7% of global cropland area. Others argue that the area of cropland going out of use because of degradation is in the order of 5–6 million ha every year (UNEP, 1997). It is estimated that soil degradation is severely affecting 15% of the earth's cropland area, and in Europe alone 16% of the soils are prone to soil degradation (Holland, 2004). UNEP (1999) estimated that 500 million ha of land in Africa have been affected by soil degradation since about 1950, including as much as 65% of agricultural land.

The degradation and loss of agricultural land arises mainly from soil erosion, salinization, waterlogging, and urbanization. In addition nutrient depletion, overcultivation, overgrazing and soil compaction contributes to the deterioration of soil fertility. Many of these processes are caused by agricultural management practices. Soil erosion is considered the single most serious cause of arable land degradation, and the major cause is poor agricultural practices that leave the soil without vegetative cover or mulch to protect it against water and wind erosion. In developing countries, the degradation is worsened by low inputs, partly due to lack of credits and partly because available crop residues and dung are used for fuel. This reduces soil nutrients and intensifies soil erosion.

The global nitrogen cycle

Nitrogen is one of the most abundant chemical elements in the atmosphere and biosphere. However, more than 99% of the nitrogen is present as molecular nitrogen, which is not available to most organisms. Only a small proportion of the nitrogen is thus present as reactive nitrogen, which includes inorganic forms (NH_3, N_2O, NO, NO_2 and NO_3) and organic compounds (urea, amines, proteins and nucleic acids).

In the pre-industrial world, creation of reactive nitrogen occurred primarily from lightning and biological nitrogen fixation, and the denitrification process balanced the input of reactive nitrogen. However, in the industrialized world reactive nitrogen is accumulating in the environment at all spatial scales (Galloway *et al.*, 2003). During the past few decades, reactive nitrogen has been accumulating in the environment (Figure 1.5), primarily due to the industrialized production of fertilizer nitrogen by the Haber-Bosch process, which converts non-reactive N_2 to reactive NH_3.

The remarkable change in the global N cycle caused by the higher inputs of reactive N has had both positive and negative consequences for people and ecosystems. A large proportion of the global population is sustained because reactive nitrogen is provided as fertilizer nitrogen or by cultivation introduced biological nitrogen fixation (Smil, 2002). However, nitrogen is accumulating in the environment, because the rate of input is much larger than the removal by denitrification, and this accumulation is projected to continue to increase as human population increases and per capita resource use increases. The accumulation of reactive nitrogen in the environment contributes to a number of local and global environmental problems (Galloway *et al.*, 2003):

- Increases in reactive nitrogen in the atmosphere leads to production of tropospheric ozone and aerosols that induce respiratory disease, cancer and cardiac disease in humans (Wolfe and Patz, 2002).
- Increases in nitrate contents of groundwater, which have potential health effects (Jenkinson, 2001).
- Productivity of terrestrial systems (e.g. grasslands and forests) is affected with loss of biodiversity in oligotrophic ecosystems.
- Reactive nitrogen contributes to acidification and biodiversity loss in lakes and streams in many parts of the world (Vitousek *et al.*, 1997). There are several examples of streams and lakes, where recent reductions in fertilizer inputs have led to reduced N concentrations (Iital *et al.*, 2005).
- Reactive nitrogen is responsible for eutrophication, hypoxia, biodiversity loss and habitat degradation in coastal ecosystems (Howarth *et al.*, 2000). This environmental problem appears to be increasing globally (Burkart and James, 1999; EEA, 2003).
- Reactive nitrogen contributes to global climate change and stratospheric ozone depletion, both of which have impacts on human and ecosystem health (Mosier, 2002).

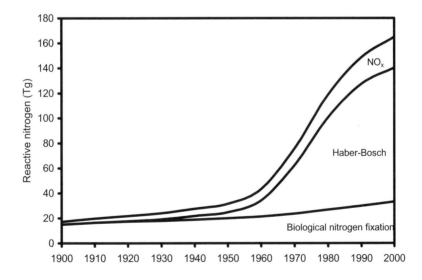

Figure 1.5. Global input for reactive nitrogen through biological nitrogen fixation, the industrial Haber-Bosch process and NO_x (based on Galloway *et al.*, 2003).

Intensively managed agro-ecosystems are the primary drivers of the changes that have occurred in the global nitrogen cycle. About 75% of the reactive nitrogen generated globally by humans is added to agro-ecosystems to sustain production or food and fibre. About 70% of this input comes from the Haber-Bosch process and about 30% from biological nitrogen fixation. There is only a small net residence of nitrogen in the agro-ecosystem, and most of the reactive nitrogen that is input to the system in a given year is lost again, either through consumption by humans or as losses to the environment.

On a global basis, about 120 Tg (1 Tg = 10^{12} g = 10^6 t) N from new reactive N (fertilizer and biologically fixed N) and about 50 Tg N from previously created N (manure, crop residues etc.) is added annually to global agro-ecosystems (Figure 1.6). Only about a third of this N input is converted into crop yield, whereas the rest is lost, primarily to the environment (Raun and Johnson, 1999). Animals consume about 33 Tg N per year of crop produce and humans consume about 15 Tg per year. Of the nitrogen input consumed by animals, only about 15% is converted to food used by humans. Of the 120 Tg N per year in new reactive nitrogen, only 21 Tg N per year is converted to food for humans (Figure 1.6). Since the change in soil nitrogen storage is very small, the rest is lost to the environment. On a global basis 6 to 12% of the added active nitrogen is denitrified to N_2 (Smil, 2002). The remaining losses of nitrogen occur as NO_3, NO_x, NH_3 and N_2O, and all of these emissions can cascade through natural ecosystems, where they alter their dynamics and in many cases reduce ecosystem services.

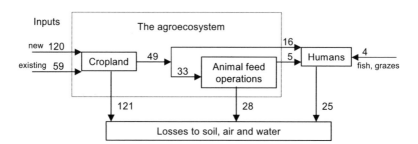

Figure 1.6. Major reactive nitrogen flows in crop and animal production components of the global agro-ecosystem (Tg N). Inputs represent new reactive nitrogen created through the Haber-Bosch process and through biological nitrogen fixation, and existing reactive nitrogen in crop residues, manure, atmospheric deposition, irrigation water and seeds. Portions of the lost reactive nitrogen may be reintroduced into the cropland component (modified after Galloway *et al.*, 2003 who refer to Smil, 2002).

Since the 1970s extensive leaching of nitrate from soils into surface and groundwater has become an issue in almost all industrial countries (OECD, 2001b). OECD (2001a) estimated that agriculture accounts for around two-thirds of nitrogen emissions into surface and marine waters and about one-third for phosphorus. In the EU countries, there is a large nitrogen surplus in the agricultural soils that can potentially pollute both surface and groundwater (Nixon *et al.*, 2003). Nitrate concentrations in rivers are highest in those Western European countries where agriculture is most intensive, but has during the 1990s been stabilized (Nixon *et al.*, 2003). Nitrate drinking water limit values (50 mg per litre) have been exceeded in around one-third of the groundwater bodies in the EU (EEA, 2003). In general, there has been no substantial improvement in the nitrate situation in European groundwater and hence nitrate pollution of groundwater remains a significant problem (EEA, 2003). Total nitrogen loading to the environment (air, soil and water) from livestock production in OECD regions is expected to increase by about 30% between 1995 and 2020 with particular large increases in Central and Eastern Europe and levels in Western Europe actually declining (OECD, 2001b). The problem of nitrate pollution of groundwater is now also serious in parts of China and India and a number of other developing countries and will presumably get worse (Zhang *et al.*, 1996). Nitrogen and phosphate enrichment of lakes, reservoirs and ponds can lead to eutrophication, resulting in high fish mortality and algae blooms, which may in the future be potentially more serious in warmer developing countries with more intense sunshine (Gross, 1998; FAO, 2003).

Pesticides

In most industrialized countries pesticides with serious toxic effects to vertebrates have been at least partially phased out. However, globally serious intoxications and incidences due to misuse of organophosphorous pesticides continue to be a problem (Satoh and Hosokawa, 2000). Both the intoxication rates and the fatality rates are highest in developing countries. UNEP (1999) estimated that global pesticide use results in 3.5–5 million acute poisonings each year.

Pesticides enter surface and groundwater from point source contamination following spillage events and from diffuse sources following their application to crops. They can be toxic to aquatic organisms and some are potentially carcinogenic (Cartwright *et al.*, 1991). In aquatic environments the leaching of pesticides into rivers, lakes and coastal waters is known to cause damage to aquatic biodiversity (OECD, 2001a).

Direct measurements of pesticides in surface or groundwater are not widely available across OECD countries, mainly because of the high costs of chemical analysis. Furthermore, many pesticides are not found in water bodies simply because they are not searched for, although when they are looked for they are frequently detected (OECD, 2001a). While the use of pesticides has fallen in many OECD countries since the mid 1980s, the long time lag between use and their detection in groundwater means that, as with nitrates, the situation could deteriorate before it starts to improve (OECD, 2001b). According to a survey of pesticide pollution of waters in the USA, in agricultural areas more than 80% of sampled rivers and fish contained one, or more often, several pesticides. Pesticides found in rivers were primarily those that are currently used, whereas in fish, organochlorine insecticides, such as DDT (now prohibited), which were used decades ago, were detected (USGS, 1999). The US survey also revealed that nearly 60% of wells sampled in agricultural areas contained one or more pesticides. The results of pesticide sampling in groundwater across a number of European Union countries, found a considerable number of sites with pesticide concentrations >0.1 µg per litre, which is the maximum admissible concentration of pesticides specified in the EU Drinking Water Directive (EEA, 1998). Finally, a French study found excessive quantities in the water environment, with surface waters being most affected where only 3% of the monitoring points showed no pesticides were present, and groundwater being better protected with 52% of all monitoring points considered to be unaffected (IFEN, 1998 cf. OECD, 2001a). Pesticide pollution is now appearing in developing countries as well and is likely to grow more rapidly than in developed countries (FAO, 2003).

The use of pesticides also affects terrestrial flora and fauna (OECD, 2001a). Herbicides are known to give rise to a decline in the flora of arable cropping systems (Andreasen *et al.*, 1996). The flora of farming systems are particularly diverse along the field margins, where herbicide uses also reduce biodiversity by removing or reducing the first step (plants) in the food web for e.g. birds and mammals (Chiverton and Sotherton, 1991). Farmland bird populations in the EU countries have fallen substantially in recent decades (EEA, 2004). The herbicide usages have been reported to have direct and knock-on effects on invertebrate abundance and species diversity (Moreby *et al.*, 1994). Broad-spectrum insecticides can cause substantial damage to populations of beneficial invertebrates and honeybees (Grieg-Smith *et al.*, 1995). Hence loss of biodiversity is not limited to the land clearing stage, but continues long afterwards.

Socio-economic impacts

Developments in agriculture and food systems such as industrialization and globalization have had socio-economic impacts all over the world, both for the millions who are engaged in farming and for the urban populations, as illustrated in Figure 1.1. More details on socio-economic impacts are discussed below.

The present phase of globalization, characterized chiefly by the proliferation of wireless communications, satellite television and the Internet, may be seen as the final outcome of a process that began in the mid-19th century with the first network technologies; the railroads and the telegraph. Beginning with these two early agents of mass transport and mass communication, the 20th century could well be characterized as the coming into being of a global mass society. Social, economic and political life has become increasingly dominated by the rise and spread of technologies of mass production and mass transport that are highly intensive in the use of energy, minerals and capital. With the accompanying trends of urbanization and rapid population growth, the impacts on agriculture and rural communities have been enormous worldwide (The Ecologist, 1993).

Industrialized countries

Agricultural modernization in the 20th century has brought major changes in socio-economic conditions in the industrialized countries of Western Europe, Oceania and North America. Along with the increases in agricultural production (see 'Trends in agricultural production'), smaller farms have been consolidated into larger ones and there has been a dramatic decline in the percentage of the population engaged in agricultural activities (FAO, 2000). Thus, in the USA, the number of farms has shrunk from about 6 million in 1950, to about 2 million today (Pretty, 2002). With the shift of agriculture, from small- and medium-scale farms serving local needs to a mass-production industry aiming at global markets, has come the growth of international competition for selling surplus agricultural produce, and the constant pressures to lower costs. Agricultural modernization has thus resulted in an abundance of raw and processed foods in national and international markets, with declining food prices (FAO, 2003). Cheaper food allows consumers in industrialized countries to spend only a small percentage of their household disposable income on food (10% for American consumers in 2003). Furthermore, a larger variety of food, especially fruit and vegetables, independent of season, can presumably be beneficial for public health and may help to revive the cultivation of some marginalized crops, such as certain millets and legumes. Despite the falling commodity prices of agricultural produce such as maize and soybean, the price of food has continued to rise with inflation (FAO, 2003), an increase attributed to the marketing costs of agribusiness and food companies, such as transportation, packaging etc. Declining real prices of agricultural produce also implies that governments in the industrialized countries have had to constantly prop up their small rural populations engaged in high external inputs agriculture with large subsidies and other incentives. These farms, in turn, have been forced to consolidate into ever larger operations and enter into contracts with large agribusiness corporations in order to remain economically viable. Thus, in the USA, about 60 to 90% of all wheat, maize and rice is marketed by only six transnational companies; and about 90% of poultry production is controlled by just ten companies (Pretty, 2002). Trends in Western Europe have been similar over the past few decades (see e.g. Mies and Bennholdt-Thomsen, 1999).

Developing countries

Farm sizes in many developing countries are typically small (often less than 1 or 2 ha). In addition there is often a substantial rural population of landless households. Therefore, on-farm mechanization of agricultural activities has not occurred to the same extent as in industrialized countries. However, many trends of modern agriculture (often hailed by many agricultural scientists, governments and international donor agencies as the 'Green Revolution') have also been witnessed by most developing countries over the past three decades. With their large rural populations and small land-holdings, the arrival of high-input agriculture has brought sweeping socio-economic impacts upon tens of millions of families in Asia, Latin America and, lately, Africa as well. Certainly, some parts of the rural population have benefited greatly from better irrigation facilities and access to subsidized diesel and electricity for pumping water from canals or deep aquifers. But, the vast majority of rural households in developing countries, especially sub-Saharan Africa (SSA), lack the ecological resources or the financial means to shift to intensive modern agricultural practices.

Integration into the global markets can be a two-edged sword for farmers in developing countries (FAO, 2000). With declining real prices of agricultural produce, farmers in developing countries tend to focus on cash crops such as cotton, paddy, sugarcane and groundnuts to take advantage of the widening access to external trade, and are forced to adopt many modern practices such as the increased use of chemical fertilizers and pesticides. This entails significant increase in the costs of agricultural inputs such as high-yielding seeds, chemical fertilizers and pesticides. The socio-economic impacts of this have become plainly visible in South Asia, with its large population of small farmers and landless labourers (Shiva, 1991). Lacking sufficient access to financial institutions (e.g. microfinance and rural credit), small farmers and labourers tend to borrow from local moneylenders at exorbitant rates of interest, which they are often unable to repay due to the vagaries of weather or unfavourable market conditions. This results in a deepening of the economic problems for small farmers in developing countries (see e.g. Sainath, 1996). The farmers are thus obliged to concentrate their efforts on short-term returns and to neglect the maintenance of the cultivated ecosystem, leading to fertility decline (FAO, 2000). This process of impoverishment and exclusion is affecting primarily the most deprived, small farmers who are especially numerous in resource-poor regions and constituting the bulk density (three-quarters) of the undernourished people in the world (FAO, 2000).

Focusing on cash crops leads furthermore to a decline in local food production and increased dependency of food imports (FAO, 2000). Developing countries have become increasingly dependent on agricultural imports. A rapid growth in imports of temperate-zone commodities (especially meat) has been seen and is expected to continue far into the 21st century (FAO, 2003). Some regions have remained sheltered for a long time from the cheap imports of cereals and other staple foods from the more advantaged regions and countries, being able to maintain their production systems longer than others. However, as soon as these regions are penetrated by the advance of motorized transport and commerce, they also find themselves caught up in interregional trade, exposed to low-cost imports of cereals and other food commodities (FAO, 2000).

Food security

For the past few decades, global food production has generally been adequate to meet human nutritional demands, and has kept pace with the rapid growth in human population. Food security has been substantially increased for some developing countries over the last decades, whereas other countries such as sub-Saharan Africa have seen no improvements. With the socio-economic disparities and political asymmetries that continue to exist, nearly 800 million people remain undernourished (see Chapter 10), where the vast majority of this undernourished population lives in rural areas and urban shanties of South Asia and sub-Saharan Africa (FAO, 2000). Thus on one hand, increases in agricultural productivity and falling real prices of produce benefit global food buyers and even raise the economic status of the urban poor in developing countries, helping to reduce food insecurity for many. On the other hand, a combination of specialization, industrialization and increased price competition, accompanied by negative environmental externalities, holds the risk of marginalizing a large number of small agricultural producers in developing countries. Exacerbating this problem is the underdevelopment of regional food storage and distribution systems linking small producers to local and regional markets. Even where such systems exist (such as the public distribution system in India), small producers in developing countries are often unable to take advantage of them due to socio-economic inequities and political imbalances that exist in many rural areas.

On balance, the socio-economic implications of agricultural trends and the larger impacts of globalization are twofold. Based on the problems described above and the principles of organic farming, it is interesting to discuss the potential of organic farming for contributing to a solution to some of the issues. These opportunities, if utilized well, may reverse some of the ill-effects of modern agriculture witnessed in the 20th century as discussed in the (following) sections. This includes both the environmental problems in intensive agriculture and the problem that there does not appear to be sufficient safeguards and policies to ensure that small producers in developing countries can benefit from the present phase of globalization. A broad range of initiatives to foster sustainable land, energy and water use practices, and social equity policies at the regional, national and international levels will be required if global trends toward organic agriculture and renewable energy, for example, are to prove beneficial to small agricultural producers in developing countries. Additionally, the role of non-governmental organizations (regional, national and international) in helping address issues of smallholder farms can be critical if these producers are to benefit from the global trends toward organic agriculture.

Box 1.1. Case study on increasing Roundup Ready soybean export from Argentina.

Increasing Roundup Ready soybean export from Argentina
by Walter Pengue

The soybean production area in Argentina has shown a remarkable increase within the last decade caused by an increasing global demand for soybeans for the pig and poultry industry, an open market and a strong campaign of technological change to Roundup Ready (RR) soybeans among other things. Concurrently with the expansion of the RR soybean production in Argentina, the use of glyphosate has showed a remarkable increase too. However, excessive reliance on a single agricultural technology, like RR soybeans and glyphosate, can set the stage for pest and environmental problems that can erode systems performance and profitability. In the following a case study on the expanding soybean production in Argentina will be presented, focusing on the agricultural and environmental sustainability.

Expanding soybean export from Argentina
Over the last decade, soybean has become the most important crop in Argentina. The majority of the expanding soybean production in Argentina is exported to world markets for animal protein supplement and vegetable oil (Benbrook, 2005). Increasing demand for meat has increased the demand for fodder for e.g. the pig and poultry industry in Europe. At the same time globalization has expanded global markets for agricultural commodities and enabled production to be separated from consumption in geographical terms.

Argentina is the world's leading exporter of cake of soybeans, followed by Brazil (FAO, 2005a). Since 1997, the export of cake of soybean from Argentina has increased dramatically from 8 million t to 18.5 million t in 2003 (FAO, 2005a). The importing countries are primarily European countries, such as Spain, Italy, The Netherlands and Denmark (Figure 1.7). The majority (82%) of the cake of soybean imported to Denmark in 2003 came from Argentina (FAO, 2005a). Denmark is the world's leading exporter of pig meat and the cake of soybean is primarily used for the pig production (FAO, 2005b).

Figure 1.7. Export of cake of soybeans from Argentina in 2003 (18,476,000 t) (FAO, 2005a). The lines show the export of cake of soybeans to different countries, where Spain, The Netherlands, Italy and Denmark are the major importers of cake of soybean from Argentina.

Rapid adoption of RR soybeans and expanding soybean areas
The dramatic growth of the soybean industry in Argentina was made possible by the combination of two technologies – no tillage system and transgenic Roundup Ready (RR) soybeans. Since 1996, the area devoted to soybean production increased a remarkable 2.4-fold, from 6 million ha to 14.2 million ha in 2004 (Figure 1.8). Of the land devoted to major crops, approximately 50% was grown with soybeans in 2003. Over a 4-year period from 1997–2001, the adoption rate of transgenic RR soybeans rose dramatically from 6 to 90%.

The increase in the soybean area and the rapid adoption of transgenic soybean were a direct consequence of globalization in commodity trade, an open market and a strong campaign on technological changes. For the farmers, RR soybean came up with a solution for one of the main problems in the farm management, namely weed control. A cost reduction in the herbicide price, less fossil energy consumption and simple application made the offer of the technical package very attractive. For the private pesticide and seed production sector, it opened a unique possibility to concentrate and rearrange the business of production and commercialization of insecticides and herbicides to the new biotechnological alternative.

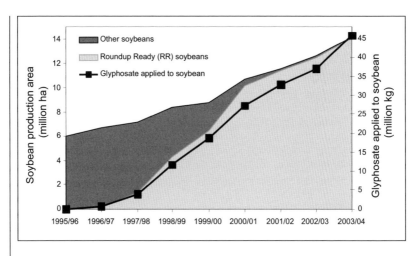

Figure 1.8. Soybean production area (million ha) and glyphosate consumption (million kg active ingredient) in Argentina from 1996 to 2004 (modified after Benbrook, 2005).

At first, soybeans were mainly produced on Pampas, one of the naturally most productive places in the world. But currently, due to the need for larger scale production, farmers are expanding the area and increasing the pressure on more environmentally sensitive areas.

During the period of expansion (1996–2004) in soybean production, the new areas needed for soybean production came from four main sources; i) approximately 25% came from conversion of cropland growing wheat, maize, sunflowers and sorghum; ii) approximately 7% came from conversion of areas growing other crops including rice, cotton, beans and oats; iii) approximately 27% came from conversion of former pastures and hay fields, and finally; iv) an estimated 41% came from conversion of wild lands, including forests and savannahs.

The Argentinean agricultural sector has set the goal of a total grain production of 100 million t by 2010, of which the soybean production is projected to be 45 million t. Achieving this goal would require an increase of the soybean planting area to about 17 million ha (Benbrook, 2005).

Increasing glyphosate consumption and resistant weeds
Given the expansion of the RR soybean hectares and the no-till systems, glyphosate herbicide usage has also risen dramatically (Figure 1.8). However, the reliance on a single herbicide year after year accelerates the emergence of genetically resistant weed phenotypes. It is predicted that continual glyphosate application for longer periods of time might lead to the development or higher increases in abundance of weeds tolerant to the herbicide (Puricelli and Tuesca, 2005). Tolerance to glyphosate in certain weeds in Argentina has already been documented (Puricelli and Tuesca, 2005; Vitta et al., 2004). Given the steady increase in the intensity of glyphosate use in Argentina, the development of resistant weeds is essentially inevitable (Benbrook,

2005). The unresolved questions include how fast will resistant weeds spread, how will farmers respond and how will the spread of resistant weeds impact weed management costs, efficacy and crop yields?

Phosphorus export and depletion of Argentinean soils
In Argentina, soybean has been cropped without fertilization, although soil phosphorus (P) contents have decreased. Areas previously considered well supplied are at present P-deficient (Scheiner *et al.*, 1996). The demand for phosphorus and depletion of natural reposition is particular important in the Pampas, where the P extraction has been increasing during the last decade (Casas, 2003).

The intensification of the production system was followed by a decline in soil fertility and increase of soil erosion (Prego, 1997). Consequently, during the last decade, fertilizer consumption stepped up from 0.3 million t in 1990 to 2.5 million t in 1999. The increase in the soybean sector in the 1990s and the increase in fertilizer use thus drove the Argentinean Pampas into a more intensive agriculture that is typical of the Northern hemisphere. Before that the nutrient budgets of the Pampas were relatively stable, with a rotation of crops and cattle being the most common production system.

Each year the country exports a considerable amount of nutrients – especially nitrogen, phosphorus and potassium, in its grains – that are not replenished, except from the part of nitrogen that is derived from N_2 fixation. Argentina annually exports around 3.5 million t of nutrients – with no recognition in the market prices, increasing the 'ecological debt' (Martinez Alier and Oliveras, 2003). Soybean, the engine of this transformation, represents around 50% of this. If the natural depletion were compensated with mineral fertilizers, Argentina will need around 1.1 million t of phosphorous fertilizers and an amount of 330 million American dollars to buy it in the international market (Pengue, 2003). Estimations for 2002 showed that around 30% of the whole soybean area was fertilized with mineral fertilizers. Ventimiglia (2003) predicts that nutrients of Argentinean soils will be consumed in 50 years with the current trend in nutrient depletion in Argentinean soils and an increasing soybean area.

Increasing soybean production – and the environmental impacts
Soybean has had and will have, an emblematic role in relation with nutrient balance, loss of quality and richness of Argentinean soils, and in marginal areas it has transformed itself into an important factor of deforestation. During the last years, advances on natural areas in Argentina have known no limits. Forest areas and marginal lands are facing the advances of agricultural borders. The campaign to increase grain production to 100 million t by 2010 will demand more land for grain crops and especially soybeans. An important part of these hectares are new land, which implies deforestation and loss of biodiversity (in terms of bioecological and sociocultural concept), replacement of other productive systems (dairy, cattle, horticulture, other grains) or an advance on marginal lands.

From an ecological economics point of view, the agricultural border expansion without environmental and territorial considerations will produce not only environmental transformations but also social and economic consequences that Argentina, and the world, is currently not considering. On the one hand, Argentina is facing an important degradation of soil and biodiversity in the country that is being promoted to solve only with the application of mineral imported fertilizers, with more environmental impacts. On the other hand the countries importing the grain and nutrients are facing problems of eutrophication and loss of habitats and biodiversity due to accumulation of especially nitrogen and phosphorus in the environment (see further 'Environmental impacts').

Box 1.2. Case study on beef trade and deforestation of the Brazilian Amazon.

Beef trade and deforestation in Brazilian Amazon

The increased globalization and demand for meat has increased Brazilian beef exports significantly during the last decade, with the EU importing a significant fraction. However, according to a recent World Bank report, medium- and large-scale cattle ranching is the key driving force behind recent deforestation in the Brazilian Amazon (Margulis, 2004). Sustainable cattle grazing is however not necessarily linked to environmental losses, but is a widely used management tool in restoration and conservation of semi-natural grasslands to e.g. reverse the decline of northern European floristic diversity.

Beef production in the EU has decreased by nearly 10% between 1999 and 2003 and a further decrease is expected (Anonymous, 2004b). For the first time in 20 years beef production was lower than consumption in 2003 in the EU and it is projected that the EU will remain a net importer of beef until at least 2011. The main reasons are a declining dairy cattle herd, the impact of the market disruptions of the 2001 BSE crisis and an expected impact of decoupling of direct payments (such as suckling cow premium and slaughter premium) from 2005 (Anonymous, 2004b).

More than 55% of the beef imported to the EU comes from Brazil (Anonymous, 2004a). Beef production in Brazil has been rapidly increasing during the last 10 years (Figure 1.9; FAO, 2005b). According to FAO (2005b), Brazil was, in 2003, the third largest exporter of boneless beef and veal in the world, in volume terms after Australia and USA. More than one-third of these exports go to the EU (Figure 1.9) and the remainder is sold primarily to Chile, Russia and Egypt (FAO, 2005a). Projections show a steady increase in beef production in Brazil (at more than 3.2% per year on average from 2004–11) (Anonymous, 2004b). Demand is expected to grow rapidly in Asia, Egypt and Russia (Anonymous, 2004b).

According to Kaimowitz *et al.* (2004), Brazilian beef exports have grown markedly mainly due to devaluation of the Brazilian currency (Cattaneo, 2002) and factors related to animal diseases. Other factors in the Amazon have also given greater force to the dynamics, such as expansion in roads, electricity, slaughterhouses etc. and very low land prices and easy illegally occupation of government land (Kaimowitz *et al.*, 2004). The overwhelming majority of the new cattle are concentrated in the Amazon states of Mato Grosso, Para and Rondonia, which are also the states with the most deforestation (Figure 1.10).

Global trends in agriculture and food systems 25

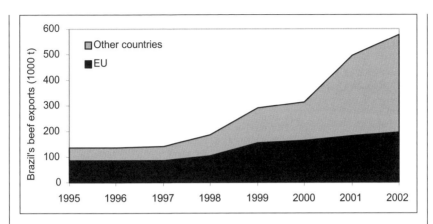

Figure 1.9. Brazil's beef exports (1000 t) to the EU and other countries (based on a table in Kaimowitz *et al.*, 2004).

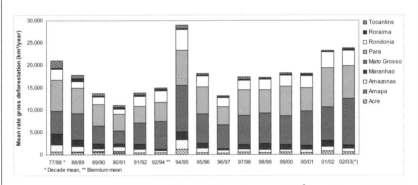

Figure 1.10. Deforestation rates in Brazilian Amazon (km^2/year) (modified after INPE, 2004).

According to a World Bank report, medium and large-scale cattle ranchers are the key driving force behind recent deforestation in Brazilian Amazon, and the overall social and economic gains are less than the environmental losses (Margulis, 2004). The expansion of the soybean cultivation into the Amazon explains only a small percentage of total deforestation according to Kaimowitz *et al.* (2004), who notes that logging is only partially responsible for deforestation, and is much less important than the growth of cattle ranching. Contrary to the occupation process in the 1970s and 1980s that was largely induced by government subsidies and policies, the dynamics of the recent occupation process gradually has become more autonomous, as indicated by the significant increase in deforestation in the 1990s despite the substantial reduction of subsidies and incentives by government. The study argues, that from a social perspective the private benefits from large-scale cattle ranching are largely exclusive, having contributed little to alleviate social and economic inequalities (Margulis, 2004).

Cattle grazing in the world however are not necessarily linked to environmental losses. Sustainable livestock grazing can enhance plant species richness and diversity of grasslands (Dupré and Diekmann, 2001; Pykälä, 2003; 2005; Rodriguez *et al.*, 2003; Pakeman, 2004) and is a widely used management tool in conservation programmes of natural grasslands (van Wieren, 1995; WallisDeVries, 1998). According to Pykälä (2003), restoration of semi-natural grasslands by cattle grazing is among the most practical options for reversing the decline of northern European floristic diversity.

Global trends in organic agriculture

Organic production and consumption has been increasing over the last decade. The organic products are not only being processed and consumed locally. Trade with organic products all over the world is a growing reality and organic products from developing countries like Uganda are being exported to e.g. Europe (see case study from Uganda in Box 1.3). However, apart from these globalization trends in organic agriculture, trends aiming at local production and consumption of organic food can also be seen (see cases from Denmark and USA in Box 1.4 and 1.5). In the following, status and developments of global organic farming will be given.

Status in global distribution of organic farming

Organic farming is practised in approximately 100 countries of the world and its share of agricultural land and farms is growing. The major part of the certified organic land is located in Australia followed by Argentina and Italy (Table 1.1). However, European countries have the highest percentage of agricultural area under organic management followed by Australia (2.5%) (Table 1.1; Willer and Yussefi, 2005).

Figure 1.11 shows the share for each continent of the total area under certified organic management. In Oceania and Latin America there are vast areas of animal pastures having a low productivity per ha, whereas the productivity per ha in European organic farming can be very high. Therefore, 1 ha in e.g. Australia cannot be directly compared to 1 ha in e.g. Denmark.

Table 1.1. 'Top ten countries worldwide' concerning percentage of agricultural area (%) or total land area (1000 ha) under organic management ranked according to highest percentage or total area (modified after Willer and Yussefi, 2005).

'Top ten worldwide' concerning land area under organic management			
Percentage organic area (%)		Total organic area (1000 ha)	
Liechtenstein	26.4	Australia	11,300
Austria	12.9	Argentina	2,800
Switzerland	10.3	Italy	1,052
Finland	7.2	USA	930
Italy	6.9	Brazil	803
Sweden	6.8	Uruguay	760
Greece	6.2	Germany	734
Denmark	6.2	Spain	725
Czech Rep.	6.0	UK	695
Slovenia	4.6	Chile	646

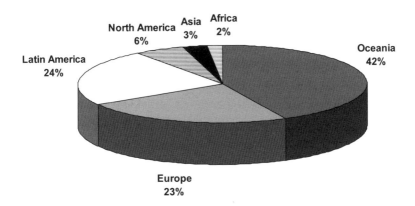

Figure 1.11. Total area under organic management – share for each continent (modified after Willer and Yussefi, 2005).

The major markets for organic food and drink are Europe and North America, which account for roughly 97% of global revenues and the markets are growing (Raynolds, 2004). Other important markets are Japan and Australia (Willer and Yussefi, 2004). Major northern markets offer good prospects for suppliers of organic products not domestically produced. These include coffee, tea, cocoa, spices, sugarcane, tropical fruits and beverages, as well as fresh produce in the off-season. Increasingly, governments in developing countries are creating conditions in support of organic export (Scialabba and Hattam, 2002). Regional markets of organic products are also expected to increase in developing countries like Brazil, China, India and South Africa along with increasing economic development and a more educated and affluent middle-class of consumers (Willer and Yussefi, 2004). Although certified organic products make up a minor share of the world food market (1–2%) it is the fastest growing segment of the food industry (Raynolds, 2004). Official interest in organic agriculture is emerging in many countries, shown by the fact that many countries have a fully implemented regulation on organic farming or are in the process of drafting regulations. Home-based certification organizations are found in 57 countries (Willer and Yussefi, 2004). The new international organic trade has two central strands, both supplying key markets in the global North. The first and largest strand is dominated by US exports to Europe and Japan, trade between European countries, and exports from Australia, New Zealand and South Africa to the top markets (Raynolds, 2004). The second strand is dominated by North–South trade and involves a growing number of production sites, most importantly in Latin America, which ship to major Northern organic markets (Raynolds, 2004). Latin America represents the hub of certified organic production in the global South, with Argentina having the greatest area and largest percentage of agricultural land under organic management (1.7%) (Willer and Yussefi, 2005). Uganda has the largest percentage of agricultural land under organic management in Africa (1.4%) (Willer and Yussefi, 2005) (see case study from Uganda in Box 1.3). A large part of African agriculture is however low external input agriculture (but not necessarily organic) where methods of the Green Revolution are risky, inappropriate or inaccessible (Willer and Yussefi, 2004). Ukraine and China are the major certified organic producers in Asia, measured by the number of certified organic hectares and enterprises, having 0.8% and 0.06% of agricultural land under certified organic management (Willer and Yussefi, 2005).

Global developments and challenges of organic farming

The organic food system has over the past two decades been transformed from a loosely coordinated local network of producers and consumers to a globalized system of formally regulated trade which links socially and spatially distant sites of production and consumption (Raynolds, 2004). Organic products were once largely produced locally, but as markets have grown, the range of organic items demanded has increased, moving beyond local seasonal products and bulk grains, to include a wide array of tropical products, counter-seasonal produce, processed foods etc. (Raynolds, 2004). Though preferences for local organic food persist, Northern countries are increasing their reliance on organic imports, particularly from the South (Raynolds, 2004). In 1998, 70% of the organic food sold in the UK was imported, 60% in Germany and The Netherlands and 25% in Denmark (Raynolds, 2004). At the same time supermarket sales of organic products have been increasing, dominating sales in the UK and Switzerland and controlling 90% of sales in Denmark. Supermarket sales comprise 20–30% of organic sales in the USA, Germany and Italy, but only 2% in The Netherlands (Raynolds, 2004).

'Conventionalization' and bifurcation in local- or export-oriented producers?

The extraordinary growth in the organic markets offers export opportunities to developing countries. At the same time the development of organic farming has led some analysts to warn that organic farming might be pushed towards the conventional farming model as agribusiness capital penetrates the organic community and its markets (Buck *et al.*, 1997; Tovey, 1997; Guthman, 2004). According to this scenario organic farming is becoming a slightly modified version of modern conventional agriculture, resulting in the same basic social, technical and economic characteristics – specialization and enlargement of farms (Milestad and Darnhofer, 2003), decreasing prices, increasing debt loads with increasing capital intensification, increased use of internal inputs and marketing becoming export-oriented rather than local (Hall and Mogyorody, 2001; Milestad and Hadatsch, 2003). Buck *et al.* (1997) are concerned that smaller alternative producers are increasingly being marginalized by larger producers who think and act like conventional producers in terms of production and marketing methods as they are forced to compete directly with larger more heavily capitalized producers within the same commodity and input markets. Although Buck *et al.* (1997) suggest that this process is leading to a bifurcation of the movement into two groups, they also argue that the alternative-oriented farmers are being pressured to adopt a number of conventional cropping, labour and marketing practices in order to survive.

In a case study of New Zealand, Coombes and Campbell (1998) found that there was some 'delocalization' in the relationship between organic producers and consumers, but due to a major growth in export-oriented organic production in New Zealand, the smaller producers were not being marginalized by the growth of larger production units or agribusiness penetration into organic agriculture. Agribusiness was focusing on converting their larger conventional growers for export-oriented markets, while the domestic markets were largely being ignored leaving the small-scale producers to continue to focus their attention on local consumers, retaining their alternative orientations and practices without any major threat or competition from agribusiness. When exporters attempted to dump certain products on the local market, there was no substantial effect on small-scale growers, as the export-oriented production was quite narrow in the range of crops, while the smaller-scale producers remained highly diversified (Coombes and Campbell, 1998). Hall and Mogyorody (2001) found little support for the idea of polarization between large export-oriented producers and small locally oriented producers in Ontario, but did find some support for the idea of 'conventionalisation' as organic field crop farmers tended to be export-oriented, large, mechanized, capitalized and specialized in cropping patterns. However, Campbell and Coombes (1999) argue that there are significant constraints and contradictions in any move to conventionalize organic farming, which creates significant space for the development of an alternative oriented organic movement. Hall and Mogyorody (2001) point out that organic farming is developing in distinct ways in different national contexts and one has to be cautious about drawing general conclusions regarding the development of organic farming. Campbell and Liepins (2001) argue that organic farming is still exceptional and provides a unique challenge to the standardizing food system. Even if it is not revolutionary, organic agriculture and food consumption highlight some ways in which the broad tendencies in food production and consumption are not linear, inevitable and uncontested – thereby providing an interesting terrain for examining the processes that are occurring at the margins of the globalizing food system. Raynolds (2004) suggests that while much of the literature on the preservation of organic movement values adopts a localist stance, these same values can be extended globally by linking small-scale peasant producers and conscientious consumers.

The development of farmers' markets, box schemes, farm gate sales, fair trade importing etc. may be seen as examples where those involved in the organic sector are attempting to develop alternative networks and patterns of control than exist in the conventional sector (La Trobe and Acott, 2000; Rigby and Bown, 2003). An example of an initially alternative trade network of organic milk in Denmark moving towards the trade patterns of the conventional sector is given in Box 1.4, where the degree of local links between food production and consumption are discussed. For some proponents of organic farming, it is exactly the potential for strengthening the local links between food production and consumption that is the promising issue. A large movement towards local production and consumption of organic food counteracts the trends of globalization in the organic sector. 'Eco-localism' is a concept presented by Curtis (2003) as an alternative economical paradigm as opposed to the global capitalist economy. The central argument is that economic sustainability is best secured by the creation of local or regional self-reliant, community economies (Curtis, 2003). An example of ongoing efforts to strengthen the local links between production and consumption of organic products in Iowa, USA, is presented in Box 1.5. The selected cases serve to illustrate the attempts to develop alternative supply networks and the problems associated with trying to 're-localize' the food chain, as the local markets have not (yet) proven adequate for sustaining a local production on a wider scale.

Environmental issues

The environmental impacts of organic farming have primarily been assessed in developed countries, pointing out however a number of benefits. Studies have shown that regarding soil biology, organic farming is usually associated with a significantly higher level of biological activity and a higher level of soil organic matter (Stolze *et al.*, 2000; Hansen *et al.*, 2001; Mâder *et al.*, 2002; Pulleman *et al.*, 2003; Oehl *et al.*, 2004), indicating a higher fertility and stability as well as moisture retention capacity (Stolze *et al.*, 2000; Scialabba and Hattam, 2002). Furthermore, Stolze *et al.* (2000) concluded that in productive areas, organic farming is currently the least detrimental farming system with respect to wildlife conservation and landscape, and a higher species diversity is generally found in organic fields (van Elsen, 2000; Pfiffner and Luka, 2003). The absence of pesticides precludes pesticide pollution and increases the number of plant species in the agricultural fields (Stoate *et al.*, 2001), which benefits natural pest control and pollinators. Organic farming furthermore reduces the risk of misuse of antibiotics (see Chapter 9).

Organic farming systems must rely on a closed nitrogen cycle and on nitrogen input via N_2 fixation by legumes. This leads to management practices that also reduce emissions of reactive nitrogen to the environment (Drinkwater *et al.*, 1998; Olesen *et al.*, 2004). The use of cover crops and mulches in organic farming also has the capacity to maintain soil fertility and reduce soil erosion. The recycling in organic farming of animal manure contributes to maintaining soil nutrients and avoiding soil degradation. Furthermore, there are indications that arable organic farming systems may reduce net greenhouse gas emissions per unit of agricultural area for arable farming systems (Robertson *et al.*, 2000).

In developing countries, organic farming has a potential of increasing natural capital, such as improved water retention in the soil, improved water tables, reduced soil erosion, improved organic matter in soils, increased biodiversity and carbon sequestration (Scialabba and Hattam, 2002; Rasul and Thapa, 2004). The potential of organic farming to enhance soil fertility and reduce soil erosion is discussed in Chapter 8. Furthermore, the risk of pesticide accidents and pollution is absent.

However, the environmental benefits of organic farming are challenged by globalization. The patterns of organic trade that are developing between North and South are to a high degree replicating those of the conventional sector. As organic produce becomes a larger part of the global food system, and as such is processed, packaged and transported more, the environmental effects become worthy of attention. 'Food miles' is one measure of this increasing transportation of organic food that captures the distance food travels from producer to consumer (Rigby and Bown, 2003). When measuring and discussing 'food miles' it can be important to distinguish between agricultural produce that can be produced locally and those that cannot. With the intensification of intra- and international transportation of organic commodities, organic agriculture systems are increasingly losing their nutrient and energy closed-system characteristic (Scialabba, 2000b) and risk encountering the same problem of nutrient transfer, depleting the production resources, as discussed in Box 1.1. The potential of closing urban–rural nutrient cycles in organic farming, especially in low-income countries, is discussed in Chapter 7. Scialabba (2000b) points to the risk that the environmental requirements of organic agriculture are becoming looser as the organic system expands and that few certification schemes explicitly mandate e.g. soil building practices, shelter for wild biodiversity and integrate animal production. This points to the need to supplement the organic farming principles with more guidelines or rules concerning e.g. ecological justice as discussed in Chapter 3.

Socio-economic issues

Organic farming in developed countries has a potential of narrowing the producer–consumer gap and enhancing local food markets (Scialabba and Hattam, 2002; see Box 1.5). Furthermore, organic farming has a potential of decreasing local food surplus and expanding employment in rural areas (Scialabba and Hattam, 2002). A better connectedness with external institutions and better access to markets has been seen through strengthened social cohesion and partnership within the organic community (Scialabba and Hattam, 2002; Box 1.4 and 1.5).

The extraordinary growing organic markets offer export opportunities to developing countries. Provided that producers of these countries are able to certify their products and access lucrative markets, returns from organic agriculture can potentially contribute to food security by increasing incomes (Scialabba, 2000b). A large number of farmers in developing countries produce for subsistence purposes and have little or no access to inputs, modern technologies and product markets. As productivity of traditional systems is often very low, organic agriculture could provide a solution to the food needs of poor farmers while relying on natural and human resources (Scialabba, 2000b). In Chapter 11, the effect of organic farming on food security will be discussed.

In developing countries, organic farming has a potential to improve social capital, such as more and stronger social organization at local level, new rules and norms for managing collective natural resources and better connectedness to external policy instruments (Scialabba and Hattam, 2002). Furthermore, improvements in human capital have been seen, such as more local capacity to experiment and solve problems, increased self-esteem in formerly marginalized groups, improved status of women, better child health and nutrition, especially from more food in dry seasons, reversed migration and more local employment (Scialabba and Hattam, 2002). It is assumed that organic agriculture in developing countries facilitates women's participation, as it does not rely on purchased inputs and thus reduces the need for credit (Scialabba and Hattam,

2002; FAO, 2003). However, insecure long-term access to the land is a major disincentive for both men and woman, since organic agriculture requires several years to improve the soil (Scialabba and Hattam, 2002). Chapter 6 illustrates the approaches of organic farming in developing countries.

Furthermore, organic agriculture has the potential to use fair trade conventions and to introduce ecological justice and the view of the theories of ecological economics. These issues are discussed in Chapters 2 and 4. Furthermore, Chapter 4 discusses the limitations of global organic trade and Box 1.3 shows a case on organic fair trade.

However, organic food and farming are challenged by globalization and development. The increasing export-orientation and supermarket domination of the organic market goes beyond the transportation effects. Supermarkets source primarily on the basis of range, quality, availability, volume and price and hence seek large volume suppliers who can supply at competitive prices all year round (Rigby and Bown, 2003). Raynolds (2000) points out that several studies suggest that due to substantial costs and risks of organic production, much of the international trade is controlled by medium and large enterprises, challenging the assumption that it is the small farms that benefit from the growing organic market. Organic farming may offer an opportunity for marginalized smallholders to improve their production without relying on external capital and inputs and to gain premium prices from trading with industrialized countries using organic production methods that have potential benefits to e.g. soil fertility and biodiversity. However, marginal organic farmers in the South are likely to be dependent on exploitative middlemen, corporate buyers and volatile prices as are conventional producers, unless they enter into fair trade networks (Raynolds, 2000). An example of organic farming as a development agent in Uganda is given in Box 1.3, where a fair trade network has been developed between organic farmers in Uganda and a Danish company. Producers, consumers and IFOAM acknowledge the convergence between the holistic social and ecological values of the fair trade and organic movements (IFOAM, 2000; Raynolds, 2004).

The certification issue is another challenge facing organic movements, especially with regard to developing countries. The term organic agriculture is backed with strict standards and rules that govern the 'organic' label of certified food found on the market. However, according to Raynolds (2004), onerous and expensive certification requirements create significant barriers to entry for poor Southern producers and encourage organic production and price premiums to be concentrated in the hands of large corporate producers. Furthermore, producers often have to comply with foreign standards not necessarily adapted to their country conditions (Scialabba, 2000b). Raynolds (2004) suggests that shifting certification costs downstream and empowering local producers to fulfil monitoring tasks should reduce barriers for small-scale producers. The issues of social justice in organic agriculture are further discussed in Chapter 2 and 3 and trade with organic products is discussed in Chapter 4.

The focus on certified organic products (and attendant costs and risks) has distracted attention on this system's potential to contribute to local food security, especially in low-potential areas in developing countries (Scialabba, 2000a). According to Scialabba (2000a), market-driven organic agricultural policies need to be complemented with organic agriculture policies that target local food security. The issues of food security are further discussed in Chapter 11.

Box 1.3. Case study on trade with organic products from Uganda.

Trade as an option of enhancing development? A case story from Uganda
by Åge Dissing and Ingelis Dissing

This case study tries to look at trade as a development tool. Most developing countries have been used to export agricultural commodities to e.g. Europe due to their former status as colonies, but often the population has hardly used the products themselves. Thus development of the products and processed produce is not incorporated in the society. In countries where agriculture is dominantly based on subsistence farming with a few cash crops for bulk export, handling of products for the market is a fairly new thing.

Scope of cooperation

Two Ugandan companies had formed a partnership with a Danish retail company in organic produce in order to supply the Danish partner with dried banana, pineapple and mango. The objectives were to process and export organic fruit in a fair trade arrangement to the Danish/European market.

The companies were both new-started, and to obtain the objectives they had to finalize and increase the infrastructure of the processing factory, including processing facilities, drying capacity and capacity-building of staff and management. On the supply side farmers should be trained in organic agriculture and be certified organic. The Danida Private Sector Development Programme has supported the cooperation.

Presentation of the Ugandan partners
Company X

The shareholder company X consists mainly of local people (like business people, teachers and agriculturists) in a middle-size town up country plus a few expatriates. The company X was initiated out of the interest of increasing agro-processing and of course of making a profit. The company was initiated as a start up trial and moved from there into a long-term cooperation of 5 years. Now 4 years after the start the company has built a factory in two stages including wet processing room, sorting/packing room, stores for fresh and dried produce and an office. Furthermore different types of dryers, a water tank, eco-toilets, a bathroom, a changing room and a store have been constructed. Factory staff and the board have been trained, and training and certification of factory and farms is an ongoing process. The monthly production is now on average 1000 kg of dried pineapple and banana, equivalent to 15,000 kg of fresh fruit from the farmers.

Company Y
The privately owned company Y has also a couple of other businesses to sustain the family, and to invest in the drying business. The company Y started like company X almost from zero like a start up trial and have reached almost the same infrastructure. However, the second stage of the factory is not finalized and fewer dryers are built. Director, staff and farmers have been trained, factory and farms are certified, but some inconsistency in the management policy has caused a high staff turnover. The monthly fruit production is now on average 400 kg of pineapple, banana and mango, equivalent to 6000 kg of fresh fruit, but the availability of fruit and the processing is very uneven over the year.

Hardships and obstacles for agro-business in Uganda
In fact there are many, but let us shortly describe the most obvious ones found in the two companies and the trade arrangement.
1. Financing a new business and especially in the countryside is almost impossible if you do not have the needed cash to invest. Banks give out only short-term loans and on very harsh conditions – at least as long as you are new in business. The main reason is that the government uses all money available to finance their part of donor investments like roads, hospitals etc. Most people do not opt for long-term investments; they still take actions from season to season, and prefer to invest in land, houses for renting or cows. A savings culture is not incorporated in society, partly due to family structures, where those who have money must give out to all relatives in need.
2. Management attitudes. It is hard to find people with a lot of management knowledge and experience, especially in agro-business. Uganda is recently recognized as a country with a very high level of entrepreneurs, but it is predominantly on very small scale like starting a tiny stall on the market or making bread on a veranda. Management experience to go into export is still difficult to find.
3. Consistency within the workforce on the factory and at farm level as well, is often very difficult and is therefore time-consuming. The inconsistency in the workforce is generally the case for both the leadership and the general staff in the factory. Industrialized working attitudes are new, as often seen in traditional agricultural societies.
4. Lack of proper logistics at all levels causes financial losses. Better logistics are needed to e.g. ensure that the needed fruit is available in time, and that the factory has all needed utensils etc. in place, not to lose too much time and money in the process.
5. Partnership and cooperation for mutual benefit is often difficult to create. At least it takes some years as farmers especially have been cheated by governmental 'cooperatives' and exploitative middlemen. Not only farmers have bad experiences of that kind, it also includes traders and other companies.
6. Cultural differences are big when African and European lifestyle and business attitudes have to find a mutual understanding. Industrialized countries have very difficult markets and are furthermore very protective and that is an additional constraint in a trade arrangement.

Conclusions

The current school system (especial secondary school) is not encouraging questions, curiosity and personal developments, on the contrary. This indicates that a boss can still handle staff in the usual feudalistic way, and thus developments are difficult, as we have seen in company Y.

A culture of subsistence farming is difficult to leave for farmers; they have to give up a lot of independence and freedom when going for commercial farming. In fact it is a very big change of lifestyle and work. Nevertheless, farmers connected to the two companies can now deliver the quantities and qualities required.

Factory work requires strict consistency in the workforce, which is not usual in countries that have not had the impact of long-term industrialized experiences. But company X has now built some capacity within the staff.

A fair trade agreement has a lot of good impact for the companies: fair prices for farmers and company, fair conditions for staff, transparency in the cooperation. It can include prepayment of the fruit, but in company X it has somehow caused delay in adjustment of the business as prepayment arrived anyway – for a period at least.

It is possible to see some good impact from these two trade arrangements. A lot of capacity building within farms, the staff and directors has taken place. In a country like Uganda it is definitely still needed, with some development supporting training and technical assistance to demonstrate success stories in international fair trade.

Box 1.4. Case study on local trade with organic milk products in Denmark.

Eco-localism and trade with organic products the case of Thise Dairy in Denmark
by Chris Kjeldsen

The dairy sector in Denmark is dominated by one big shareholder company that apart from the conventional milk also has organic shareholders and trades organic milk. However, the organic cooperative dairy Thise has successfully been established at the market through both alternative and more ordinary distribution channels. Thise Dairy was rooted in closer contact between producer and consumer and was initially a local dairy. Thise Dairy has over years increased its sales to all over Denmark.

Thise Dairy is an independent cooperative dairy, which was started in 1987, when a group of organic farmers in northern Denmark approached a privately owned dairy plant in an effort to acquire processing facilities for their production of organic milk. As a result of these negotiations, the cooperative, Thise organic dairy was formed in September 1988. The scale for the cooperative dairy was relatively small initially, reflecting the modest size of the market for organic milk at the time. There were only eight shareholders (organic and biodynamical farmers) in the cooperative, and the amount of milk weighed in at the dairy plant was only 1.6 million kg/year in the period 1989–90 (Jensen and Michelsen, 1991). The first 2 years were very costly for the cooperative, since they had to establish their own distribution network. One of the problems encountered in the early days of the dairy was that they distributed small amounts of milk over long distances to various rural locations in northern Denmark, e.g. shop owners with a sufficient dedication to organic products (Jensen and Michelsen, 1991). The main reason behind this very costly distribution strategy was that Thise could only gain access to stores without contracts with the major retail chains and their distribution networks. The problem was that individual shops within the major retail chains have only very limited autonomy regarding what to put on the shelf, since they are obliged to use centralized distribution networks.

Distribution costs for Thise Dairy were reduced by about 70% in 1990, when Thise joined a national distribution and sales organization for Danish organic dairy farmers, Dansk Naturmælk, which made it possible for Thise to sell their milk to some of the major retail chains, most notably 'FDB' (Jensen and Michelsen, 1991). The reduced distribution costs were mainly due to the fact that Thise, and with them other independent dairies, now could use the distribution network of the dominating (conventional) dairies 'MD Foods/Kløvermælk'. Due to financial and organizational problems within Dansk Naturmælk, the organization was terminated in 1992. After the termination of Dansk Naturmælk's agreement with the large retail chains in 1992, the future appeared quite bleak for Thise.

A crucial turning point for Thise Dairy happened in 1993. The Danish market for organic food expanded radically, when 'FDB' discounted organic products, which had a significant influence on Thise's sales to 'FDB'. The most important event in Thise's history took place in 1995, when Thise Dairy signed a contract with the retail chain Irma in Copenhagen. Irma has since then been the most important distribution channel for Thise Dairy. Today, around 50% of Thise's products are being sold in Irma shops in Copenhagen. Thise has more than doubled its sale, both in terms of turnover and the amount of milk weighed in at the dairy. At the same time, the number of shareholders in the cooperative has expanded to 42 (Anonymous, 2004c). In the late 1990s, Irma was bought by 'FDB', which marked an important change, since the forward sales of milk were sold in Irma's own brand. However, Thise has maintained a high degree of branding of their own name, and is often praised in the media for their high degree of innovation in developing new product types. Compared to the much larger cooperative dairy Arla with 15–16,000 shareholders (that dominates the Danish dairy market and sells both conventional and organic dairy products), Thise launches a much wider range of new products each year, and has been quite a trendsetter in the organic dairy sector.

As the map below illustrates (Figure 1.12), the most important market for Thise Dairy today is in Copenhagen, where around 50% of their dairy products are being sold, approximately 400 km from Thise Dairy. The second most important markets are export markets in England, Germany and Sweden, where up to 20% are being sold (Anonymous, 2004c). The dotted line on the map indicates the initial 'heartland' of Thise, within which most of their sales were taking place in the early 1990s. The majority of the producers are still placed within or around that perimeter. Although Copenhagen is the major market, Thise is trying to diversify its operations, since it also sells its milk via different alternative distribution channels. One example is that Thise are selling dairy products to a company called 'Anemonemælk', a web-based milk delivery scheme, which delivers milk and other dairy products to people's doorsteps in the wider Århus area. Other examples are health shops throughout the country.

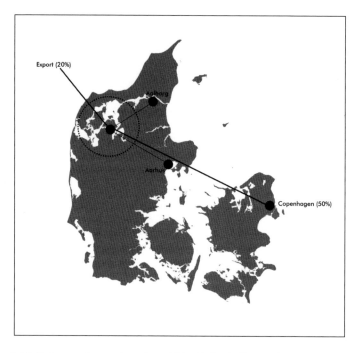

Figure 1.12. Pathways for products from Thise Dairy in Denmark.

Compared to the initial circulation of their products in a primarily rural (and regional) setting, Thise has moved beyond that context, now circulating its products in a primarily urban setting, geographically remote from the production sites in the network. One of the main reasons for this shift in direction was the inability of the local markets to support an economically viable scale of production, reflected in the fact that the shareholders for prolonged periods in the early history of Thise had to accept lower prices than at the other organic dairies, for example MD Foods (Arla).

Conclusion
Thise Dairy has moved from an alternative distribution network towards the supply patterns resembling those of the conventional sector. The initial supply pattern was characterized by a center–periphery structure dependent on place and personal relations as seen in some box schemes, farm shops etc. Thise Dairy has over time moved towards another distribution pattern characterized by standardization and regulation requiring no personal relations and less dependency of place, as seen in the supermarket distribution. The case illustrates the trend in Denmark where supermarket sales represent 90% of the sales of organic products (Raynolds, 2004). However, there are elements of the Thise Dairy, that exhibit some degree of 'regionalization' or dependency on place, as expressed in the idea of 'eco-localism'. It can be argued that Irma is a primarily regional based retail chain, and that 'Anemonemælk' is regionalizing Thise's products around Århus. Furthermore, one should of course not forget the very important regional importance of Thise in its 'home region' in terms of

local jobs. But an important issue in this regard is that each of these distinctive patterns is not spatially adjacent to each other and their interaction is primarily based on the standardization. Thise Dairy can, however, in some ways be described as expressing some of the classical virtues of the Danish cooperative dairy sector, such as producer autonomy through cooperative organization, a high degree of innovation and also an orientation towards exporting their products. The challenge of Thise is to span across different geographical and social spaces in order to recruit enough consumers to obtain a viable economic scale. It has proven a very successful market strategy, but leaves other challenges to be met, both regarding how to 'regionalize' the circulation of organic milk and how to obtain a higher degree of social integration between producers and consumers.

Box 1.5. Case study on foodsheds and eco-localism in the USA.

The development of (local) foodsheds in Iowa
by Chris Kjeldsen

The notion of foodsheds has its origin in the use of watersheds as the organizing spatial unit for integrated biophysical and social systems in bioregionalism (Wackernagel and Rees, 1996; Hansson and Wackernagel, 1999). In the same manner, the notion of foodsheds has been proposed as an organizing spatial unit for closely integrated networks between production and consumption of food (Kloppenburg *et al.*, 1996). Taken at face value, the notion of foodsheds implies a strong degree of local embeddedness. In practice though, this might not be the case, since many food networks labelled as 'sustainable' might exhibit a large scale in terms of size of their foodshed. One obvious example is fair trade networks, where producers and consumers are half a world apart and products travel over very large distances.

Initiatives in Iowa
In recent years, Iowa has seen an increase in the number of food system initiatives aiming at 're-localizing' the circuit of food between producers and consumers (Hinrichs, 2003). Historically, Iowa is in a way not the most typical place for such initiatives to appear, since Iowa appears as 'the quintessential agricultural state in the US' (Hinrichs, 2003). Compared to many other Midwestern states, Iowa has less diversity in its terrain and climatic features, making it an obvious target for agricultural development. Because of its obvious potential for agricultural use, the prairie state of Iowa was rapidly ploughed and the early white settlers drained the abundant wetlands. From the early days of settlement, Iowa agriculture was oriented towards non-local (mostly national) markets. Commodity agriculture seemed to be a strong cultural force within the agricultural community, since Iowa agriculture was rapidly modernized, in terms of specialization and integration with the agri-food industry. From the mid-20th century and onwards, the range of crops grown in Iowa has decreased significantly, as well as the number of farmers active within the sector. One example is that many labour-intensive crops such as apple or other horticultural crops vanished to a large degree from Iowa (Pirog and Tyndall, 1999), being replaced

by imports from production sites within the USA, such as Washington State, or overseas producers like China. The heavily industrialized and export oriented grain–livestock–meat systems became the most typical food system in Iowa.

The interest for re-localizing Iowa food chains is very recent. Food system localization in Iowa first took place with direct marketing initiatives such as Farmer's Markets growing from a number of 50–60 markets in the early 1980's to some 120 markets by the mid-1990s (Hinrichs, 2003). The first direct markets were mainly producer-driven, more than consumer-driven and should be seen as part of a strategy aiming at finding ways to overcome the massive farm crisis for commodity agriculture during the 1980s. Important actors in this regard were county extension officers and chambers of commerce, who initiated the first direct markets. Even though direct markets remain a focus area for food systems activists, there was a growing disquiet about their limited ability to sustain the livelihoods of many Iowa farmers. Aided by the activities of other actors, such as researchers from Iowa State University and the Leopold Centre for Sustainable Agriculture, both sited in Ames, Iowa, food systems activists started initiating other projects, which were supposed to extend the possibility to channel farm production flows. One of the significant developments, which took place during the 1990s, was the growth of Community Supported Agriculture (CSA) projects. By 1996, there were nine CSAs in Iowa, whereas this number had grown to about 50 by the year 2000 (Hinrichs, 2003).

CSA was an improvement of the alternative market channels for locally produced food, but as in the case of direct markets, small, decentralized, face-to-face direct market initiatives like CSA could not sustain many Iowa producers. Instead, food system activists and organizers have increasingly focused on changing the patterns of institutional food procurement. One of the first initiatives was a publicly funded demonstration project in 1997–98, which determined that it was possible for a university dining service, a hospital and a restaurant in north–east Iowa to purchase a significant proportion of the food needs locally (Hinrichs, 2003). Another important development was the development of a type of event called the Iowa-grown banquet meal. The first of these events was held at the Leopold Centre for Sustainable Agriculture in 1997. As both a promotional event and a celebratory enactment of local Iowa foods, the banquet meals have helped to establish a new ritual that showcases and redefines local Iowa food. Since 1997, the Iowa-grown banquet meals have spread all over the state, coordinated by a brokering office of the farmer's organization Practical Farmers of Iowa, with 57 meals at 47 different events being held in 2000 (Hinrichs, 2003). A loosely knitted network of 23 farmers has supplied the food being served at these events.

As a symbolic way of redefining and sustaining a local food culture, the Iowa-grown banquet meals have been very important. Still, the banquet meal is episodic and supplemental for any individual Iowa producer (Hinrichs, 2003), and has not been able to sustain any larger number of local farmers. Organics as an element in the localization of food chains of Iowa have until now been overshadowed by the valorisation of local produce. So in that sense the banquet meals conform to what Michael Winter has termed 'defensive localism' (Winter, 2003), where localization is the top priority for development of food systems, more than progressive social and ecological priorities. The challenge for initiatives like the Iowa-grown banquet meal seems to be to balance between defensive localism and a more receptive attitude to wider social and ecological objectives.

Conclusions

- Increasing globalization and production in agriculture has primarily benefited the industrialized countries and certain developing countries such as China that are integrated into the global markets. In those countries, food security has increased, a greater variety of food has been offered and diets have changed towards a greater share of meat and dairy products.
- However, the development hides a growing disparity among agricultural systems and population, where especially developing countries in Africa have seen very few improvements in food security and production. The vast majority of rural households in developing countries lack the ecological resources or financial means to shift into intensive modern agricultural practices as well as being integrated into the global markets.
- At the same time, intensive agriculture especially in industrialized countries has contributed to environmental problems such as pollution of surface and groundwater with nitrates and pesticides, global warming, reductions in biodiversity and soil degradation, and virtual monocultures and specialized livestock productions have spread over entire regions.
- Organic farming offers a potentially more sustainable form of production. Organic farming is practised in approximately 100 countries of the world and the area is increasing. Trade with organic products all over the world is a growing reality with the major markets being Europe and North America. These major markets offer good prospects for suppliers of organic products from developing countries.
- However, the recent development holds the risk of pushing organic farming towards the conventional farming model, with specialization and enlargement of farms, increasing capital intensification and marketing becoming export-oriented rather than local. Furthermore, as the organic products are being increasingly processed, packaged and transported long-distance, the environmental effects need to be addressed.
- Organic farming might offer good prospects for marginalized smallholders to improve their production without relying on external capital and inputs, either in the form of uncertified production for local consumption or certified export to Northern markets. However, in order to create a sustainable trade with organic products focus should be given to issues like trade and economics (Chapters 4 and 5), certification obstacles, and ecological justice and fair trade (Chapters 2 and 3). Furthermore, the implications of certified and non-certified organic farming in developing countries need to be addressed (Chapters 6 and 9) including issues on soil fertility (Chapter 8) and nutrient cycles (Chapter 7) and the contribution to food security (Chapter 10).

References

Andreasen, C., Stryhn, H. and Streibig, J.C. (1996). Decline in the flora of Danish arable fields. Journal of Applied Ecology 33: 619-626.

Anonymous (2004a). Agricultural trade statistics. Part IV: The EU15 and EU10N Main Markets by selected commodity aggregate in 2003 (by values and quantities). Online at http://europa.eu.int/comm/agriculture/agrista/tradestats/2003/part4

Anonymous (2004b). Prospects for agricultural markets and income 2004-2011 for EU-25. European Commission Directorate-General for Agriculture. December 2004. Online at http://europa.eu.int/comm/agriculture/publi/caprep/prospects2004b/fullrep.pdf, 126 pp.

Anonymous (2004c). Thise Mejeri: Økologisk landsbymejeri med internationale ambitioner (In Danish). Thise Mejeri, Salling. Online at http://www.thise.dk/Generelt/default.asp

Benbrook, C.M. (2005). Rust, resistance, run down soils, and rising costs – problems facing soybean producers in Argentina. Benbrook Consulting Sevices. Ag BioTech InfoNet. Technical Paper Number 8. 51 pp.

Bremen, H., Groot, J. and van Keulen, H. (2001). Resource limitations in Sahelian agriculture. Global Environmental Change 11: 59-68.

Brown, L. (1984). State of the world 1984. Worldwatch Institute, Washington, DC.

Buck, D., Getz, C. and Guthman, J. (1997). From farm to table: The organic vegetable commodity chain of northern California. Sociologia Ruralis 37(1): 3-20.

Burkart, M.R. and James, D.E. (1999). Agricultural-nitrogen contributions to hypoxia in the Gulf of Mexico. Journal of Environmental Quality 28: 850-859.

Byrne, J. and Glover, L. (2002). A common future or towards a future commons: Globalization and sustainable development since UNCED. International Review for Environmental Strategies 3(1): 5-25.

Campbell, H.R. and Coombes, B.L. (1999). Green protectionism and organic food exporting from New Zealand: crisis experiments in the breakdown of Fordist trade and agricultural policies. Rural Sociology 64(2): 302-319.

Campbell, H. and Liepins, R. (2001). Naming organics: understanding organic standards in New Zealand as a discursive field. Sociologia Ruralis 41(1): 21-39.

Cartwright, N., Clark, L. and Bird, P. (1991). The impact of agriculture on water quality. Outlook on Agriculture 20: 145-152.

Casas, R. (2003). Sustentabilidad de la agricultura en la región pampeana. Instituto Nacional de Tecnología Agropecuaria. INTA. Instituto de Clima y Agua. Buenos Aires, September. Online at http://www.inta.gov.ar/balcarce/info/documentos/recnat/suelos/casas.htm

Cattaneo, A. (2002). Balancing agricultural development and deforestation in the Brazilian Amazon. Research Report 129, International Food Policy Research Institute, Washington, DC.

CBD (2001). The Global Biodiversity Outlook. Secretariat on the Convention on Biological Diversity, United Nations Environment Programme. Online at http://www.biodiv.org/doc/publications/gbo/gbo-ch-01-en.pdf

Chiverton, P.A. and Sotherton, N.W. (1991). The effects on beneficial arthropods of the exclusion of herbicides from cereal crops. Journal of Applied Ecology 28: 1027-1039.

Coombes, B. and Campbell, H. (1998). Dependent reproduction of alternative modes of agriculture: organic farming in New Zealand. Sociologia Ruralis 38(2): 127-145.

Costanza, R., d'Arge, R., de Groot, R., Farber, S., Grasso, M., Hannon, B., Limburg, K., Naeem, S., O'Neill, R.V., Paruelo, J., Raskin, R.G., Sutton, P. and van der Belt, M. (1997). The value of the world's ecosystem services and natural capital. Nature 387: 253-260.
Cromwell, E., Cooper, D. and Mulvany, P. (1999). Agriculture, biodiversity and livelihoods: issues and entry points for development agencies. FAO, ITDG and ODI. Online at http://www.ukabc.org/odi_agbiod.pdf
Crosson, P. (1997). Will erosion threaten agricultural productivity? Environment 39(8): 4-12.
Curtis, F. (2003). Eco-localism and sustainability. Ecological Economics 46(1): 83-102.
Dalgaard, T., Halberg, N. and Fenger, J. (2000). Simulering af fossilt energiforbrug og emission af drivhusgasser (In Danish). FØJO-rapport nr. 5.
Drinkwater, L.E., Wagoner, P. and Sarrantonio, M. (1998). Legume-based cropping systems have reduced carbon and nitrogen losses. Nature 396: 262-265.
Dupré, C. and Diekmann, M. (2001). Differences in species richness and life-history traits between grazed and abandoned grasslands in southern Sweden. Ecography 24: 275-286.
EEA (1998). Europe's Environment: The Second Assessment, Office of Official Publications of the European Communities, Luxembourg.
EEA (2003). Europe's environment: the third assessment. Environmental assessment report No. 10. EEA, Copenhagen.
EEA (2004). EEA Signals 2004 – A European Environment Agency update on selected issues. European Environment Agency, Denmark. 31 pp.
EEA (2005). EEA core set of indicators, guide. EEA Technical report No. 1. European Environment Agency, Luxembourg, 38 pp.
FAO (1997). Report on the State of the World's Plant Genetic Resources for Food and Agriculture. Food and Agriculture Organization of the United Nations (FAO), Rome, Italy.
FAO (2000). The State of Food and Agriculture. Food and Agriculture Organization of the United Nations (FAO), Rome, Italy. Online at http://www.fao.org/docrep/ x4400e/ x4400e00.htm
FAO (2003). World Agriculture: towards 2015/2030 – an FAO perspective. Earthscan, Food and Agriculture Organization of the United Nations (FAO), Rome, Italy. Online at http://www.fao.org/documents/show_cdr.asp?url_file=/docrep/005/y4252e/ y4252e00.htm
FAO (2005a). A graphical presentation of the World's agricultural trade flows, WATF. Online at http://www.fao.org/es/ess/watf.asp. Food and Agricultural Organization of the United Nations, Economic and Social Department, The Statistics Division. (The WATF is an interactive map that displays the trade flows between countries/ territories and provide basic trade data).
FAO (2005b). Key statistics of food and agriculture external trade. Online at http://www.fao.org/es/ess/toptrade/trade.asp. Food and Agricultural Organization of the United Nations, Economic and Social Department, The Statistics Division.
FAOSTAT data (2005). Online at http://apps.fao.org/default.jsp (FAOSTAT is an on–line and multilingual database currently containing over 3 million time-series records covering international statistics in the agricultural area).
Galloway, J.N., Aber, J.D., Erisman, J.W., Seitzinger, S.P., Howarth, R.W., Cowling, E.B. and Cosby, B.J. (2003). The nitrogen cascade. BioScience 53: 341-356.

Grieg-Smith, P.W., Thompson, H.M., Hardy, A.R., Bew, M.H., Findlay, E. and Stevenson, J.H. (1995). Incidents of poisoning of honeybees (Apis mellifera) by agricultural pesticides in Great Britain 1981-1991. Crop Protection 13: 567-581.

Gross, E. (1998) Harmful algae blooms: a new international programme on global ecology and oceanography. Science International, 69.

Guthman, J. (2004). The trouble with 'organic lite' in California: a rejoinder to the 'conventionalisation' debate. Sociologia Ruralis 44(3): 301-314.

Hall, A. and Mogyorody, V. (2001). Organic farmers in Ontario: an examination of the conventionalization argument. Sociologia Ruralis 41(4): 399-422.

Hansen, B., Alrøe, H.F. and Kristensen, E.S. (2001). Approaches to assess the environmental impact of organic farming with particular regard to Denmark. Agriculture, Ecosystems and Environment 83: 11-26.

Hansson, C.B. and Wackernagel, M. (1999). Rediscovering place and accounting space: how to re-embed the human economy. Ecological Economics 29(2): 203-213.

Heuer, O.E. and Larsen, P.B. (2003). DANMAP 2003 – Use of antimicrobial agents and occurence of antimicrobial resistance in bacteria from food animals, foods and humans in Denmark. Statens Serum Institut, Danish Veterinary and Food Administration, Danish Medicines Agency and Danish Institute for Food and Veterinary Research, Denmark. Online at http://www.dfvf.dk/Files/Filer/Zoonosecentret/Publikationer/Danmap/Danmap_2003.pdf

Hinrichs, C.C. (2003). The practice and politics of food system localization. Journal of Rural Studies 19(1): 33-45.

Holland, J.M. (2004). The environmental consequences of adopting conservation tillage in Europe: reviewing the evidence. Agriculture. Ecosystems and Environment 103: 1-25.

Howarth, R.W., Anderson, D., Cloern, J., Elfring, C., Hopkinson, C., Lapointe, B., Malone, T., Marcus, N., McGlathery, K., Sharpley, A. and Walker, D. (2000). Nutrient pollution in coastal rivers, bays and seas. Issues in Ecology 7: 1-15.

Houghton, J.T., Ding, Y., Griggs, D.J., Noguer, M., van der Linden, P.J. and Xiaosu, D. (2001). Climate Change 2001 – The Scientific Basis. Contribution of Working Group I to the Third Assessment Report of the Intergovernmental Panel on Climate Change. Intergovernmental Panel of Climate Change (IPCC), Cambridge University Press, Cambridge. Online at http://www.grida.no/climate/ipcc_tar/wg1/index.htm

IFEN (1998). Les pesticides dans les eaux (only in French 'Pesticides in Water'). Etudes et Traveaux n 19, IFEN (Institut francais de l'environnement), Orleans, France.

Iital, A., Stålnacke, P., Deelstra, J., Loigu, E. and Pihlak, M. (2005). Effects of large-scale changes in emissions on nutrient concentrations in Estonian rivers in the Lake Peipsi drainage basin. Journal of Hydrology 304: 361-273.

IFOAM (2000). Organic Agriculture and Fair Trade two concepts based on the same holistic principal. Online at http://www.ifoam.org, see 'Social Justice' and next 'Organic Agriculture and Fair Trade'.

INPE (2004). Monitoring of the Brazilian Amazonian Forest by Satellite 'Projeto Probe'. Brazil's National Institute of Space Research (Instituto Nacional de Pesquisas Espaciais, INPE). Online at http://www.obt.inpe.br/prodes/prodes_1988_2003.htm

Jacobs, M. (1995). Sustainable development from broad rhetoric to local reality. Conference Proceedings from Agenda 21 in Cheshire, 1 December 1994, Cheshire County Council, Document No. 493.

Jenkinson, D.S. (2001). The impact of humans on the nitrogen cycle, with focus on temperate arable agriculture. Plant and Soil 228: 3-15.

Jensen, E. and Michelsen, J. (1991). Afsætning af økologiske mælkeprodukter. Kooperativ Forskning Notat 23/91. Esbjerg, Sydjysk Universitetscenter, Institut for Samfunds- og Erhvervsudvikling: 112.

JETACAR (1999). The use of antibiotics in food-producing animals: antibiotic-resistant bacteria in animals and humans. Report of the Joint Expert Advisory Committee on Antibiotic Resistance (JETACAR). Commonwealth Department of Health and Aged Care. Commonwealth Department of Agriculture, Fisheries and Forestry, Australia.

Kaimowitz, D., Mertens, B., Wunder, S. and Pacheco, P. (2004). Hamburger connection fuels Amazon destruction – cattle ranching and deforestation in Brazil's Amazon. Center for International Forestry Research (CIFOR), Indonesia. Online at http://www.cifor.cgiar.org/publications/pdf_files/media/Amazon.pdf, 10 pp.

Kindall, H.W. and Pimentel, D. (1994). Constraints on the expansion of the global food supply. Ambio 23: 198-205.

Kloppenburg, J.J., Hendrickson, J. and Stevenson, G.W. (1996). Coming into the foodshed. Agriculture and Human Values 13(3): 33-42.

La Trobe, H.L. and Acott, T.G. (2000). Localising the global food system. International Journal of Sustainable Development and World Ecology 7(4): 309-320.

Lindert, P. (2000). Shifting ground: the changing agricultural soils of China and Indonesia. MIT Press, USA.

Mâder, P., Fliessbach, A., Dubois, D., Gunst, L., Fried, P. and Niggli, U. (2002). Soil fertility and biodiversity in organic farming. Science 296, 1694-1697.

Margulis, S. (2004). Causes of deforestation of the Brazilian Amazon. World Bank Working Paper No. 22. The World Bank, Washington DC, USA, 71 pp.

Martinez Alier, J. and Oliveras, A. (2003). Deuda Ecológica y Deuda Externa. Quién debe a quién? Icaria, Barcelona.

Mazzucato, V. and Niemeijer, D. (2001). Overestimating land degradation, underestimating farmers in the Sahel. IIED Issue Paper No. 101. May.

Mellon, M., Benbrook, C. and Benbrook, K.L. (2001). Hogging it: Estimates of Antimicrobial Abuse in Livestock. Union of Concerned Scientists.

Mies, M. and Bennholdt-Thomsen, V. (1999). *The Subsistence Perspective*: *Beyond the Globalized Economy*. Zed Books (London and New York).

Milestad, R. and Darnhofer, I. (2003). Building farm resilience: The prospects and challenges of organic farming. Journal of Sustainable Agriculture 22(3): 81-97.

Milestad, R. and Hadatsch, S. (2003). Growing out of the niche – can organic agriculture keep its promises? A study of two Austrian cases. American Journal of Alternative Agriculture 18(3): 155-163.

Moreby, S.J., Aebischer, N.J., Southway, S.E. and Sotherton, N.W. (1994). A comparison of the flora and arthropod fauna of organically and conventionally grown winter wheat in southern England. Annals of Applied Biology 125: 12-27.

Mosier, A.R. (2002). Environmental challenges associated with needed increases in global nitrogen fixation. Nutrient Cycling in Agroecosystems 63: 101-116.

Nixon, S., Trent, Z., Marcuello, C. and Lallana C. (2003). Europe's water: An indicator-based assessment. EEA Topic report 1, Copenhagen, 95 pp.

Novotny, V. (2005). The next step – Incorporating diffuse pollution abatement into watershed management. Water Science Technology 51: 1-9.

OECD (2000). Environmental Indicators for Agriculture: Methods and Results Executive Summary, OECD, Paris.

OECD (2001a). Environmental indicators for Agriculture. Volume 3. Methods and results. Agriculture and Food. OECD Publications Service, France. 400 pp.

OECD (2001b). OECD Environmental Outlook. Environment. OECD Publications Service, France. 309 pp.
Oehl, F., Sieverding, E., Mäder, P., Dubois, D., Ineichen, K., Boller, T. and Wiemken, A. (2004). Impact of long-term conventional and organic farming on the diversity of arbuscular mycorrhizal fungi. Oecologia 138, 574-583.
Olesen, J.E., Jørgensen, L.N., Petersen, J. and Mortensen, J.V. (2003). Effects of rates and timing of nitrogen fertiliser on disease control by fungicides in winter wheat. 1. Crop yield and nitrogen uptake. Journal of Agricultural Science 140: 1-13.
Olesen, J.E., Sørensen, P., Thomsen, I.K., Eriksen, J., Thomsen, A.G. and Berntsen, J. (2004). Integrated nitrogen input systems in Denmark. In: Mosier, A.R., Syers, J.K. and Freney, J.R. (eds) Agriculture and the nitrogen cycle: assessing the impacts of fertilizer use on food production and the environment. SCOPE 65, Island Press, pp. 129-140.
Pakeman, R.J. (2004). Consistency of plant species and trait responses to grazing along a productivity gradient: a multi-site analysis. Journal of Ecology 92: 893-905.
Pengue, W.A. (2003). La economía y los subsidios ambientales: Una Deuda Ecológica en la Pampa Argentina. Buenos Aires. Fronteras 2: 7-8.
Pfiffner, L. and Luka, H. (2003). Effects of low-input farming systems on carabids and epigeal spiders – a paired farm approach. Basic and Applied Ecology 4, 117-127.
Pimentel, D., Harvey, C., Resosudarmo, P., Sinclair, K., Kurz, D., McNair, M., Christ, L., Shpritz, L., Fitton, L., Saffouri, R. and Blair, R. (1995). Environmental and economic costs of soil erosion and conservation benefits. Science 267: 1117-1123.
Pirog, R. and Tyndall, J. (1999). Comparing apples to apples: An Iowa perspective on local food systems. Leopold Center for Sustainable Agriculture, Ames, Iowa. Online at http://www.leopold.iastate.edu/pubs/staff/apples/applepaper.pdf
Prego, J. (1997). El deterioro del ambiente en la Argentina (suelo-agua-vegetación-fauna). 3th Edition. Fundación para la Educación, la Ciencia y la Cultura. Buenos Aires.
Pretty, J. (2002). Agri Culture. (2002). Earthscan Publications, London.
Pulleman, M., Jongmans, A., Marinissen, J. and Bouma, J. (2003). Effects of organic versus conventional arable farming on soil structure and organic matter dynamics in a marine loam in the Netherlands. Soil Use and Management 19: 157-165.
Puricelli, E. and Tuesca, D. (2005). Weed density and diversity under glyphosate-resistant crop sequences. Crop Protection 24(6): 533-542.
Pykälä, J. (2003). Effects of restoration with cattle grazing on plant species composition and richness of semi-natural grasslands. Biodiversity and Conservation 12(11): 2211-2226.
Pykälä, J. (2005). Cattle grazing increases plant species richness of most species trait groups in mesic semi-natural grasslands. Plant Ecology 175(2): 217-226.
Rasul, G. and Thapa, G.B. (2004). Sustainability of ecological and conventional agricultural systems in Bangladesh: an assessment based on environmental, economic and social perspectives. Agricultural Systems 79: 327-351.
Raun, W.R. and Johnson, G.V. (1999). Improving nitrogen use efficiency for cereal production. Agronomy Journal 91: 357-363.
Raynolds, L.T. (2000). Re-embedding global agriculture: The international organic and fair trade movements. Agriculture and Human Values 17: 297-309.
Raynolds, L.T. (2004). The globalization of the organic agro-food networks. World development 32(5): 725-743.
Rigby, D. and Bown, S. (2003). Organic food and global trade: Is the market delivering agricultural sustainability? Discussion Paper, ESEE Frontiers II Conference. 21 pp.

Rigby, D. and Cáceres, D. (2001). Organic farming and the sustainability of agricultural systems. Agricultural Systems 68: 21-40.

Robertson, G.P., Paul, E.A. and Harwood, R.R. (2000). Greenhouse gases in intensive agriculture: Contributions of individual gases to the radiative forcing of the atmosphere. Science 289: 1922-1925.

Rodriguez, C., Leoni, E., Lezama, F. and Altesor, A. (2003). Temporal trends in species composition and plant traits in natural grasslands of Uruguay. Journal of Vegetation Science 14(3): 433-440.

Sainath, P. (1996). Everybody Loves a Good Drought: Stories from India's Poorest Districts. Penguin Books India Ltd., New Delhi.

Satoh, T. and Hosokawa, M. (2000). Organophosphates and their impact on the global environmental. Neurotoxicology 21: 223–227.

Scheiner, J.D., Lavado, R.S. and Alvarez, R. (1996). Difficulties in recommending phosphorus fertilizers for soybeans in Argentina. Communications in Soil Science and Plant Analysis 27(3and4): 521-530.

Scherr, S.J. (1999). Soil degradation. A Threat to Developing-Country Food Security by 2020. Food, Agriculture and the Environment Discussion Paper No. 27. IFPRI. Available online www.IFPRI.org

Scialabba, N. (2000a). Factors influencing organic agriculture policies with a focus on developing countries. Proceedings from the 13th IFOAM Scientific Conference, Basel, Switzerland, 28–31 August 2000.

Scialabba, N. (2000b). FAO perspectives on future challenges for the organic agriculture movement. Proceedings from the 13th IFOAM Scientific Conference, Basel, Switzerland, 28–31 August 2000.

Scialabba, N.E. and Hattam, C. (2002). Organic agriculture, environment and food security. Environment and Natural Resources Series 4. Food and Agriculture Organization of the United Nation (FAO).

Shiva, V. (1991). The Violence of the Green Revolution: Third World Agriculture, Ecology and Politics. Zed Books, Atlantic Highlands, New Jersey.

Smil, V. (2002). Nitrogen and food production: proteins for human diets. Ambio 31: 126-131.

Stoate, C., Boatman, N.D., Borralho, R.J., Carvalho, C.R., de Snoo, G.R. and Eden, P. (2001). Ecological impacts of arable intensification in Europe. Journal of Environmental Management 63: 337-365.

Stolze, M., Piorr, A., Häring, A. and Dabbert, S. (2000). The environmental impact of organic farming in Europe. Organic Farming in Europe: Economics and Policy, vol. 6. University of Hohenheim, Germany.

The Ecologist (1993). Whose Common Future? Reclaiming the Commons. New Society Publishers (Philadelphia, PA and Gabriola Island, BC).

Thompson, P.B. (1996). Sustainability as a norm. Techné: Journal of the Society for Philosophy and Technology 2(2): 75-94. Online at http://scholar.lib.vt.edu/ejournals/SPT/v2n2/pdf/thompson.pdf

Tovey, H. (1997). Food, environmentalism and rural sociology: on the organic farming movement in Ireland. Sociologia Ruralis 37(1): 21-37.

UNEP (1997). Global environment outlook. UNEP and Oxford University Press.

UNEP (1999). Global environment outlook 2000. Earthscan Publications Ltd., London.

USGS (United States Geological Survey) (1999). The quality of our nation's waters – nutrients and pesticides. USGS Circular 1225, Washington DC, USA.

van Elsen, T. (2000). Species diversity as a task for organic agriculture in Europe. Agriculture, Ecosystems and Environment 77: 101-109.
van Wieren, S.E. (1995). The potential role of large herbivores in nature conservation and extensive land use in Europe. Biological Journal of the Linnean Society 56: 11-23.
Ventimiglia, L. (2003). El suelo, una caja de ahorros que puede quedar sin fondos. La Nación. Suplemento Campo. Pag. 7. Buenos Aires. October 18.
Vitousek, P.M., Howarth, R.W., Likens, G.E., Matson, P.A., Schindler, D., Schlesinger, W.H. and Tilman, G.D. (1997). Human alteration of the global nitrogen cycle: causes and consequences. Issues in Ecology 13: 1-17.
Vitta, J.I., Tuesca, D. and Puricelli, E. (2004). Widespread use of glyphosate tolerant soybean and weed community richness in Argentina. Agriculture, Ecosystems and Environment 103(3): 621-624.
Von Braun, J. (2003). Overview of the world food situation – food security: new risks and new opportunities. Paper for Annual General meeting of the Consultative Group on International Agricultural Research, Nairobi, 29 October 2003. International Food Policy Research Institute (IFPRI).
Wackernagel, M. and Rees, W. (1996). Our ecological footprint: Reducing human impact on earth. New Society Publishers, Gabriola Island, British Columbia.
WallisDeVries, M.F. (1998). Large herbivores as key factors for nature conservation. In: WallisDeVries, M.F., Bakker, J.P. and Van Wieren, S.E. (eds) Grazing and Conservation Management. Kluwer Academic, Dordrecht, The Netherlands, pp. 1-20
Willer, H. and Yussefi, M. (2004). The world of organic agriculture – statistics and emerging trends 2004. International Federation of Organic Agriculture Movements (IFOAM), Bonn, Germany, 6th revised edition, 170 pp. Available online at http://www.soel.de/oekolandbau/weltweit_grafiken.html
Willer, H. and Yussefi, M. (2005). The world of organic agriculture – statistics and emerging trends 2005. International Federation of Organic Agriculture Movements (IFOAM), Bonn, Germany, 7th revised edition, 197 pp. Available online at http://www.soel.de/oekolandbau/weltweit.html
Winter, M. (2003). Embeddedness, the new food economy and defensive localism. Journal of Rural Studies 19(1): 23-32.
Wolfe, A. and Patz, J.A. (2002). Nitrogen and human health: direct and indirect impacts. Ambio 31: 120-125.
Zhang, W., Tian, Z., Zhang, N. and Li, Z. (1996) Nitrate pollution of groundwater in northern China. Agriculture, Ecosystems and Environment 59: 223-231.

2
Globalization and sustainable development: a political ecology strategy to realize ecological justice

John Byrne, Leigh Glover and Hugo F. Alrøe*

Introduction	50
Organic farming and the challenge of sustainability	51
Political ecology as one approach to globalization and sustainable development	53
Growth without borders	54
Growth within limits	54
Growth and ecological injustice	55
Commons as the basis of ecological justice	55
Defining commons in the contemporary era	56
State and corporate solutions to commons protection	59
From commons to commodity	61
Ecological commons	62
Overcoming commodification	63
Reclaiming the commons idea	63
Globalization and trade	64
Free trade versus fair trade	65
Traditional and indigenous agriculture in developing nations	67
Putting ecological justice into practice: guidelines for policy	68
A role for 'fair trade'	68
The 'nearness' principle	69
Identifying organic production and produce	69
Sustainability targets	70
Non-certified organic agriculture	70
Ecological justice assessment	70
Conclusions	71

---

* Corresponding author: Center for Energy and Environmental Policy, University of Delaware, Newark, DE 19716-7301, USA. E-mail: jbyrne@udel.edu

Summary

Organic agriculture is, like mainstream agriculture, faced with the challenges of globalization and sustainable development. Ecological justice, the fair distribution of livelihoods and environments, has emerged as a key concept in efforts, on the one hand, to resist negative consequences of globalization and ecological modernization and, on the other to propose new agenda and institutional arrangements. This chapter investigates the role that ecological justice as a political ecology strategy may have in addressing the present problems of organic agriculture in a global political economy. The investigation has two interacting elements, a theoretical analysis of the political, economic and ecological aspects of ecological justice and a discussion of how its key concepts can be put into practice. The political basis of ecological justice is the idea of shared responsibility for livelihoods and environments, or what we have termed commons-based governance. Typically, ecological justice positions social and ecological interests ahead of market liberalism and economic growth. Therefore it may suggest ways to resist the pressures of globalization and associated structural and technological developments. The concepts of commons and ecological justice when joined, define a post-globalist pattern of governance that may facilitate the spread of organic agriculture and other socio-ecological practices that thrive on cooperative, sustainability-focused relations.

Introduction

Release of the United Nations Millennium Ecosystem Assessment, in early 2005, revealed the parlous condition of the global environment. In an era when environmental awareness is high and there are unprecedented international efforts to create global environmental governance, nearly all major indicators of the world's ecological health are in decline. Global economic growth and industrialization, under the influence of the forces of globalization, are increasing natural resource consumption, drawing down non-renewable resources, stressing ecosystem processes, and generating unprecedented amounts of wasted nature. As Chapter 1 describes, modern agriculture has become part of the problem.

Farming in industrialized nations and increasingly in the developing world bears many of the hallmarks of industrialism and of the goals of modernity. Indeed, the recommended path for feeding the world by globalization's proponents typically features the following elements:

- Greater mechanization, standardization (including production techniques, varieties and breeds, and monocultural production), 'factory farming' and increasing scale of production;
- Rising inputs of fossil fuel energy, fertilizers, pesticides and GMOs; and

- Integration into a network of transnational and transcontinental markets shaped by conglomerate 'agribusinesses' and highly complex technology.

Industrial agriculture is inextricably woven onto the modern world through its techniques of production, its market ideology and its technology. Further, globalization has ensured that the demands, preferences and practices of the developed nations are being diffused throughout the world, connecting developing nations to the markets of the developed world.

Modern agri-food production presents an array of environmental concerns associated with intensive water and fossil energy consumption, rising greenhouse gas emissions, increasing application of artificial fertilizers and biocides, and the uncertain effects of biotechnology. Addressing the goal of global ecological sustainability therefore, necessarily challenges the assumptions and practices of industrial agriculture. In this chapter, the role for ecological justice as a political ecology strategy in developing and guiding organic agriculture along a pathway of sustainable development in a globalized world is explored.

Organic farming and the challenge of sustainability

For several reasons, organic farming is providing a sustainable form of agriculture in this era of globalization, at least for industrial nations. Organic production, processing, distribution and sales have grown immensely in size and efficiency in the past two decades, and the movement can no longer be regarded as merely a niche activity serving the needs of a normatively motivated wealthy few. The International Federation for Organic Agriculture Movements (IFOAM) epitomizes a 'coming of age' for the initiative with an adopted worldwide goal of ecologically, socially and economically sound food production (IFOAM, 2004). But, like mainstream agriculture, organic farming is faced with the trends of globalization and the ensuing challenges of sustainable development (see Byrne and Glover, 2002, for a discussion of the general problem).

Yet the case for promoting organic agriculture as ecologically sustainable is complex. Organic farming cannot be considered entirely free of the grip of industrial agricultural practices. Adhering to the standards of organic farming can secure more sustainable development in specific areas, such as regulation of fertilizer, pesticide use, cautions about genetic engineering, opposition to additives and calls for the protection of animal welfare. But for other aspects of agricultural production, the pathway of organic farming is not as clear and its contribution to sustainability has still to be addressed. For example, how will organic farming interface with the following attributes of the modern food regime:

- Large-scale production;

- Processing and marketing through large conventional food companies;
- Sale through supermarkets, sometimes using supermarket brands;
- Trade of feed, seed and other inputs through conventional companies; and
- Global trade.

Successful partnership of the movement with non-organic actors has been an important factor in the recent growth of organic production and expansion of organic food markets. On the other hand, this development can, in itself, lead to unwanted social and environmental impacts, by way of reduced landscape diversity, increases in 'food miles', greater distance between producers and consumers, and unfair competition from large players. Further, partnership can and has put pressure on the integrity of the organic agro-ecological production systems by imposing constraints on the selection and diversity of crops, varieties and breeds.

Globalization and ecological modernization together constitute the mainstream approach to sustainable development (Byrne and Glover, 2002). Globalization is here understood as 'the erosion of the barriers of time and space that constrain human activity across the earth and the increasing social awareness of these changes' (Byrne and Glover, 2002). It embodies a normative interest in modernity's technological, economic and political architecture. Specifically, globalization seeks to remove barriers to state- and market-based organization of society. Its politics privileges ideals of rationality, efficiency, objectivity and competitiveness.

Sustainability was placed on the global agenda in a large consensus-building work under the World Commission on Environment and Development, which gave an often quoted description of sustainable development: 'Humanity has the ability to make development sustainable – to ensure that it meets the needs of the present without compromising the ability of future generations to meet their own needs' (WCED, 1987: 8). The Commission pointed out that sustainable development implies limits – limitations imposed by the existing technological and social development – in the form of environmental resources and the abilities of the biosphere to absorb the effects of human activities. But they also stated that humanity has the ability to create a sustainable future through a marriage of economy and ecology which is today known as 'ecological modernization' – a reform of economics, technologies, and social institutions.

While globalization and ecological modernization constitute mainstream approaches today, they have also generated great resistance from many stakeholders, most noticeably developing nations, local communities, advocates of civil society, and environmentalists. Although diverse, there is a general philosophical theme that unites this resistance, that of the cause of 'ecological justice' (Low and Gleeson, 1998; Byrne *et al.*, 2002a, b). Ecological justice seeks to promote justice in relation to the environment for both present and future generations. In this sense it extends the more familiar concept of environmental justice through a broadening of the ambit of political concern to include future generations and to ecological interests (both living beings and ecological processes). To give a first impression of what this means, some examples of ecological injustice are shown in Box 2.1.

Box 2.1. Examples of ecological injustice.

A large Coca-Cola factory in Plachimada (a hamlet in the state of Kerala, south India) pumps large amounts of groundwater daily for use in producing the famed soda. The pumpage has been shown to deplete groundwater in the area, and polluting the local basin (AIPRF, 2002; India Resource Centre, 2004). While urban consumers far from the plant enjoy the beverage at a relatively modest price, the health and livelihoods of people in the local communities who depend on local natural resources are put at risk.

The construction of China's Three Gorges Dam (CNN, 2001) and India's Narmada Dam (Wagle, 2002) has disrupted the lives of millions of peasant farmers, inundating villages, settlements and agricultural lands, causing great social upheavals, and creating great ecological losses through habitat loss, changes to streamflows and other hydrological effects. Distant cities and downstream communities will benefit from the electricity and flood control created by both projects, but at substantial cost to the rural lives and ecologies of the disrupted valleys.

The corporate dominated world banana industry is characterized by ecologically and socially destructive practices. Chiquita and Dole operate huge Latin American plantations, monocropping bananas over thousands of acres using heavy applications of fungicides, insecticides and other chemicals. This has fuelled significant environmental and health problems, including deforestation, soil erosion, water pollution, and pesticide poisonings (Murray and Raynolds, 2000).

Anticipated changes in climate are caused by the industrialized, high-income countries (Byrne and Inniss, 2002). In general, these changes will have their greatest impact on those that have the fewest resources available to respond (Byrne *et al.*, 2004). In particular, rising sea levels will have major consequences for low-income, lowland countries like Bangledesh and many small ocean states (Byrne and Inniss, 2002). Because anthropogenic releases of carbon to the atmosphere will remain for up to 250 years, the inequality wrought by climate change will continue into the 22nd century (Byrne *et al.*, 2002b).

Political ecology as one approach to globalization and sustainable development

Sustainable development as described by the World Commission emphasizes the possibility for a new era of economic growth through better technologies and social organization (WCED, 1987). But the complex and interdependent relationships between globalization, economic growth, sustainability, and ecological limits have become contested questions. These relationships lie at the core for the discussion of the role of organic agriculture in a global perspective.

Elsewhere, two of the authors of this chapter (Byrne and Glover, 2002) identify three basic positions with regard to globalization and sustainable development:

- Growth and free trade without ecological borders (market liberalism);
- Growth and free trade within certain limits (ecological economy);
- Opposition to growth and free trade on the grounds of ecological injustice (political ecology).

Growth without borders

From a neoliberal economic perspective, globalization does not present a problem. On the contrary, globalization is seen as an improvement of the possibilities for free market forces to allocate resources, which in this view is economically and socially ideal and a prerequisite for liberal democracy (Byrne and Yun, 1999). The solution to world poverty and environmental problems lies in growth and open markets, according to advocates, because growing wealth will furnish more than enough capital to repair whatever damage the growth may have caused.

This position presupposes an independent, always growing economic system as well as well-distributed benefits from the system. So called 'environmental economics'[1] recognizes that there are market failures with respect to the environment and advocates institutions to internalize external costs, so that markets can settle on 'optimal' levels of pollution and ecological losses. From the neoliberal perspective, sustainable development is measured by a single economic indicator: growth in the value of society's collected capital. The price for this simplicity is an assumption of substitutability – that all natural resources and environmental goods can be replaced with produced goods or, in other words, that there is no critical natural capital.

Growth within limits

Market liberalism can be characterized as having a 'weak' conception of sustainability (e.g. Ayres *et al.*, 1998; Neumayer, 1999). Other economic

---

[1] Environmental economics is a relatively new extension of neo-classical economics that applies neoclassical principles to environmental problems (see, especially, Coase 1960). Ecological economics is a broader, transdisciplinary field of study that includes contributions from institutional economics and ecology, as well as from several of the social sciences, the humanities, and the natural and engineering sciences. See, e.g. Söderbaum, 2000: 9, 19.

perspectives endorse stronger conceptions of sustainability. For example, many believe that the economic system is dependent on a finite, vulnerable, ecological system and that there are only limited possibilities of substituting natural capital with manufactured capital (Hawken et al., 1999; Daly and Farley, 2003).

'Ecological economics' is a pluralistic, transdisciplinary alternative to market liberalism that considers ecological limits and the scale of the material and energy flows to which the economical processes connect.[2] A key argument from the ecological economics perspective is that sustainable scale, just distribution, and an efficient allocation are three distinct, but interdependent, problems requiring different policy instruments (Daly and Farley, 2003). Sustainable scale here implies that the throughput associated with economic activities remains within the natural capacity of the ecosystem to absorb wastes and regenerate resources.

Growth and ecological injustice

As a third position, Byrne and Glover argue for a perspective of political ecology, which does not see development and efficiency as solutions, but as the primary sources of social and ecological problems. Political ecology opposes both globalization and ecological modernization because both presume trade is essentially an economic issue. Political ecology, on the other hand, situates trade within a political frame as a contest between resources taken as 'commodities' and taken as 'commons', a contest, in essence, of ecological justice. From this perspective, sustainable development in the form of ecological modernization has primarily been the agenda of the wealthy. Relatedly, sustainable development is seen not as a remedy for problems created by globalization, but a reform programme that currently tends to advance a globalization agenda. Together, globalization and sustainable development spur a replacement of commons valuation with commodity valuation that benefits multinational corporations and exploitive commodity interests, while simultaneously undermining sustainable commons systems and community governance.

Commons as the basis of ecological justice

Ecological justice is founded on the principle that an environment is fundamentally shared. The environment constitutes a 'commons' from a societal perspective, since all human interaction depends upon impacts and is impacted by nature.

---

[2] On the concept of scale in ecological economics, see, e.g. Gibson *et al.* (2000) and Jordan and Fortin (2002).

For organic farming, an ecological justice perspective highlights a number of distinct issues. Organic agriculture is more dependent on the environmental characteristics of the site of production than conventional industrial agriculture, because it bases agricultural production on a close interaction with natural systems and processes and because it has fewer technological remedies available to counteract depletions of these systems. Organic farming in industrial nations represents an effort to move beyond industrial farming because it strives to align its practices with a set of societal, political and ecological principles that cannot be satisfied by conventional farming. Furthermore, organic agriculture may well have unconventional ideas about what can be considered as commons, due to its integrated ecological view of nature, a matter that will now be explored.

Defining commons in the contemporary era

Commons are long-standing social institutions serving diverse cultures throughout human history in their need to share efforts to sustain daily life and in their need to organize shared resources. In modern life, shared effort and shared resources sometimes seem less compelling concerns as we rely on markets, technology and scientific knowledge to solve problems in ways that make 'sharing' apparently unnecessary. Issues of ecological justice, however, can re-establish the importance of commons institutions.

Broadly two conceptually relevant dimensions of commons can be identified; in one depiction, a commons refers to a natural resource ecosystem or spatial area that is regarded as having certain characteristics which enable or encourage common social usage; secondly, a commons can be rendered as a social system or organization that intends to recognize social and/or natural phenomena, processes or areas as common resources, and leads to the formation of informal and possibly formal institutions that govern social relations in support of the intended commons (see Figure 2.1).

Spatial extension
Local: e.g. common lands
Global: e.g. atmosphere
Non-spatial: e.g. knowledge

Provided and reproduced by
Nature: natural commons
Society: social/political/economic commons

Usage characteristics
Reusable
Renewable
Concurrent use
Multifunctional

Ownership and usage regime
Common property: commons regime (no exclusive owner)
State property: state ownership
Private property: individual, exclusive owners

Figure 2.1 Features of commons.

Community, political and scholarly interest in commons has increased in recent years in the wealthy and developing nations. In European history, the concept is familiar from the case of medieval 'common land', in which local communities had traditional rights to common grazing, planting, etc. Commons, however, have come to be recognized as an indispensable feature of social life, and can be identified across all cultures and peoples and from prehistoric times to the contemporary period (Ostrom, 1990). When we consider commons in the form of language and culture, for example, we find that the institution precedes many of the social formations now considered essential, such as governments, nations and corporations. In the absence of effective commons or when commons fail, individual and community welfare is reduced, and in some cases human survival can be problematic (a contemporary case of this last point might be sudden climate change).

Environmental and natural resource issues are at the centre of much of the current interest in commons and have invigorated inquiries into the wide array of new and ancient commons. Part of this interest has been prompted by the search for commons approaches to novel and emerging environmental issues. The international relations literature has recognized that many environmental challenges supersede controls of specific nations, making various forms of international agreements and policy 'regimes' necessary to address them. Byrne and Glover (2002), Volger (1995) and Buck (1998), for example, have explored this new category of international initiatives (known as 'global commons') that concern problems such as ozone-depleting emissions, climate change, biodiversity loss, international toxic waste trade, international endangered species trade, and degradation of the high seas and the polar regions. Noteworthy is the number of global commons issues that are essentially environmental problems.

The concept of commons is also used in another sphere of life, the 'intellectual commons', which includes art, music, fiction and research. Commons arising in this context encompass rituals, language, culture and the store of knowledge generally. Intellectual commons are recognized in law and norms for public activity, through new forms of copyright that expand usage rights and through commitments to open access of publicly funded science (e.g. Suber, 2004). The recent focus on intellectual commons is due partly to the rise of the Internet, and partly to new technologies of digitalization that harbour options for unconstrained reproduction of digital resources. In this respect, the Internet itself exemplifies a commons institution.

Generally, commons are created to govern social interactions with resources, processes, services and other phenomena that are potentially reusable, renewable or sustainable in some sense. A condition for the creation of a commons regime is the feasibility of common and continued use – in terms of how a resource, service or process can be used and what is available for use. The availability for common usage, by different social actors can take different forms – successive use (e.g. a well or spring), concurrent use (e.g. common grazing areas, the sea and the atmosphere) and multifunctional use (e.g. the use of trees for fruit, fodder and firewood). The availability for continued use can depend on inexhaustibility or durability, allowing for re-use of the same resource (such as the physical landscape and space being used for motion and transportation), or on renewability, recycling, reproduction, etc., or processes that replenish the resources. Use is a relation, so the availability for use will not only depend on the resource but also on the users, and the possibilities that different individuals and groups have for using a common resource will depend on their abilities to do so. In the same way, the options for re-use and renewability of a resource will always be relative to the cumulative and technological abilities that are put into the use of the resource. This is why the question of sustainability often comes up in relation to new technological abilities for utilizing natural resources.

Considered as social institutions, commons present an ancient and venerable solution to the problems of resources that need to be shared and governed as such. Commons have been an essential feature of human life since the formation of social groups. The provision of food, water, fibre, shelter and social cohesion has involved commons. Two aspects appear to be critical, firstly, management of the commons resources such that they provide the stream of benefits sought by a community, and secondly, social governance so that shared effort and sharing of resources are sustained. In effect, commons institutions exercise a political role in two senses: i) the creation of common resources; and ii) the evolution of a regime of governance which serves to protect the commons, and also the community's interests that are using the resource. Commons governance of natural resources does not axiomatically result in the protection of environmental values or in the assurance of just access, distribution or disposition of these resources. But a number of reasons make the consideration of environmental issues in a commons context attractive for environmental protection and ecological justice.

As human societies have evolved, new forms and techniques of governance, technology and social institutions have emerged, such as capitalism, mechanization and liberal democratic governance, which address historic commons problems. Changes in technology and the ever-increasing demand for resources by industrial societies have resulted in a continual expansion of resource harvesting and their inclusion in the global economy of production and consumption. Wastes and by-products from industrial society are also accumulating and resultant pollution problems have continued to worsen. Commons feature prominently in this system, providing many of the resources for consumption and the sinks for waste outputs. Rather than being static, therefore, commons are dynamic – being created and lost, as a result of changing circumstances (see Ostrom *et al.*, 2002).

Regardless of the specific characteristics of the shared resources, commons regimes must address the relation between different aspects of common use – different users and usages can conflict in various ways. Furthermore, commons are multidimensional and the impact of one kind of use on other kinds of use (multifunctionality) also needs to be considered in connection with ecological justice. Although the allocation of natural resources usually evokes concepts of conflict, the history of commons finds an expression of social cooperation in a multitude of forms for the successful resolution of these problems.

State and corporate solutions to commons protection

Inherent in the concept of commons is the idea that human interaction with natural resources and ecological services can be governed so as to meet human needs in perpetuity, in other words, to provide for their sustainable use. The degradation and depletion of commons through over-use has been the topic of the 'tragedy' discourse that followed from the influential article 'The tragedy of the commons' by Garrett Hardin (1968). Hardin presumes a state of unrestricted usage of a common grazing area by selfish, rational herdsmen and shows how this will inevitably lead to overgrazing. The tragedy of 'the tragedy of the commons' is that it has been taken as a demonstration of the inability of 'common property' regimes to manage commons (McCay and Jentoft, 1998). Hardin's argument, however, is hardly about a commons. Rather he conceives a regime of free usage in which private gain is paramount, resembling (in this respect) more a commodity approach found in capitalist systems, than a commons approach in a cooperatively organized economy. This is not a proper commons regime as there is no governing social institution where resource users cooperate and follow instituted rules for resource use (e.g. The Ecologist, 1993). Unfortunately, Hardin set in train a widespread misconception through his assumption that commons were open access regimes, thereby promoting the view of the modern impossibility of community governance of commons resources.

Having ignored community governance – historically, one of the most prevalent forms of economic governance (see, e.g. Ostrom *et al.*, 2002), Hardin reduced the question to a choice between two options: privatization or state regulation (nationalization). These options are the signature approaches to commons governance by modern industrial societies. Both capitalist and socialist nation states have sought access to natural resources to promote industrial growth, and many of these resources were originally commons – organized before their identification by the state as constituting state or entrepreneurial property. Oftentimes, the role of the state in capitalist societies has been to make these resources available for private ownership. This can be readily identified with regard to many of the major natural resources used during the formative stages of industrialization (timber from public lands, leases on mining and grazing lands, sale of water rights, sale of transport corridors, sale of broadcast rights and so on). Historically, industrial nations have also used the process of colonization to extend the realm of commons resources to which access could be gained, a process in which state and corporate interests were often joined.

In socialist nations, commons regimes have often been supplanted by central planning in which natural resources and ecosystem services are conceived as available inputs for meeting collective social needs. Community interests are presumed to be represented by the state planning apparatus. As with their capitalist counterparts, development goals have been the principal forms of socialist regimes, although differences can arise with regard to such concerns as equity and democratic participation.

Overall, the response to commons in the global economic system has been one of commodification.[3] As manifested by the global environmental crisis and the multitude of local environmental problems besetting contemporary life, the routine functioning of industrial nation states – socialist and capitalist – has produced well-documented patterns of ecological injustice.[4] Industrial societies have lived unsustainably for more than a century with the effects of their unsustainability being disproportionably borne by the poor and disadvantaged. Moreover, ecological processes have been harmed, and the effects of these changes will be experienced by future generations.

Industrial societies have responded to these crises with strategies of ecological modernization and sustainable development. These strategies seek solutions to environmental problems from within the array of state and market powers (as described above) and have sought to bring remaining commons into state or market control. In this manner, the polluted commons are now regulated by governments, or by corporations working with governments to devise approaches that accommodate both parties' interests. Accordingly, the environment is protected for economic use, but the extent to which the goals of ecological justice are served is less certain.

Under the rationale that commons are best handled by being converted into private or state property, the modern world has struggled to protect societies and ecosystems from the problems of over-use, degradation, and pollution that accompany industrial development. Under globalization, the rate of resource consumption and waste generation continues to increase. Ecological modernization has attempted to use science and state powers to regulate environmental problems without undue disruption to routine industrial activities, but this has emerged as, at most, a partial solution. Governance approaches that conceive of commons as commodity resources are therefore deficient in their

---

[3] Byrne and Rich (1992: 271, footnote 1): 'Commodification is defined as a development orientation pursued by societies in which progress is determined by increased social capacities to produce and purchase goods and services. Under this orientation, the physical environment is valued either directly as a commodity in the form of energy, raw materials and resources extracted for social use; or indirectly as a 'least-cost' means of disposing of wastes (thereby improving the efficiency of commodity production and use)'.

[4] The *World Resources* reports regularly issued by the World Resources Institute and the *State of the World* annuals of the Worldwatch Institute record empirical trends of ecological injustice. See, for example, World Resources Institute (2003) and Worldwatch Institute (2004).

ability to protect the environmental and social values sought under ecological justice.

From commons to commodity

The concept of ecological justice includes a systemic, political relation that property regimes such as those classically described by Bromley and Cernea (1989) cannot reproduce. Property rights regimes treat land and other natural resources as commodities whose benefit streams can only be maximized if enforceable rules of exclusive access are imposed (e.g. Coase, 1960). But maximization will itself result in patterns of ecological injustice that cannot be corrected except by extraordinary means; this is the gist of the ecological modernization proposal (Brown, 2002; Bell, 2003). Similarly, collectivization mobilizes natural resources to maximize socialist development, which may differ (or may not) in its distributive efforts, but does not prioritize ecological justice over development. In a 'commons regime', humans rely on their environment as a (multifunctional) 'lifeworld' for realizing livelihoods – they depend on the land, the waters and the atmosphere, as life support systems. In this sense, life and well-being depend upon the environment; no separation or dualism of 'nature' and 'society' exists (e.g. Byrne *et al.*, 2002b). This is expressed in the social welfare concept of environmental justice and is extended to other living organisms, with the idea of ecological justice.

In the case of global commons, it might seem that there is little use for a broader concept of ownership than the common property regime described above, since the group is basically the whole human population. But there is still the question of who in the group has property rights and the consideration of fairness towards other members, human and non-human. The pre-eminence of property rights in this instance remains a barrier to ecological justice if the challenge is conceived as a problem of political economy (or more extensively, political ecology – see Byrne and Glover, 2002; Byrne *et al.*, 2002b).

Private enclosure or state appropriation effectively operates in two ways. First, as described above, both partition a commons and turn it into a commodity for the purposes of development, often under the rationale that this best serves society's interest (usually meaning efficiency is served, which in turn is conceived as the rational norm for any social allocation). Secondly, enclosure or state appropriation prevents social access into the realm of governance, so that no community institution can exercise its judgements in the governance of the commons. Protecting ecological values becomes difficult because the market or state systems focus on development at the expense of other values (including ecological and social values) and because communities cannot offer an alternative set of views and values.

Commons governance emerges as an activity most likely to protect social and environmental values when commons social institutions are involved, and less likely when commons are treated only as a resource. Governments frequently realize this fact and often re-introduce community involvement into management of public assets through community advisory committees and the like, but having first established and de-limited the powers and authorities of such groups, making them (i.e. communities) creatures of the market or state. In this respect, enclosure or state appropriation not only alters access and use of heretofore socially organized commons, both also undermine political voice in governance. However badly or incompletely community voices may have been previously recognized in commons regimes, privatization and collectivization appear to have caused an acute weakening of community governance. With the ideas of shared resources and common effort in retreat, the ecological justice problems of commodity regimes magnify in the present circumstance.

Ecological commons

Renewed interest in commons governance approaches can be attributed partly to mounting problems of ecological injustice. Discussion of a paradigm shift needed to redress these problems has sparked investigation of existing and earlier commons regimes (Buck, 1998; Ostrom *et al.*, 2002). But there are, as well, empirical reasons since the commons proposition rests on solid ground, as the most highly successful, efficient and long-lived resource management systems are those based on commons (Ostrom, 1990; Ostrom *et al.*, 2002).

Certainly commons governance has shown in practice that approaches can be designed to ensure long-term environmental protection and supply of resources and services in perpetuity. Local resource management practices are able to employ proven, often experience-based knowledge firmly grounded in local cultural norms. Further, we find that historically commons production is typically oriented towards local consumption, rather than for surplus (which is the aim of industrial development), so that the demands made on natural resources tend to be lower than when surpluses are sought. Extraction of resources and waste production are usually conducted with an awareness of local social and environmental implications. Communities exploiting local resources have a vested interest in minimizing the harmful effects of economic activity on local communities and environmental values. From the perspective of creating and maintaining the institutional aspects of local political governance, family and communal relations tend to be reinforced by commons regimes.

It is possible to apply an ecological commons approach to an array of agricultural issues, as can be demonstrated by using the example of soils. Although soils are rarely considered as parts of a global environmental commons, it can be instructive to conceive of them, and their degradation, in this light. Little needs to be said about the role of soil condition for agricultural production, but less well articulated are those connections between the processes of globalization and the corresponding influences on local soils.[5] Globalization of agriculture can influence local soils through the importation of new organisms, including GMOs, and by diffusing new farming practices and technologies. Socioeconomic influences include those brought about by changes in global markets and the demands for certain products, and (often collaterally) by shifting ownership and management regimes.

Soils are influenced by the spread of modern agriculture under the influence of globalization, and associated effects brought about by mechanization, especially fossil fuel-powered equipment and the application of fertilizers and biocides. Using modern agricultural practices, 'feeding the world' has the effect of contributing to the commodification of the earth's mantle. Treating that mantle as an ecological commons, both in the global and local context, can facilitate an understanding of needed policy, institutional and social changes in order to restore values such as nearness, equity and sustainability that would be key to an ecological justice strategy for agricultural practice.

Overcoming commodification

With socialist strategies in decline, globalization offers an unfettered opportunity for neoliberal design of the international order. Neoliberal economics can be characterized as the art of externalizing costs, and private property as a way of internalizing social and ecological benefits. By contrast, ecological justice can be seen as a political strategy for reinstating political voice and elevating the interests of sustainability and social justice above those of neoliberal development. This section examines possible linkages between ecological justice and organic agriculture, and how the pressures of globalization can be resisted.

Reclaiming the commons idea

Social and environmental costs associated with agriculture, such as biodiversity loss and pollution, often stem from practices shaped by the economics of surplus production. That is, modes of agricultural production that require large and continually growing surpluses for sale in markets as the basis for profitable operation can be expected to rely increasingly upon chemical inputs, irrigation and biocides and to farm by mechanical means large, continuous tracts of land in order to raise yields and lower unit costs. Resulting social and environmental impacts, in principle, are to be externalized as the necessary costs of efficient, high-yield agriculture. The externalization of costs becomes a key ingredient for

---

[5] A foundation for such an approach is the classic work of Blaikie (1985), *The Political Economy of Soil Erosion in Developing Countries*, London, New York: Longman. See also the recent McNeill and Winiwarter (2004) article.

success under this model. Ecological modernization proposes to address environment impacts of modern agriculture's progress by regulating the scale and seriousness of these impacts. Its counterpart, social modernization, promises to compensate 'losers' from the revenues of the commonwealth or special fees levied against agricultural wealth. In either case, the source of the problem is unaddressed since doing so would undermine modern, 'efficient' development.

A commons regime, by contrast, traces the problem to commodification and seeks redress by valorizing globalization's external costs and assigning them in a manner that discourages harmful practices. Of particular interest for organic agriculture, certification procedures and North–South agricultural partnerships can be employed to reveal externalities and to promote nearness, sustainability and equity in agricultural practice. These tools can be readily employed in a commons regime, while they are exceptionally difficult to apply in neoliberal commodification contexts.

Globalization and trade

Conventional agriculture is indivisible from the global economy. There is a multitude of ways in which conventional agricultural practices and outputs are shaped by external factors, such as technology, markets, international transport and the activities of multinational corporations. Central to these influences is the role of international trade as an agent that promotes commodification of social and environmental values, resources and services.

Alrøe and Kristensen (2005) identify two problematic trade issues relevant to organic products. Firstly, there are trade barriers and other economic impediments that organic products must overcome in order to compete fairly with conventional agriculture. Of particular concern are state subsidies for conventional agriculture which provide products from these nations with a competitive advantage over organic ones. Secondly, conventional agricultural products are offered at prices that do not reflect the local and global environmental and social costs entailed in their production, so that often environments and communities of Southern countries are forced to bear the burdens of unsustainable production while 'low-cost' foods are enjoyed in the North.

Global trade has the effect of obscuring or effectively eliminating the connections between production and consumption. Where production and consumption are closely linked, the costs and impacts of production are part of the awareness of most consumers, and the effects of local social values and regulations influence consumption. But when foods are sold at a great distance from their sites of production, the social and environmental costs of production are less likely to be known and less likely to influence choices.

Placing organic products into the global market has a number of implications. Global markets are characterized by the strong role played by corporations in transport, handling, distribution, marketing and sales. Entering into the same markets as conventional agricultural products is likely to result in organic produce being subject to the same economic conditions that have shaped conventional agriculture and made sustainable practices unattractive. Organic producers competing in existing global markets will face economic incentives likely to erode the principles of organic farming. An emerging issue of potentially great concern is challenges brought against nations whose trading preferences run counter to such groups as the World Trade Organization. Entry into global markets may offer grounds on which to challenge national subsidies for conventional agriculture, but retaliatory challenges against organic farming are likely. A further concern is that global markets are uncertain and often volatile, which have the effect of reducing the security of farming enterprises and can be added to the economic incentives for larger-scale enterprises.

Free trade versus fair trade

Central to the argument for economic globalization is the advocacy of free trade. Long established as one of the tenets of neo–liberalism, free trade seeks the unencumbered movement of goods, services, labour and capital between markets with minimum state interference, such as in the form of regulations, tariffs and restrictions on capital flows. Free trade is supported by claims that it best produces economic growth and that markets without state restrictions are the most efficient. Neoliberalism's ideal role for government is to provide national security and the rule of law, but intervention in markets is supposed to be minimal. International agencies, notably the Bretton Woods institutions (i.e. the International Monetary Fund and World Bank) and the World Trade Organization, now promote free trade strongly. National governments, especially those of the OECD, have similarly espoused the principles of free trade.[6] Measured by an array of indices, such as annual global trade or resource consumption, the process of economic globalization continues to expand (see, e.g. Held *et al.*, 1999).

Free trade has long been controversial on geopolitical, human rights and environmental grounds. While promoted as an economic goal that produces desirable social outcomes, in practice free trade economics cannot be isolated from questions of politics and history. Disputations over the theory and practice of free trade typically entail a broad range of issues. In agriculture, the issues concerning international free trade are especially complicated, but a few stand out. Global markets provide economic advantage to the more powerful economic states and corporations, so that integration into global markets often produces

---

[6] However, OECD countries have only sporadically moved national policies toward this ideal, and existing agricultural policies often are defended as requiring exceptions of one kind or another (OECD, 2004).

local hardships for producers as prices are depressed. Production can be guided by global markets, rather than local needs, and as farming communities become increasingly oriented towards 'cash crops', they increase their reliance on distant markets and reduce their self-sufficiency. Global commodity markets are frequently unstable, making local producer incomes more uncertain and less secure. Processes of modernization are accelerated under the influence of the global economy, thereby increasing the use of unsustainable methods of production, expanding energy and resource consumption rates, and causing higher ecological costs.

A high-profile effort to resist globalization has emerged in the agro-food network's creation of an alternative market system known as 'fair trade'. This system arose from the alternative trade networks started in the 1960s and 1970s that sought to find and create markets for neglected developing world goods, as sponsored by organizations such as Oxfam in the UK and Equal Exchange in the USA. Principally dealing with coffee, tea, and handicrafts, this movement began its own stores, run as cooperatives. In 1990, a collaborative organization of 11 fair trade organizations in nine European nations was formed: the European Fair Trade Association (EFTA) (see www.eftafairtrade.org). Collectively, EFTA now imports products from some 400 rural communities in Africa, Asia and Latin America, with a turnover of 150m Euro in 2001 (www.eftafairtrade.org). There is also a network of alternative trade groups, the International Federation for Alternative Trade, comprising around 220 member organizations from 59 nations (see www.ifat.org).

'Fair trade' began as a labelling initiative by an NGO in the late 1980s in The Netherlands for marketing coffee from a Mexican cooperative attempting to break through a strong oligopoly (Renard, 2003). This initiative evolved into several fair trade labels in many nations, 17 of which were eventually brought under an umbrella group, the Fair Trade Labelling Organization (FTLO) (see www.fairtrade.org), responsible for certification, standards and labelling. Dozens of products are covered by fair trade labels, notably coffee, tea, rice, bananas, mangoes, cocoa, sugar, honey and fruit juices.

Renard (2003: 90) summarizes the general fair trade criteria involved. Buyers are to meet these conditions: direct purchase, a price covering the costs of production and a social premium, advance payments to prevent smallholder indebtedness, and contracts that allow for long-term planning. Certification requires of the growers: smallholders can participate in a democratic organization, plantation and factory workers can participate in trade unions, no forced or child labour, and programmes to improve environmental sustainability.

Fair trade has been a success, as measured by the growth in sales of its products. FTLO reports that in 2003 it sold 83,480 million t (a 42% increase over the previous year) (www.fairtrade.net/sites/impact/facts.html). This group represents 389 certified producer organizations and over 800,000 families of farmers and workers in 48 countries and selling to consumers in 19 nations.

Jaffee *et al.* (2004) offer that the concept of fair trade can be applied to initiatives within developed nations, in addition to its well-known North–South usage.

The fair trade movement offers a strategy consistent with the promotion of commons regimes. Political voice and social and environmental values take precedence in this movement over questions of efficiency and economic growth. Linking the two could strengthen the interest of ecological justice while offering effective, practical resistance to globalization.

Traditional and indigenous agriculture in developing nations

In many respects, traditional and indigenous farming practices offer an effective foil against commodification and there are several lessons to be drawn for the organic farming movement. Clear distinctions need to be drawn, however, between (certified) organic agriculture, traditional/indigenous agriculture and industrial agriculture. Organic farming in many respects draws on and is popularly identified with older farming traditions and practices and may therefore appear radical (i.e. 'returning to its roots', as it were). However, it is perhaps more accurately understood as a development of modern agriculture, arising from farmers and consumers in industrial societies disenchanted with conventional industrial farming. 'Certification' itself denotes a modern process characterized by objective standards, measurement and assessment, monitoring, performance evaluation, authoritarian control and other activities. Certified organic agriculture in developed nations typically incurs higher costs, which are largely successfully passed onto consumers in the form of premium prices, thereby ensuring economic viability of the organic farming enterprise as a whole. Agricultural products bearing organic certification thereby compete with often lower-priced conventional agricultural products derived from local and distant sources. A relatively small volume of agricultural trade from developing nations is certified organic produce destined for developed nation markets.

Traditional/indigenous agriculture may well satisfy the requirements of certified organic agriculture (especially where there is an absence of use of artificial fertilizers and biocides), yet farmers relying on these long-established methods may be unable to afford or unwilling to commit the time needed to secure certification. Hence, a category for 'non-certified' organic agriculture in developing countries might be warranted in which the organic farming movement promotes smallholder farming in developing nations. We can imagine a scenario where non-certified organic farming of this type is advocated as an alternative response to problems of food security (see Chapters 6 and 10 of this volume). This approach can avoid the problematic effects on soil fertility and biodiversity that solutions based on high external inputs cannot promise.

The question of how this might be done is an important issue, however, for organic agriculture in a global perspective. The pursuit of 'non-certified' organic agriculture might need to be tempered by an awareness of the frequently negative experience of integrating traditional and indigenous farming with modern agriculture and the global economy in general. Concern continues to mount that organic agriculture could evolve towards conventional systems or in ways that are similar to conventional systems (particularly, through involvement with supermarketing and lengthy transport of products to serve organic food demand). Such conventionalization would move organic farming into direct competition with traditional/indigenous agriculture and could result in the organic food movement influencing developing country farming in ways that conventionalize its integration into globalized production, with destructive consequences for rural livelihoods. For this reason, we believe that a note of caution is in order for proposals of this kind (for a detailed discussion of the issues, please see Chapters 3 and 6 of this volume).

Putting ecological justice into practice: guidelines for policy

There is a wide variety of means to incorporate the principles of ecological justice into practice. Here, a number of suggestions are offered that are intended to address how ecological justice can be operationalized in relation to organic agricultural production and trade.

A role for 'fair trade'

As discussed above, the concept of fair trade applies equally to exchanges between North and South nations and within nations of the North and South. Organically grown foods, in the North, have benefited from labelling, standards and marketing systems because the values embodied in the production of these products finds a clear resonance in communities and among individuals who seek to restore a commons idea of food production and consumption. Southern farmers and communities may be less served by these strategies, but efforts to support non-certified organic farming may be applicable. Ecological justice is promoted because the restored sense of commons relations builds social and ecological values into the decision process. Further, the political character of the decision process is explicitly recognized (rather than being muddled by the rhetoric of free trade and efficiency). Unfettered economic globalization cannot realize these things and this finding suggests that a closer alignment of the organic foods movement, and the social and ecological values it reflects, with that of fair trade will benefit a more systemic process of change.

The 'nearness' principle

Globalization encourages the movements of goods, services and capital and the erosion of local identity where this does not add obvious market value. Throughout the process of globalization, the global movement of produce has continued to accelerate and the concept of global markets now exists for all major forms of agro-produce. As argued in this chapter, there are a number of ecological and social implications from this trend, including an increasing mechanization of the food production system, greater transport costs, higher energy inputs, greater application of preservatives and food storage technologies, loss of farmer independence, greater corporate involvement, incentives for unsustainable production and the dominance of cash cropping. Of particular concern to organic farmers is the goal of resisting the global market, wherein the factors influencing market prices become increasingly remote and market relationships become more volatile. Incorporating a principle of 'nearness' into the agro-production system could promote the consumption of local and regional produce over goods imported from afar. Confirmation of the principle would aid in the identification of the local and broader commons interests. Coordination with the fair trade movement would be necessary so as not to intentionally harm farmers in developing nations.

Identifying organic production and produce

Identification of organic produce serves several goals simultaneously. A community is established through the system to devise and administer the organic identity of produce, which in turn reinforces a sense of community among identified organic producers. Such an identity allows consumers to express their preferences and can spur the formation of an alternative market for farm produce. Wider educative benefits for the community become possible because markets now express a broader range of social and ecological values, so that communities and individuals can demand specific goals that stand in opposition to those of conventional agriculture. The ascendance of social and ecological values can have the direct effect of reducing the role of the global economy. Establishing such an identity is complex and there is the risk of creating a technocratic system that repeats the same undesirable effects of conventional industrial agriculture. A desirable outcome is a system of identification reflecting both community and farming interests and values and supporting a diversity of political voices.

Sustainability targets

Environmental and social justice goals can be transformed into specific indicators and applied to farming activities in order to assess the extent to which sustainability and justice targets are being met. To some degree, such measures could be an extension of the broad set of environmental and social goals currently found in organic farming and fair trade standards. Sustainability and justice targets could apply to various inputs to the farming process and to overall measures, such as the 'ecological footprint' approaches (Wackernagel and Rees, 1996) and measures of socio-ecological performance that are built on interlinked principles of equity and sustainability (see, e.g. Byrne *et al.*, 1998, 2004).

Targets can ensure that organic food production uses ecologically sustainable and socially equitable tools to reach long-term goals. By communicating to society that organic produce meets 'green' and fair objectives, the appeal of organics can be broadened. Further, such identification would highlight the social and environmental failings of conventional agriculture and could lead to increased pressures from civil society on behalf of an agenda of justice and sustainability. Because organic farming focuses on local circumstances, the setting of targets would have to consider the extent to which local (e.g. nutrient inputs) and more global concerns (e.g. greenhouse gas emissions or impacts on Southern farming) are included.

Non-certified organic agriculture

Certification of organic produce is ideally suited to production in the North but poses difficulties for the South, and indeed its application in the developing world is potentially harmful to the interests of smallholder farming communities. Recognition of the organic farming approaches of Southern farmers is needed but operationalized in a manner that avoids imposing the burden of Western-style certification. Here the basic approach could involve local decision-making to promote sustainability and fairness objectives based on local and regional conditions. Consumers and producers in Southern nations should be able to benefit from knowing whether agricultural produce is contributing to the goals of ecological justice. At the same time, a system of imports from the South to the North can ensure the commitment of Northern resources to redress problems created for agriculture by economic globalization.

Ecological justice assessment

If the preceding initiatives are considered collectively, a nascent assessment strategy on behalf of ecological justice can be defined. The assessment process would involve the creation of a series of social institutions seeking to revitalize commons-centred agricultural production and consumption. These commons regimes would be capable of taking into consideration not only local assessments of fairness and sustainability, but also would reflect global sustainability and justice goals. To this end, ecological justice goals could be established to assist and guide the organic farming community in this activity. These goals could consider some of the key components of ecological justice, including the extent of sustainability, the effects on future generations, the effects on non-human species, the pursuit of fair trade, the practice of the nearness principle and the extent to which social justice goals are served.

Conclusions

There are potentially strong links between organic farming and explicit strategies to pursue the values of ecological justice. Organic farming already exhibits a commitment to social and ecological values that conform with principles of ecological justice, including protecting the productive capacities of farming systems, meeting local needs, contributing to local community development, and considering the interests of future generations. However, the forces of economic globalization offer a number of challenges to the spread of organic agriculture and increase the incentives for the organic food systems to become more like conventional food schemes. Alternatives to economic globalization are available and can be organized around the concept of commons-centred organic agriculture. In this way, organic farming may well play a vital role in the quest for an ecologically just and sustainable future.

References

AIPRF (All India People's Resistance Forum) (2002). Struggle against Coca Cola in Kerala, ZNet September 10, 2002. Available online at http://www.zmag.org/content/showarticle.cfm?SectionID=32andItemID=2316 (accessed 28 September 2004).

Alrøe, H.F. and Kristensen, E.S. (2005). Organic agriculture in a global perspective. In: Wulfhorst, J.D. and Haugestad, A.K. (eds) Building Sustainable Communities: Ecological Justice and Global Citizenship, Rodopi (forthcoming). Available online at http://orgprints.org/3855

Ayres, R.U., van den Bergh, J.C.J.M. and Gowdy, J.M. (1998). Viewpoint: Weak versus Strong Sustainability, Tinbergen Institute Discussion Papers no. 98-103/3. Available online at http://ideas.repec.org/p/dgr/uvatin/19980103.html

Bell, D.R. (2003). Political liberalism and ecological justice. Paper presented at ECPR General conference, Environmental politics section, 19th September 2003. Available online at: http://www.essex.ac.uk/ECPR/events/generalconference/marburg/show_panel.asp?panelID=46

Blaikie, P. (1985). The Political Economy of Soil Erosion in Developing Countries. Longman, London, New York.

Bromley, D.W. and Cernea, M.M. (1989). The Management of Common Property Natural Resources: Some Conceptual and Operational Fallacies, World Bank Discussion Paper 57. The World Bank, Washington, DC. Available online at http://www.wds.worldbank.org/servlet/WDS_IBank_Servlet?pcont=detailsandeid=000178830_9810190357252

Brown, L.R. (2002). Restructuring the economy for sustainable development. International Review for Environmental Strategies 3(1): 73-80.

Buck, S.J. (1998). The Global Commons: An Introduction. Island Press, Washington, DC.

Byrne, J. and Glover, L. (2002). A common future or towards a future commons: Globalization and sustainable development since UNCED. International Review for Environmental Strategies 3(1): 5-25.

Byrne. J. and Inniss, V. (2002). Island Sustainability and Sustainable Development in the Contest of Climate Change. In: Hsiao, M. (ed.) Sustainable Development for Island Societies. Academia Sinica, Taipei, Taiwan pp. 3-29. Available online at ceep.udel.edu/publications/globalenvironments/publications/2002/ge_island_sustainability.pdf

Byrne, J. and Rich, D. (1992). Toward a Political Economy of Global Change: Energy, Environment and Development in the Greenhouse. In: Byrne, J. and Rich, D. (eds), Energy and the Environment: The Policy Challenge. Transaction Publishers, New Brunswick, NJ and London, pp. 269-302.

Byrne, J. and Yun, S.-J. (1999). Efficient Global Warming: Contradictions in Liberal Democratic Responses to Global Environmental Problems. Bulletin of Science, Technology and Society 19/6: 493-500.

Byrne, J., Wang, Y.-D., Lee, H. and Kim, J.-D. (1998). An Equity- and Sustainability-Based Policy Response to Global Climate Change. Energy Policy 26(4): 335-343.

Byrne, J., Glover, L. and Martinez, C. (2002a). Environmental Justice: Discourses in International Political Economy. Transaction Publishers, New Brunswick, NJ and London.

Byrne, J., Glover, L. and Martinez, C. (2002b). The Production of Unequal Nature. In: Byrne, J., Glover, L. and Martinez, C. (eds). Environmental Justice: Discourses in International Political Economy. Transaction Publishers, New Brunswick, NJ and London, pp. 261-291.

Byrne, J., Glover, L., Inniss, V., Kulkarni, J., Mun, Y.-M., Toly, N. and Wang, Y.-D. (2004). Reclaiming the atmospheric commons: Beyond Kyoto. In: Grover, V. (ed.) Climate Change: Perspectives Five Years After Kyoto. Science Publishers, Plymouth, UK, pp. 429-452.

CNN (2001). China's Three Gorges Dam (includes interview with this chapter's co-author John Byrne). Available online at http.www.cnn.com/SPECIALS/1999/China.50/asian.superpower/three.gorges

Coase, R. (1960). The Problem of Social Costs. Journal of Law and Economics 3: 1-44.

Daly, H. and Farley, J. (2003). Ecological Economics: Principles and Applications. Island Press, Washington, DC.

Gibson, C.C., Ostrom, E. and Ahn, T.K. (2000). The concept of scale and the human dimensions of global change: a survey. Ecological Economics 32(2), 217-239. Available online (restricted access) at http://dx.doi.org/10.1016/S0921-8009(99)00092-0

Hardin, G. (1968). The tragedy of the commons. Science 162 (December): 1243-48.

Hawken, P., Lovins, A. and Lovins, L.H. (1999). Natural Capitalism: Creating the Next Industrial Revolution. Little, Brown and Co., Boston, MA.

Held, D., McGrew, A., Goldblatt, D. and Perraton, J. (1999). Global Transformations: Politics, Economics and Culture. Stanford University Press, Stanford, CA.

India Resource Centre (2004). Campaign to Hold Coca-Cola Accountable. Available online at http://www.indiaresource.org/campaigns/coke (accessed 28 September 2004).

IFOAM (2004). International Federation of Organic Agriculture Movements. Online at http://ifoam.org (accessed 30 March 2005).

Jaffee, D., Kloppenburg, J.R. and Monroy, M.B. (2004). Bringing the 'moral charge' home: Fair trade within the North and within the South. Rural Sociology 69(2): 169-196.

Jordan, G.J. and Fortin, M.-J. (2002). Scale and topology in the ecological economics sustainability paradigm. Ecological Economics, 41(2): 361-366. Available online at http://www.zoo.utoronto.ca/fortin/Jordan2002.pdf

Low, N. and Gleeson, B. (1998). Justice, Society and Nature: An Exploration of Political Ecology. Routledge, London and New York.

McCay, B.J. and Jentoft, S. (1998). Market or community failure? Critical perspectives on common property research. Human Organization 57(1): 21-29.

McNeill, J.R. and Winiwarter, V. (2004). Breaking the Sod: Humankind, History and Soil. Science 304 (June 11): 1627-1629.

Murray, D.L. and Raynolds, L.T. (2000). Alternative trade in bananas: Obstacles and opportunities for progressive social change in the global economy. Agriculture and Human Values 17: 65-74.

Neumayer, E. (1999). Weak versus Strong Sustainability: Exploring the Limits of Two Opposing Paradigms. Edward Elgar, Cheltenham.

OECD (2004). Agricultural Policies in OECD Countries: At a Glance – 2004 Edition. OECD, Paris, France.

Ostrom, E. (1990). Governing the Commons: The Evolution of Institutions for Collective Action. Cambridge University Press, New York.

Ostrom, E., Dietz, T., Dolsak, N., Stern, P.C., Stonich, S. and Weber, E.U. (eds) (2002). The Drama of the Commons. National Academy Press, Washington, DC.

Renard, M.-C. (2003). Fair trade: quality, market and conventions. Journal of Rural Studies 19(1): 87-96.

Söderbaum, P. (2000). Ecological Economics: A Political Economics Approach to Environment and Development. Earthscan, London.

Suber, P. (2004). Creating an intellectual commons through open access. Presented at the Workshop on Scholarly Communication as a Commons, Workshop in Political Theory and Policy Analysis, Indiana University, Bloomington, IN, March 31–April 2, 2004. Available online at http://dlc.dlib.indiana.edu/archive/00001246

The Ecologist (1993). Whose Common Future? Reclaiming the Commons. Earthscan Publications, London.

Volger, J. (1995). The Global Commons: A Regime Analysis. John Wiley and Sons, Chichester, UK.

Wackernagle, M. and Rees, W. (1996). Our Ecological Footprint: Reducing Human Impact on the Earth. New Society Publishers, Philadelphia.

Wagle, S. (2002). The Long March for Livelihoods: Struggle Against the Narmada Dam in India. In: Byrne, J., Glover, L. and Martinez, C. (eds) Environmental Justice: Discourses in International Political Economy. Transaction Publishers, New Brunswick, NJ and London, pp. 71-96.

WCED (World Commission on Environment and Development) (1987). Our Common Future. Oxford University Press, New York.

World Resources Institute (2003). World Resources 2002-2004: Decisions for the Earth, Balance, Voice, and Power. World Resources Institute, Washington, DC.

Worldwatch Institute (2004). State of the World 2004 Special Focus: The Consumer Society. W.W. Norton and Company, New York.

3
Organic agriculture and ecological justice: ethics and practice

Hugo F. Alrøe,* John Byrne and Leigh Glover

Introduction	76
Scope and purpose of the chapter	79
Sustainability, globalization and organic agriculture	79
Dimensions of sustainability	80
Different meanings of globalization and sustainability	82
Sustainability and organic agriculture	83
The ethics and justice of ecological justice	84
Ecological justice as an ethical concept	85
The justice of ecological justice	87
Summing up	89
Challenges for organic agriculture	90
Commodification of commons	91
How to address externalities	92
Growing distances	94
Putting ecological justice into organic practice	97
The way of certified organic agriculture	98
The way of non-certified organic agriculture	102
Organic agriculture as an alternative example	106
Conclusions	108

Summary

Ecological justice is a challenging concept in relation to the current development of agriculture, because it positions social and ecological interests against market liberalism and economic growth. Ecological justice concerns fairness with regard to the common environment based on the idea that environments are fundamentally shared. This chapter investigates the role that ecological justice may have in relation to the global challenges of organic agriculture. We perform a philoso-

---

* Corresponding author: Danish Research Centre for Organic Food and Farming, Research Centre Foulum, Blichers Allé 20, P.O. Box 50, DK-8830 Tjele, Denmark. Hugo.Alroe@agrsci.dk

phical analysis of the ethics of ecological justice and the relation to sustainability and globalization. On this basis, we discuss the challenges that this important concept poses to organic agriculture and how it can be put into organic practice. Organic agriculture is in an advanced position with regard to ecological justice, since it aims to interact in a positive way with the environment. But ecological justice also poses significant challenges to organic agriculture. The three main challenges are: the commodification of hitherto commons; external environmental and social costs that are not accounted for in the market; and growing distances in form of distant trade and ownership in the organic food systems. We conclude that the ideas of ecological justice can be promoted in three ways by means of organic agriculture: by implementing ecological justice more fully in the organic certification standards through incorporating a measure of 'nearness' and developing a fair organic trade; by promoting non-certified agriculture based on the organic principles as an alternative development strategy for local sustainable communities and food security; and by organic agriculture serving as an alternative example for the broader implementation of ecological justice in agriculture and society.

Introduction

Organic production, processing, distribution and sale have grown immensely in size and efficiency in the past two decades, and organics has become a global player. The International Federation for Organic Agriculture Movements (IFOAM, www.ifoam.org) states that its goal is the worldwide adoption of ecologically, socially and economically sound systems that are based on the principles of organic agriculture. But at the same time, like mainstream agriculture, organic agriculture is faced with the all-pervading trends of globalization and the ensuing challenges of sustainable development.

The current trends in mainstream agriculture have implications for social and environmental values, and most trends are to some degree shared by organic agriculture. The organic standards do secure a more sustainable development in the areas that they address, such as the regulations on fertilizers, pesticides, genetic engineering, additives and animal welfare. But on areas that are not, at present, covered by regulations, organic agriculture tends to follow the mainstream path. Some characteristic features of modern organic agriculture are thus:

- Large-scale efficient productions, incorporating modern technologies.
- Trade of feed, seed and other inputs through conventional companies.
- Global trade with organic feed and food products.
- Processing and marketing through large conventional food companies.
- Sale through supermarkets, sometimes using supermarket brands.

This market based 'modernization' and 'conventionalization' of organic food systems and the involvement of non-organic actors have been important factors in the recent growth of organic production and trade. On the other hand, this development can, in itself, lead to unwanted social and environmental impacts (Rigby and Bown, 2003), by way of reduced landscape diversity, increases in food miles, greater distance between producers and consumers, unfair competition from large players, reduced food diversity, etc. And it can also put pressure on the local adaptation and integrity of the organic production systems by imposing constraints on the selection and diversity of crops, varieties, farm animals and breeds.

In accordance with the strategy of the organic movement to operate both in and against the market, Alrøe and Kristensen (2005) identify two problematic issues relevant to the trade of organic products: how to remove unfair obstacles to free trade with organic products, and how to avoid negative effects from free, global trade.

There are trade barriers and other economic impediments that organic products from low-income countries must overcome in order to compete fairly with similar conventional and organic products. Of particular concern are state subsidies for conventional agriculture. Subsidies may also distort the competition between organic products from different regions. Moreover, conventional agricultural products are offered at prices that do not reflect the environmental and social costs entailed in their production as well as organic products do, and thereby local environments and communities are forced to bear the burdens of externalities from the production. Finally, the organic standards and control systems themselves can be a barrier that hinders the potential growth and spread of organic farming (e.g. Fuchshofen and Fuchshofen, 2000; Haen, 2000). Global uniform standards are likely to be unfair to some, because they do not attach importance to the different cultural and natural conditions in different regions. The issue of free trade with organic products is treated further in Chapter 5.

With regard to the second issue, the identity of organic farming must be broadened and strengthened to avoid negative environmental and social consequences from free, global trade with organic products. Distant trade may conceal complex systemic costs connected to organic production processes and transportation. In particular, while the present organic certification schemes do promote soil fertility and to a large degree prevent environmental degradation, they do not consider issues such as: commodification of hitherto commons like soil, water and land; social impacts and consequences for agricultural and natural biodiversity of globalized organic productions (such as when large corporate organic operations establish themselves in low-income areas and productions for self-sufficiency are replaced with organic cash crops); environmental costs connected to international transportation; and unfair prices and profits in the organic food systems.

Reflections on the current trends in the development of modern organic food systems have led to a new and renewed interest in values and principles of organic farming that can guide the future development of organic agriculture (DARCOF, 2000; Lund, 2002; Alrøe and Kristensen, 2004). With this in view, IFOAM is currently rewriting the principles of organic agriculture (see Box 3.1). All the principles have something to say in relation to the trends of globalization, but the principle of fairness speaks most directly. It says: 'Organic agriculture should build on relationships that ensure fairness with regard to the common environment and life opportunities'. This principle refers to the concept of ecological justice, which in recent decades has been subject to a fair amount of interest (e.g. Low and Gleeson, 1998; Byrne *et al.*, 2002a; Baxter, 2005; see also Chapter 2). Based on the idea that environments are fundamentally shared, *ecological justice concerns fairness with regard to the common environment*.

Box 3.1. The proposed Principles of Organic Agriculture.

The hitherto 'Principal Aims of Organic Production and Processing' are being rewritten by IFOAM as the 'Principles of Organic Agriculture' (Luttikholt, 2004). These are the proposed principles as of June 2005:

Principle of health
Organic Agriculture should sustain and enhance the health of soil, plant, animal and human as one and indivisible.

Principle of ecology
Organic Agriculture should be based on living ecological systems and cycles, work with them, emulate them and help sustain them.

Principle of fairness
Organic Agriculture should build on relationships that ensure fairness with regard to the common environment and life opportunities.

Principle of care
Organic Agriculture should be managed in a precautionary and responsible manner to protect the health and well-being of current and future generations and the environment.

The proposed principles have been presented and discussed at several occasions in and outside the organic movement in 2004–05 and there has been a comprehensive hearing process with IFOAM membership and other stakeholders. In September 2005 the IFOAM general assembly will vote on the principles. All drafts of proposed principles and details of the hearing process, including questionnaires, feedback and minutes of Task Force meetings, are available at an open website (http://ecowiki.org/IfoamPrinciples) and on the IFOAM website (http://www.ifoam.org/organic_facts/principles). The new principles are also to be used as a basis for future revisions of the EU regulation on organic agriculture, according to the plans of the EU-financed targeted research project 'Organic Revision' (http://organic-revision.org).

As concluded in Chapter 2, there are potentially strong links between organic agriculture and explicit strategies to pursue the values of ecological justice. Organic agriculture has social and ecological goals that conform with the principles of ecological justice, including protecting the productive capacities of farming systems, meeting local needs, contributing to local community development and considering the interests of future generations. However, the forces of economic globalization offer a number of challenges to the ideas of organic agriculture and increase the incentives for the organic food systems to become more like conventional food systems. If the global development of organic agriculture is to succeed, the need is urgent to investigate what the concept of ecological justice means with regard to the development of organic food systems issues and what challenges and promises it holds.

Scope and purpose of the chapter

The present chapter investigates the role of ecological justice as a key ethical principle in relation to organic agriculture, globalization and sustainability. Ecological justice is a challenging concept in relation to the current development of organic agriculture, because it places social and ecological interests against market liberalism and economic growth, and it may suggest ways to resist the pressures of market globalization and current structural and technological developments. This chapter investigates the role that ecological justice may have in relation to the present challenges for the global development of organic agriculture. The main questions are: What is the meaning and context of ecological justice? How can these ideas help resist the pressures of globalization? How can ecological justice be implemented in relation to organic production and trade? And how can organic agriculture contribute to ecological justice in a global perspective?

The investigation has two interacting elements, a philosophical analysis of ecological justice in relation to other relevant concepts and a discussion of how the concept can be put into practice to meet the present challenges.

Sustainability, globalization and organic agriculture

The World Commission on Environment and Development raised sustainability on the global, political agenda. They stated that poverty, which is an evil in itself, but also makes the world prone to ecological and other catastrophes, is no longer inevitable (WCED, 1987). Technology and social organization can be managed and improved to make way for a new era of economic growth. This approach is now the main approach to sustainable development, often called 'ecological modernization' (e.g. Hajer, 1995). The commission further stated that the overall sustainability goal of meeting the essential needs of the present requires an assurance that those poor get their fair share of the resources required to sustain economic growth. Unfortunately, current policy makers generally emphasize the overall goal of economic growth through economic globalization and neglect the goals on poverty reduction, fair access to resources, and the needs of future generations. Therefore, while ecological modernization, globalization of markets and promotion of free trade constitute mainstream approaches today, they have also generated great resistance from many stakeholders, most noticeably developing nations, local communities, advocates of civil society, and environmentalists, and a call for social, ecological and environmental justice (e.g. Bond, 2002).

The important issues today with regard to sustainability and globalization are thus not questions of sustainability versus globalization, but of different understandings of sustainability versus each other and different understandings of globalization versus each other. For instance, as pointed out by Christoff (1996), the term ecological modernization has been employed in a range of ways, bearing quite different values. 'Consequently there is a need to identify the normative dimensions of these uses as either weak or strong, depending on whether or not such ecological modernisation is part of the problem or part of the solution for the ecological crises' (Christoff, 1996: 497). Byrne and Glover (2002) conclude that the goal of ecological justice is needed to effectively resolve the world's problems with environmental decline and social deterioration – and that this is a more controversial and problematic goal than that of sustainable development. In this section we will look at different understandings of sustainability and globalization in order to indicate the relation with the rising discourse of ecological justice, and in the next we perform a normative analyses of ecological justice in order to clarify the meaning and the values of this important concept.

Dimensions of sustainability

It is common to speak of three dimensions of sustainability: ecological, economic and social. But this distinction is not very helpful in the present context. Even though discussions on sustainability and globalization with regard to agriculture should be seen in the context of the more general discussions on these issues, agriculture also brings in new perspectives. In particular, agriculture makes the relationship between man and nature very explicit. From the perspective of organic agriculture, agriculture is an ancient and very intimate relationship between human and nature that involves both ecological and social systems – man is not separate from nature, human and nature are in many ways an integrated whole. Speaking of ecological, economic and social sustainability tends to remove focus from the relations between the three and thereby counteract the insights of organic agriculture. Moreover, it does not capture the really significant differences in how sustainability is understood and used.

Joachim Spangenberg (e.g. 2002) has formulated a broader framework of sustainability that includes a fourth dimension, institutional sustainability (which was introduced by the UN Division for Sustainable Development in 1995). This framework has been depicted as a sustainability prism (see Figure 3.1) with four dimensions/aspects/subsystems/imperatives (they are described in different terms) placed in the corners with six 'interlinkages' between them: justice, burden sharing, democracy, eco-efficiency, care, and access (e.g. Valentin and Spangenberg, 2000; Spangenberg, 2002).

This richer picture of sustainability is more useful in this context due to the interlinkages between different dimensions of sustainability, which are in accordance with a focus on the relations between human and nature, social and ecological. 'Access', the interlinkage between the environmental and social aspects, is thus described in ways that resemble ecological justice. This may be of some importance in relation to putting ecological justice into practice, since the sustainability prism is being used as a framework for development of indicators and since it provides an eye and a space for ecological justice in this type of work. However, while the sustainability prism does offer a rich view that opens up for discussions of ecological justice, there is still a need to look at the differences in how the concept of sustainability is used in different discourses and the meanings and values inherent in these differences.

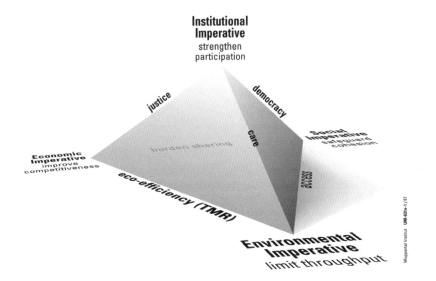

Figure 3.1. The sustainability prism (Valentin and Spangenberg, 2000).

Different meanings of globalization and sustainability

Globalization means that technological, institutional and social changes enforce global communications and interactions. There is absolutely no consensus on whether globalization as such is good or bad, but globalization seems to have accentuated both the positive and negative aspects of the global changes at the same time (Halle, 2002). The concept of globalization therefore should not be used in an unqualified way for analysis of globalization processes.

Ritzer (2003) suggests that the conventional opposition between globalization and the local is of little use – 'it is increasingly difficult to find anything in the world untouched by globalization'. Instead, Ritzer proposes that it is more useful to look at the conflict between glocalization (the interpenetration of globalizing processes and local heterogeneous conditions) and grobalization (the imperialistic ambitions of nations, corporations and organizations to global growth in power, influence and profits) as a key dynamic under the broad heading of globalization. The global sale of local fair trade products and the local adaptations of the general principles of organic agriculture can both be seen as examples of glocalization. But the tag of glocalization may also be used as a thin disguise a to mask the ambitions of grobalization – as when McDonald's uses the figure of Asterix instead of Ronald McDonald in France.

Byrne and Glover (2002) identify three different positions that harbour different perspectives on globalization and sustainable development in relation to trade and environment (see also Chapter 2). The first position endorses the goals of growth and free trade and finds that sustainable development is best sought solely by means of the market. It harbours a 'weak' conception of sustainability. A characteristic approach to address environmental problems within this position is environmental economics. The second position shares the same goals, but deems that there are ecological limits that need to be considered separately and thereby harbours a 'strong' conception of sustainability. This is a characteristic perspective within ecological economics (see further in Chapter 4). The third position, on the contrary, sees growth and free trade as a recipe for ecological injustice and therefore opposes both globalization and ecological modernization. This perspective is characteristic of political ecology.

These three positions show that the perspective from which one observes the issues of globalization and sustainability strongly influences what one sees. All three perspectives can be useful, but in relation to organic agriculture it is important to consider how the chosen perspective relates to the basic aims and values of the organic movement. And it is important to realize that what can be seen from one perspective may not be visible from another. The issues of ecological justice can thus be observed from the third perspective, but may be more or less hidden from the other perspectives.

Sustainability and organic agriculture

Within the context of agriculture, Gordon Douglass (1984) described a distinction between three dominant visions of agricultural sustainability that are used by different groups with different views and values (a distinction that resembles the above one by Byrne and Glover, 2002). *Sustainability as food sufficiency* looks at population growth and speaks of sustainability in terms of sufficient food production, with the necessary use of technology and resources. Agriculture is an instrument for feeding the world and economic cost-benefit analysis is the instruction, which guides application of that instrument. In this group we find the defenders of the modern 'conventional', industrialized agriculture. *Sustainability as stewardship* is concerned with the ecological balance and the biophysical limits to agricultural production. From the ecological point of view, sustainability constrains the production and determines desirable human population levels. This is a diverse group of 'environmentalists', often with a concern for the limits to growth in a finite global environment. *Sustainability as community* shares the concern for ecological balance, but with special interest in promoting vital, coherent rural cultures. Cultural practices are taken to be as important as the products of science to sustainability, and the values of stewardship, self-reliance, humility and holism are encouraged. In this group we find the 'alternative' forms of agriculture, and the modern organic farming has originated from within the community group.

From a philosophical point of view, Paul Thompson (1996) suggested that there were only two different meanings of agricultural sustainability: resource sufficiency and functional integrity. *Resource sufficiency* is an 'accounting' approach that focuses on how to fulfil present and future human needs for food, and on how we can measure and calculate the proper balance between present resource use and future needs based on the relation between input and output from the system, seen from without. Environment and nature is considered a resource that is separate from humans and society. *Functional integrity*, on the other hand, sees humans as an integrated part of nature based on an ecological view of nature (Tybirk *et al.*, 2004). Humans and nature form vulnerable socio-ecological systems that have crucial elements, such as soil, crops, livestock, ecosystems, cultural values, and social institutions, which must be regenerated and reproduced over time. (This does not mean that functional integrity determines cultural values or social institutions, only that they need to perform certain functions for the system to survive.)

Functional integrity emphasizes resilience and recognizes the limits of human knowledge and the possible risks connected to new technologies, thereby incorporating the concept of precaution. Precaution does not denigrate scientific knowledge, but uses it as far as it can within the general context of uncertainty and ignorance. The distinction made by Thompson therefore reveals a close connection between different conceptions of sustainability and different views of the scope and limits of human knowledge (for a fuller treatment of the ethical basis of sustainability and precaution, see Alrøe and Kristensen, 2003).

Thompson uses the two meanings of sustainability in analysing different case examples, showing how resource sufficiency and functional integrity each order our priorities, when we look for signs of sustainability or unsustainability.

> This means that certain kinds of values will inevitably be served in adopting one approach or the other, and in defining the system boundaries for articulating a conception of functional integrity. ... It may be impossible to arrive at consensus on these value questions, but informed interdisciplinary research will be possible only when participants have a clear sense of where they stand with respect to one another.
> (Thompson, 1996: 92)

The views of the organic movement lean towards the more radical, systemic understanding of sustainability as functional integrity (as indicated by the principles of health, ecology and care in Box 3.1). However, from this perspective functional integrity may be seen as an extension of resource sufficiency – a more comprehensive perspective that can utilize the views and tools of resource sufficiency as far as their powers go, while putting them in a larger and deeper context.

In relation to ecological justice, functional integrity concerns the workings of the system as a whole, while ecological justice concerns the individuals in the system and their relation to the system. We will now turn to the meaning of ecological justice.

The ethics and justice of ecological justice

The roots of ecological justice are in the concept of environmental justice that arose from grass-root resistance movements in the United States in the 1980's – in particular the antitoxics movement, which focused on environmental health threats from waste dumps and pollution in local communities, and the movement against environmental racism, which focused on the disproportionate environmental risks to poor and coloured communities (Byrne et al., 2002b; Schlosberg, 2003). Environmental justice is mainly concerned with the fair distribution of environmental ills among human communities.

Since then, these concepts have been treated theoretically by several authors in relation to environmental politics, justice and ethics (Low and Gleeson, 1998; Baxter, 1999, 2005; Shrader-Frechette, 2002; Bell, 2003, 2004; Schlosberg, 2003). Low and Gleeson (1998: 2) coined the term 'ecological justice' which broadens the scope of environmental justice to include the justice of the relations between humans and the rest of the natural world and between present and future generations (see also Box 3.2).

> **Box 3.2.** Environmental justice and ecological justice
>
> '*Environmental* justice is about the fair distribution of good and bad environments to *humans*. *Ecological* justice is about fair distribution of environments among *all the inhabitants of the planet*. To speak of 'environmental' or 'ecological' justice means to recognise the values that an environment has for all creatures. An environment is comprised not only of people, but also nonhuman nature in all its abundance and diversity: animals and plants, landscapes and ecologies. An environment is not divisible like property but is fundamentally shared.'
> Low and Gleeson, 1998: [emphasis added]
>
> This quote illustrates the difference between environmental and ecological justice. Note, though, that the reference to *distribution* only is expanded below to include recognition and participation as well.

For many of the arguments in this chapter, it will make little difference whether we speak of environmental or ecological justice. Protecting disadvantaged people and protecting the natural environment are not at odds; they tie in with each other (e.g. Shrader-Frechette, 2002). However, the discourses of the organic movement seem to be more compatible with the broader scope of ecological justice than with the more narrowly anthropocentric concerns of environmental justice. Therefore, we use the term 'ecological justice' as the common designation for environmental and ecological justice in the present chapter and only distinguish the two where there is a need to do so. In distinction from social justice, which has generally focused on inequalities in relation to the labour market, income and wealth, the distribution of goods and burdens by society, and human rights, ecological justice concerns fairness with regard to shared environments.

This section investigates the meaning of ecological justice as an ethical principle with reference to environmental ethics and liberal ideas of justice and, in particular, what the justice of ecological justice means.

Ecological justice as an ethical concept

As ethical concepts, environmental and ecological justice are placed squarely across the fields of *environmental ethics*, which considers the extension of moral considerability beyond humans or persons (e.g. Goodpaster, 1979), and *liberal theories of justice*, which focus on fairness to persons within human societies (e.g. Rawls, 1971). The discursive force of the concepts of environmental and ecological justice therefore depends on whether they can be successfully grounded in these two well-established theoretical bodies. We cannot attempt to fully accomplish this theoretical grounding here, but for the purposes of the present chapter we will briefly consider some key issues of these two conceptual sources. (For a broader treatment of the philosophy of justice and environmental politics in relation to ecological justice, see Low and Gleeson, 1998).

With regard to environmental ethics, both environmental and ecological justice entail a systemic conception of ethics, where the moral concern for other individuals includes a concern for the parts of their environment that they depend on for their life and well-being (Alrøe and Kristensen, 2003). Though they have a common focus on justice and fairness in relation to the environment, 'environmental' and 'ecological' justice differ in the extension of the moral concern for fairness. 'Environmental' justice limits the moral concern to humans, whereas 'ecological' justice has a broader concern that entails moral concern for non-human nature (Low and Gleeson, 1998: 21, 133).

Ecological justice extends moral considerability to animals and other living organisms and to ecological communities and systems. From the perspective of environmental ethics, limiting moral considerability to humans is arguably a chauvinistic view (Singer 1979), whereas limiting moral concerns to persons is logically consistent, but morally unsatisfactory to most ('persons' in this ethical context designate self-conscious individuals that are thus capable of moral acting – so small children and mentally disabled people are not persons in this sense). The anthropocentric position is therefore not as unproblematic as its predominance might suggest. We will not, however, consider the issue of the proper extension of moral considerability further in this chapter (see instead Low and Gleeson, 1998: Chapter 6; Alrøe and Kristensen, 2003; Baxter, 2005: part 2), apart from two brief remarks.

First, in agreement with the broader moral scope of ecological justice, we note that this concept has implications for animal welfare as well. The fair distribution of environments to animal husbandry speaks to support the concern for the possibilities for expression of natural behaviour that characterizes organic agriculture (Alrøe et al., 2001; Lund et al., 2004).

Second, it is important to note that the extension of (equal) moral considerability to include animals, living organisms and ecosystems does not imply that these are as morally significant as humans or persons. Moral 'ecologism' does not necessarily imply environmental fascism as one might otherwise conclude from the well-known critiques of deep ecologists and ecological holists (e.g. Ferry, 1995). Justice in the Aristotelian sense means proportional treatment where like instances are treated alike and relevant differences are taken into account. Treating plants and pigs alike is unjust if sentience is morally relevant, and treating pigs and persons alike is unjust if self-awareness is morally relevant (Alrøe and Kristensen, 2003: 75).

> All life forms deserve certain rights to the fullness of their natural existence but a biospherical egalitarianism cannot be sustained logically or practically
> (Low and Gleeson, 1998: 157).

The second conceptual source of ecological justice besides environmental ethics is the influential liberal ideas of justice in the Kantian and non-utilitarian tradition of Rawls (1971). These are mainly concerned with the lives of individual humans and the issues of social justice. Hence, on a first look, it seems like ecological justice is incompatible with liberalism. But the two may be reconciled if liberalism can be extended like the extension within environmental ethics that has been described above. *Environmental* justice can without too much effort be understood as Rawlsian liberalism with a special concern for the environment and its implications for the opportunities and limitations of the individual (Low and Gleeson, 1998: 89; Bell, 2004). But this first, anthropocentric, path to an extension of liberalism is not sufficient in the present context, because it is not compatible with the more comprehensive views of organic agriculture. *Ecological* justice seems more difficult to reconcile with liberalism, because the moral extension beyond persons, which is the hallmark of a genuinely ecologic ethics, goes against the reciprocity of Rawls' political conception of justice as fairness, which is based on the idea of a cooperative democratic society of citizens acting as responsible persons. But Bell in fact challenges the incompatibility between ecologism and liberalism and argues 'there is nothing in Rawls' political liberalism to exclude the possibility of liberal ecologism' (2003: 2, see also Low and Gleeson, 1998: 84-90, 199-205; Baxter, 1999: chapter 8, 2005: chapter 7). Bell's arguments are: (a) that Rawls considers the further extension of justice as fairness to animals and the rest of nature (besides his extensions to future generations, international justice and health care) and leaves open the possibility of justice to nonhuman nature; and (b) that liberal ecologism must reject biospherical egalitarianism and be substantively biased towards humans (citizens), and that most ecologists do this.

The above considerations have clarified the distinction between environmental and ecological justice, decided the focus on ecological justice here, and argued the basic coherence of this concept. But what does the justice of ecological justice mean? A more detailed understanding of this will be helpful when putting the concept into practice.

The justice of ecological justice

In political theory in general, justice has been defined almost exclusively as equal distribution of social goods. Baxter (2005: 8) remains focused on distributive justice while extending it to non-human life forms, whereas Low and Gleeson (1998: 133) argue that ecological justice is different from environmental justice in that we here have to consider our moral relationship with the non-human world in a deeper sense. As we shall see, this deeper moral understanding of justice is pertinent for both human and non-human relations.

In an analysis of the justice of environmental justice, Schlosberg (2003) describes three conceptions of justice in form of equitable distribution, equal recognition and participative procedures. Within environmental justice, *the distributive notion of justice* focuses not on wealth or money, but on the distribution of environmental qualities, be they 'bads' in form of risks and costs or 'goods' in form of access and opportunities in relation to environments (Low and Gleeson, 1998: chapter 5). *Justice as recognition* concerns equal rights and ownership to environments and the recognition of connections between community and place. Injustice is here based on a lack of recognition of identity or equalness or a lack of recognition of difference, uniqueness and heterogeneity of views, values and interests. *Justice as participation* entails that communities and persons 'have a say' in environmental matters that concern them, and that there are democratic procedures for participation and representation in relation to ecological injustices and decisions on environmental matters. Related notions are citizen sovereignty and food sovereignty. Schlosberg emphasizes that these three conceptions of justice are not competing, contradictory, or antithetical. Environmental justice requires more than an understanding of unjust distribution and lack of recognition; it requires an understanding of the way the two are tied together in political and social processes. 'The combination of misrecognition and a lack of participation creates a situation of inequity in the distribution of environmental dangers' (Schlosberg, 2003: 98).

The three-fold understanding of justice can be applied as well to ecological justice. But from an ethical point of view, there are a number of important points to make. The first point is that moral responsibility is constrained to self-aware beings. Within environmental ethics it is common to distinguish between moral agents, who are capable of acting morally and taking on moral responsibility, and moral objects, which are taken into moral consideration by others. Animals and ecosystems can not be moral agents, while persons, organizations, companies and states can. The capacity for moral responsibility works both ways. This means that the capacity of moral agents to take responsibility for their actions should be respected by involving them in democratic participatory decision-making processes, either directly or by way of representation (Bell, 2003). And it also means that demands can be made on them to act morally responsible in accordance with their capacities for doing so. Furthermore, powerful, knowledgeable agents must take on larger responsibilities than those without much power and knowledge, because the moral responsibility for ones actions relates to action ability as well as to awareness (Alrøe and Kristensen, 2003). That moral responsibility is a correlate of power seems crucial in questions of ecological injustice where large differences in power and action abilities are common. In relation to Schlosberg's analysis of justice, it is clear that the three conceptions of justice relate differently to the distinction between moral agents and objects. The participatory processes of justice are only open to moral agents, while distribution and recognition concerns all moral objects (some of who may be moral agents as well).

The second important point is that there is more to ecological justice than the distribution of known risks and options. First of all, ecologism entails that morally considerable entities should have the opportunity to exist, flourish and develop in accordance with their natures (e.g. Baxter, 1999: 95). The recognition of such identities (equally considerable rights to freedom) and differences (in accordance with their natures) is quite different from distributional ideas of justice. Second, the diversity of morally considerable entities makes rational distributional policies quite unmanageable. More generally, the application area of ecological justice – global markets, global social and ecological systems, immense heterogeneity of moral objects and moral agents – indicates that there is an obvious need to be able to address ignorance, unknown consequences and unknown impacts.

In ethical terms, there is a need for new moral reasons beyond the intentions, virtues and duties in non-consequentialist ethics and the rational calculations of consequentialist ethics. Alrøe and Kristensen (2003) suggest that this new moral ground must be based on self-reflexivity, and refer to the precautionary principle as a well-known example of this development of ethics. Ecological justice as recognition must involve similar reflexive attitudes towards the limitations of knowledge and rationality.

Summing up

Summing up, ecological justice implies a necessary bias in relation to moral responsibility and participation in decision–making processes on environmental matters, since only self-reflexive beings (such as persons and some kinds of social systems) can be moral agents; an arguable absence of bias with regard to moral considerability, extending justice and fairness to animals and other living beings and systems; but also an arguable bias in moral significance based on persons, animals and plants being different kinds of entities with different types of capacities and senses, and which should therefore be treated differently. Moreover, fairness with regard to shared environments is not just about distribution of environmental goods and bads; more fundamentally it concerns recognition and participation based on a universal right to freedom and with an eye for ignorance and uncertainties in decision–making processes.

Ecological justice is not an entirely new and different response to the problematic trends of global development – it has much in common with other concepts and reactions such as sustainable development, functional integrity and social justice. But it does have its own very specific angle, which defines the problematic in an importantly different way.

Challenges for organic agriculture: commodification, externalities and distant trade

Ecological justice is a challenging concept in relation to the current globalization and structural and technological development of organic agriculture, because it positions social and ecological interests against market liberalism and economic growth. Therefore, it may suggest new ways to look at the challenges and promises connected to the global development of organic agriculture.

In particular, ecological justice can be applied to three, related, aspects of the current trends: the commodification of hitherto commons, the externalization of environmental and social costs, and the growing distances of trade and ownership due to globalization.

- *Commodification* is the transformation of non-commercial relationships into relationships of buying and selling, based on the concept of private property. Commodification of commons brings common goods, such as land or water, into the market by way of enclosure and exclusion of others from the benefit stream.
- The term *externalities* is an economic term that refers to production costs that are not paid within the market. There is an empirical aspect of this, concerning what the costs connected to the production actually are and how they might be reduced, and a normative aspect, concerning whether the costs are to be reduced or internalized by compensations.
- The *growing distances*, inherent in the globalization processes, between those who pay the costs and those who enjoy the benefits aggravate the problems of both commodification and externalities. Distance can create problems of transparency and democratic participation in relation to ownership and trade (though globalization can also benefit transparency) and problems of externalities connected to transport.

Somewhat caricatured, market economics can be characterized as the art of externalizing social and ecological costs, and private property as a way of internalizing social and ecological benefits. Ethics, in contrast, based on the principle of responsibility (Jonas, 1984), can be seen as the art of internalizing social and ecological costs, and ecological justice as a way of externalizing social and ecological benefits. In this section we will discuss the three challenges of commodification, externalities and growing distances in relation to organic agriculture.

Commodification of commons

As argued in Chapter 2, the concept of ecological justice is closely connected to the general idea of 'commons'. The language of commons brings us to focus on the question of what aspects of the environment are or should be shared and in what respects, and what that means for ecological justice. The scope of ecological justice then depends on what rights or claims individuals and communities have or should have on these aspects of their environments. Commodification of commons can lead to unsustainable exploitation (e.g. in form of ranching, logging, mining) and ecological injustice by undermining sustainable commons systems and community governance and negatively influencing the life opportunities of those that hitherto used the commons. Issues of externalities and distant trade may add to problems of commodification, but they may also be problematic in themselves.

The idea of commons is traditionally found in relation to common lands where the use by local people for grazing or gathering is managed according to traditional rights and rules, and debates on the commodification of these common lands by way of enclosure and private property. But it is now used in a broader sense to include forests, freshwater supplies, inshore fishing grounds, etc. (The Ecologist, 1993). There are furthermore explorations of a new category of international initiatives known as 'global commons', which concern such problems as ozone-depleting emissions, climate change, biodiversity protection, international toxic waste trade, international endangered species trade, and the use of the high seas and the polar regions (Volger, 1995; Buck, 1998; Byrne and Glover, 2002). In relation to organic agriculture, soil as a production resource may also be considered a global commons (see Chapter 2).

Organic agriculture is more dependent on the environment than conventional agriculture, because the production is based on close cooperation with natural ecological systems and processes, it has fewer technological remedies available to counteract depletions and malfunctions of these systems, and there is a special focus on maintaining the local resources for production such as soil fertility. What we may call 'ecological commons' therefore have a special importance in organic agriculture. Nature plays a key role in the provision and reproduction of ecological commons whereas public goods (or public commons), such as roads, libraries and systems of justice, are produced by human actors. This distinction is important to keep in mind in relation to organic agriculture because the provisions by nature tend to be overlooked in policy analyses directed towards the challenges of globalization (e.g. Kaul *et al.*, 2003).

The question of whether something is to be considered as a commons (and thereby whether its commodification is problematic) is determined by ethical and political criteria, not by empirical criteria such as the ones found in economic textbooks: whether the benefits from the resource are excludable (whether they can be withheld from others, e.g. through the enclosure of land and water supplies) or rival (whether they are depleted when used). Technological and structural developments keep shifting the ground for such empirical criteria, and technically and economically excludable resources may well be considered commons from the ethical perspective of ecological justice.

The concepts of commons and ecological justice can be put into practice in different ways that institutionalize the fair usage of common environments. Examples are sustainable production methods, local community institutions of co-management and cooperative food networks; certification and labels that involve the consumers as a responsible actor; state or supra-state regulations of the market and environmental impacts; and global institutions under the mantle of the United Nations. Organic agriculture has little direct influence on the latter, but it can play a key role in the first. Further below, we look in more detail at how certified and non-certified organic agriculture, respectively, may promote ecological justice.

How to address externalities

Externalities are costs and benefits connected to the processes of production, processing and distribution, which are not accounted for and which do not enter into market transactions. With regard to ecological justice, externalities can appear in form of localized impacts on the living and working environment from production activities, in form of deliberate localization of environmental bads (placing of waste dumps and harmful industries, export of waste, etc.) to the disadvantage of local communities near such places, or in form of globalization of environmental bads (climate changes, ozone depletion, pollution of the global environment with heavy metals and other persistent harmful substances, etc.) to the disadvantage of those communities that are most vulnerable to such global changes.

Commodification of food systems is a frequent source of externalities. External costs in agriculture, such as biodiversity impacts and pollution, often stem from agricultural practices shaped by the economics of surplus production. That is, modes of agricultural production that require large and continually growing surpluses for sale in markets as the basis for profitable operation can find externalization of costs a key ingredient for successful development.

Today, we can find examples and suggestions of different ways to address externalities, which directly regulate the sources of externalities within the production or valorize external costs and assign their incurrence in a manner that discourages harmful practices and ecological injustices: governments and supranational institutions like the EU enforce environmental laws that regulate agricultural productions to avoid or reduce externalities. Global institutions and international agreement like the Kyoto Protocol to the UN Framework Convention on Climate Change may lead to changes in production that reduce externalities on a global scale. Low and Gleeson (1998: 199ff) propose new global institutions under the mantle of the United Nations, the World Environment Council and World Environment Court, to instigate ecological justice. Developments within the WTO also harbour possibilities for regulating global trade in a more sustainable direction (see Chapter 5). Brown (2002) lists four other forms of market regulation for sustainable development: eco-labelling, tax shifting, subsidy shifting and tradable permits. The three latter are economic means to internalize the external costs in the market, which can to some degree reduce externalities, and which may also create a revenue that can, in principle, be used to pay compensations for external costs.

The certification of environmentally friendly agricultural production and processing, which is a form of eco-labelling in Brown's sense, may be seen as a way to realize ecological justice within a distant, non-localized food system that works across national and regional borders based on certification standards that describe the rules for how to use environmental commons. Such alternative ways of production based on certification are immediately realizable by pioneer groups – they do not (at least in principle) depend on national or international regulations. Ideally, they are competitive within the mainstream market system due to consumer preferences for socially and environmentally friendly food products. Organic agriculture is a prominent example of eco-labelling, though it remains to be clarified how the current certification standards fare with regard to ecological justice and how they might be improved (see below). However, the competitiveness of such an alternative may be hampered by subsidy structures, and if the alternative is not supported by societal actions the responsibility for the commons is placed solely in the hands of the individual consumer and their daily consumer choices. Such non-localized institutions for ecological justice, which work only by way of certification and consumer preferences, will therefore have a hard time growing to be a dominant influence on global commons.

The three positions on globalization and sustainable development that we described above (see also Chapter 2), show different approaches to address externalities. Environmental economics focuses on how to internalize external costs, ecological economics focuses on identifying overall ecological limits to economic growth and the associated externalities, and political ecology focus on ecological justice and the way externalities inflict on different communities. The above examples of ways to limit, avoid and compensate externalities will fall out differently if they are analysed in relation to these distinctions, and they will have different potentials for taking on the different approaches. This is not the place to perform such a general analysis (some aspects are addressed in The Ecologist, 1993: 117–121). But there is an aspect that seems important to, at least briefly, point out: the limitations of knowledge and the associated impotence of compensation.

We typically only have limited knowledge of externalities and limited means of identifying them. This goes for the nature of the external consequences and impacts as well as for who suffers from the impact, and thereby who are to be compensated for what impacts. This is even worse when non-human beings are taken into consideration, and it is aggravated in cases of long term, systemic effects. Furthermore, there are no adequate means of compensation for severe and irreversible impacts. This is what has motivated the inclusion of a precautionary principle in environmental regulations, a principle that formulates deliberate strategies for handling ignorance and uncertainty (O'Riordan and Cameron, 1994; Raffensperger and Tickner, 1999). The precautionary principle requires preventive actions before conclusive scientific evidence of severe and irreversible externalities has been established (e.g. by saying no to unpredictable technological activities), and in addition it supports the development of society's capacity for early detection of dangers through comprehensive research and the promotion of cleaner technologies.

Growing distances

Trade is an inherent aspect of commodification, but the concept of distant trade brings up two important issues with regard to ecological justice: transport and transparency. The transport issue is pretty straight forward, since the physical exchange of the commodity and the money exchange are understood as essential processes in a market system, while the externalities connected to different transport means, the options for limiting them by more local trade and the resulting consequences for ecological justice are mostly not a factor in market transactions. The issue of transparency is somewhat more intricate, since there are very different motivations and interests (profit, branding, market domination, public health, competition and choice, transparency, consumer needs and preferences) involved in the communication of knowledge about products in market systems – and growing distances can help reduce transparency for those who wish so – but on the other hand globalization also entails better possibilities for transparency due to new communication technologies. Both transport and transparency influence the options for democratic participation in decisions on issues of commodification and externalities from the production.

So, the growing distances in food systems, which were illustrated in Chapter 1, aggravate the problems of commodification and externalities that were discussed above. Furthermore, the idea of the local has played, and still plays, a characteristic role in the organic movement due to the emphasis on working in closed systems and drawing on local resources (e.g. Woodward *et al.*, 1996: 262). We will therefore take a closer look at nearness and distance in food systems.

The pioneer farming initiatives that eventually led to modern organic farming, were mainly localized systems that focused on the living soil and its importance for agricultural production. Localized agricultural systems are characterized by close relations where owners, workers and consumers share a local environment with other local residents. In Figure 3.2 the monetary and non-monetary exchanges of a localized food system are illustrated as flows of value (commodity and money exchanges, external costs and benefits) between the production and local stakeholders. In very localized systems that mainly make for community self-sufficiency, the owners, workers, consumers and residents may even be more or less the same persons. The value flows are often not well known and difficult to identify (see also Chapter 4).

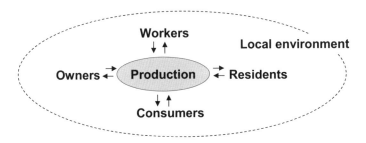

Figure 3.2. A localized food system. Arrows denote flows of value (commodity and money exchanges, external costs and benefits). Owners, workers, consumers and residents may be the same persons.

In terms of nearness and distance there are a number of steps from the very localized system towards a globalized system. Commodification and market exchange is a first step, though the markets may be very localized, because nearness is to be understood not only as physical distance, but also in terms of knowledge, communication and awareness. And market exchange in itself does introduce a distance in this regard, since the market provides strong incentives not to disclose disfavourable information about external social and ecological costs and to manufacture fictional good stories about the products instead. This is often of little consequence in a very localized market system, because people are well aware of how the local productions take place: they can see and experience the productions directly and they have other available channels of communication than the market; and they belong to the same local ecological community (and in some sense localized ecological unit, such as a watershed). Examples of such local food systems still abound in less-industrialized countries, whereas in highly industrialized countries they are found almost only as counter-reactions to the mainstream food systems, such as 'community supported agriculture' systems (e.g. Cone and Myhre, 2001) and the 'food-shed' and 'eco-localism' movement (see Chapter 1).

The erosion of barriers to distant trade and ownership, inherent in globalization, leads to increasingly non-localized systems characterized by distant relations and value flows (Figure 3.3). In global trade, the consumers are physically very far away from the production; in vertically integrated corporate businesses the owners are often far away and intermediary products are transported between different production facilities across the globe; and modern agriculture influences the global environment as well as local living and working environments due to the development of technologies and the increase in inputs of (e.g. fossil fuel) and production levels.

As such, long distance trading is nothing new; it has existed for several thousand years (e.g. spices). But the level of long distance agricultural trading is rising, as a key aspect of globalization. (The average distance of trade, and thus the *share* of global trading in relation to regional trade, seems not to be growing, however, see e.g. Davidson and Agudelo, 2004). (See Chapter 1 for an overview of the actual development of global trading in agriculture and organic agriculture). Conventional agriculture is inseparable from the global economy. There are a multitude of ways in which conventional agricultural practices and outputs are shaped by factors related to globalization, such as technology, markets, international transport, and the activities of multinational corporations. Central to these influences is the role of international trade as an agent that promotes commodification of social and environmental values, resources and services.

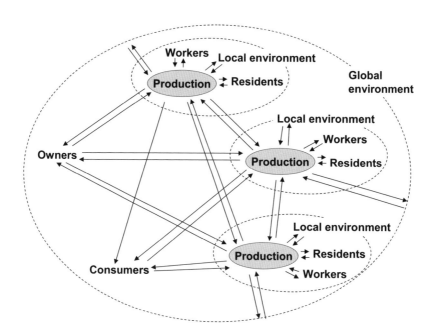

Figure 3.3. A non-localized food system with distant flows of value. Arrows denote commodity and money exchanges as well as external costs and benefits to local communities and environments and to the global environment.

Distant trade has the effect of obscuring or effectively eliminating the connections between production and consumption and thereby hampering transparency and the options for democratic influence on how the production takes place. Where production and consumption are closely linked, the costs and impacts of production are part of the awareness of most consumers, and the effects of local social values and regulations influence consumption. Similarly, in democratic countries with well-developed institutions, there is a good chance that the consequences connected to home production will come to the attention of consumers, citizens and authorities, so that they can take action in terms of market choices or societal regulations. But when products are sold at a great distance from the site of production, the social and environmental costs of production are less likely to be known and less likely to influence market choices.

The same mechanisms will work in alternative trade movements, such as certified organic agriculture, unless something in the certification standards prevents them from doing so.

Putting ecological justice into organic practice

There are several different ways in which the organic movement can implement the idea of ecological justice in relation to organic production and organic trade. This section discusses how the previous considerations can be put into practice and provides a background for a closer investigation of these issues. Three main ways are described, the ways of certified and non-certified organic agriculture (see definition in Box 3.3) and the way of organic agriculture as an alternative example for agriculture, research and society.

Box 3.3. Certified and 'non-certified' organic farming.

When assessing the potential benefits and problems of the global development of organic agriculture in relation to the principle of ecological justice, there is a need to distinguish between certified and non-certified organic farming (in line with Scialabba and Hattam, 2002). Certified organic productions compete with conventional products in regional and global markets, even though the organic production levels are usually lower than conventional, based on consumer preferences and premium prices. 'Non-certified organic farming', on the other hand, is a term for farming systems that are based on principles and practices similar to 'branded' organic agriculture, but which are targeted at local consumption based on close relations and not at the distant sale of certified products.

The way of certified organic agriculture

As discussed briefly in the previous section, certified organic agriculture is an example of an 'eco-labelling' type of market regulation for sustainable development (Brown, 2002). This solution incorporates the production process context into the market by way of elaborated certification procedures and extensive standards of production and processing that provide the foundation for an alternative way of trade, located (to large degree) within the ordinary market structures. Hence, the existing organic system already shows some promises with regard to the implementation of ecological justice in so far as the certification standards do indeed promote ecological justice by working against commodification and externalization of costs.

Organic certification

As argued above, certified organic production and trade can be seen as an example of a type of institution that may secure ecological justice across a distant, non-localized food system. The identification of organic produce by way of certification and labelling allows for the formation of alternatives on the market for farm produce and for consumers to express their preferences.

There are, however, a number of important challenges with regard to the implementation of ecological justice through certified organic agriculture. First of all the organic trade needs recognition within the World Trade Organization that organics products are different from similar conventional products (see Chapter 5). The idea that conventional and organic products of the same kind are 'like products' in the WTO sense, and therefore cannot be discriminated in the free market, runs counter to the recognition of the importance of the processes behind the products, which are of central concern in relation to social and environmental impacts and ecological justice.

Global markets are characterized by the strong role played by large, multinational corporations in transport, handling, distribution, marketing and sales. When the organic products compete on the market with conventional products they experience the economic conditions and pressures that are typical of a free market system, and which have shaped conventional agriculture and made sustainable practices unattractive. In particular, there are pressures to conceal information on the production process and possible ecological injustices and to manufacture attractive stories with no real background; pressures that will work against the goals of transparency and authenticity inherent in certified trade. These mechanisms are supported by the international goals of free competition of products without concern for the production process. There is a real threat that these pressures will erode the principles of organic farming. A further concern is that global markets are uncertain and often volatile, which has the effect of reducing the security of farming enterprises, and which adds to the economic incentives for larger-scale enterprises to replace small-scale farms.

To resolve these challenges the organic movement must create and maintain a system that reflects both community and farming interests and values and democratic objectives. And for the non-localized system to function across distant markets, ecological justice goals must be implemented clearly in the organic certification standards.

The implementation of ecological justice in the organic certification standards must consider the issues that have been outlined in the previous sections. Possible injustices connected to the commodification of commons, such as the use of land for large-scale (organic) cash crops instead of local sustenance and nature areas (in terms, also, of justice to wildlife and biodiversity), have so far not been implemented in the standards, though the key goal of maintaining soil fertility certainly implements an aspect of justice to future generations. Certain types of production systems that make extensive use of large, non-enclosed areas and natural resources, such as pastoral systems, fisheries and wild harvesting, entail special issues in this regard. Environmental externalities connected to agricultural production have to a large extent been covered in the existing standards to the benefits of local and, to some degree, more distant communities and to biodiversity in general. This has been a key concern in modern organic agriculture in line with the understanding of sustainability as functional integrity, even though the movement until now has had to compromise its goals on some areas such as the use of fossil fuel. The principles and standards on agricultural production also implement aspects of ecological justice to local residents, livestock, and biodiversity and wildlife in the area. Environmental externalities connected to processing and, in particular, distribution are less well covered in the standards. A key question in relation to the globalization of agriculture, the (distant) transport of organic certified products and intermediary products and the externalities connected to this, is not covered at all (see further below).

However, it is not possible to guide the development of organic agriculture solely by way of standards due its heterogeneous and dynamic nature (Alrøe and Kristensen, 2004). The ideals of ecological justice therefore also need to be adapted explicitly into the principles of the organic movement (as it has already been proposed, see Box 3.1). First of all to guide the development of certification standards in the directions outlined above, but also to enable regionalization of standards in consideration of the need of organic productions to adapt to different local climatic, edaphic, and cultural conditions and to serve directly as a guide for organic practices where standards are hard to define. In the same way, principles that clearly express the ideals of ecological justice can guide the practices of non-certified organic agriculture.

Organic and fair trade

The implementation of ecological justice in organic standards also includes the question of commodification of commons – which is of course a real challenge for a market-based system – and more generally justice considered as distribution, recognition and participation with regard to shared environments (as discussed above). There is a widespread recognition of the claims of husbandry to a fitting and more natural environment within organic agriculture, and some recognition of the claims to life and space for other living beings, formulated as requirements on biodiversity, for instance. For humans, social considerations are to a certain degree covered in some organic standards, such as IFOAM's standards, but not in all. Some inspiration for the implementation of these issues can be gained from fair trade, as indicated by the following, strong statement:

> Most Latin American organic bananas are grown on plantations. For example, Dole Food Corporation – which controls 25 percent of the conventional banana trade and a significant share of the US organic sector – has in recent years become a major organic banana supplier. Some Dole banana plantations might be able to pass IFOAM's relatively weak social standards; outside of IFOAM they can be certified as organic irrespective of even gross labor violations. Without the strict social standards and restrictions on eligible producers found in fair trade, organic production clearly risks being transformed from a form of alternative agriculture to a segment of the traditional corporate dominated global agro-export trade.
>
> (Raynolds, 2000: 303)

Fair trade certification is a second well-known example of an alternative form of trade that has the potential to work across globalized food networks in distant trade relations, and which goes some way towards meeting the principle of ecological justice (see also Chapter 2). But both organic and fair trade fall short of the target in some respects. Fair trade goes further in specifying the social conditions and costs of production, but is lacking in ecological considerations. Organic trade, on the other hand, goes further in detailing the ecological conditions and costs of production, but is lacking in social considerations (e.g. Raynolds, 2000). We may therefore think that organic and fair trade movements can simply combine forces to meet their ecological and social ideals and that their standards can complement each other to fulfil the promises of ecological justice. However, both standards omit, for instance, considerations on distant transport. Furthermore, fair trade has focused more on the traditional aspects of social justice and it does therefore not provide all those complementary social aspects with regard to ecological justice that focus on fairness with regard to shared environments. Some of the ideas within fair trade will presumably be elements in a future 'fair organic trade' certification that can more fully promote ecological justice. Major challenges are to secure ecological justice to those outside the trade network as well as those within, and to resolve the potential conflict between the benefits of fair global trade to low-income areas and the inherent disadvantages of distant trading.

Heterogeneity and transport

More generally, two main challenges to implementing ecological justice in organic standards are heterogeneity in the natural and cultural conditions for agriculture and (long-distance) transport. In a global perspective, the conditions for organic production and processing are extremely varied as the present book illustrates. In the pursuit of fairness and to support the local development of organic practices, there is a need to elaborate different rules for different regions on the basis of the common values and ethical principles of organic agriculture. Since regional rules might be misused to unfairly diminish the demands on organic production, there is a need for investigations of what regional differences in natural and cultural conditions can fairly require regional differences in the organic rules.

If and when the values of organic agriculture, including those of ecological justice, are fully implemented in localized production and processing practices, the only remaining issue is that of the long-distance transport of the products. One and the same product (such as wheat, soya, etc.) can be sold locally or in another part of the world, and there it may compete with a quite similar organic product that is produced locally. In other words, the local product can be substituted with imported products from far away. The challenge is how to handle this. If the total external environmental and social costs connected to the transport could be estimated, then these could be internalized and added to the price. If, as one might suspect, this approach is not feasible, rules of 'substitutability' – whether a similar product can, and should, be produced and traded more locally – could be enforced to promote a principle of localism in such cases. A less rigid, but probably also less efficient, solution could be to leave the choice to the consumer by requiring that importers put information on the origin(s) of the product on the product label. Similar rules could be implemented for information on the origin(s) of feed and other inputs to the organic production. But, as can be imagined, this can quickly become overwhelmingly complex in a non-localized food system. Common to the latter solutions (those that do not include an estimate of the external costs) are that they would treat different means of transport the same, which seems unfair. And adding information on means of transportation would add to the complexity to be communicated. In this respect, transparency and communicational barriers are important aspects. In all cases it would be necessary to balance the issues of transport with the fair access to global markets for farmers in low-income countries.

In general, the different forms of alternative, certified trade put the responsibility for ecological justice on the consumer (the so-called 'political consumer' or 'ethical consumer'). This is good in the sense that it enables any consumer to participate in decisions that concern commons and ecological justice in relation to agriculture and food. But the question remains to what degree the consumers can bear such responsibility in a situation of cheap conventional goods that are subsidized and do not carry their own environmental and social costs, and under the economic constraints of everyday purchases.

The way of non-certified organic agriculture

In large parts of the low-income countries food production is based on localized systems with low-yielding agriculture, subsistence farming, and local food markets. Here 'non-certified organic agriculture', which accords with the ideas and principles of organic agriculture without being certified, has the potential to give higher and more stable yields than the existing agriculture, based only on local natural resources and inputs of knowledge and extension services. Non-certified organic farming may therefore be promoted as an alternative solution to food security problems that is more ecologically just.

Not all traditional farming systems that do not use artificial pesticides and fertilizers are 'non-certified organic' by default, because they may very well be unsustainable due to for example soil degradation. On the other hand, non-certified organic food systems may be more in line with the organic values and principles than certified systems, because the latter face direct pressures of market competition and globalization that threaten to move organic food systems towards conventional systems, or in ways that are similar to conventional systems, and away from its original values and principles (e.g. Rigby and Bown, 2003).

As documented elsewhere in this book (Chapter 6), agricultural approaches that are based on the values and principles of organic farming, or more generally on the ideas of sustainable low external input agriculture (LEISA), but which are not certified organic, remain valid alternatives to high-input, commercialized, 'green revolution' type developments with respect to food security and sustainability. The low input alternatives show more promise in terms of ecological justice than high-input solutions, since the latter carry new risks of new external environmental costs and new, unfavourable dependencies on sources of finance and large, multinational agricultural corporations (e.g. Scialabba and Hattam, 2002: Chapter 4). And they may even be more in congruence with the organic principles than certified organic agriculture, because they are not in the same way subject to the pressures of globalization (see Box 3.4).

There is therefore a separate line of development open to organic agriculture with the promise of promoting ecological justice, the development of 'ecological communities' in the form of non-certified, community-based organic agriculture. There are a number of possible localized food system models, such as self-sufficient family or community farms, local community networks, local markets, and local participatory guarantee systems (as described by Alcântara and Alcântara, 2004). The realization of this promise in form of a variety of local practices requires support for the development of participatory research and extension services that incorporate the goals of sustainability and ecological justice. The involvement of the global organic movement is needed to guide such a development. But the main challenge will be to gain understanding and support for this development strategy within development organizations and connected research institutions for the value of sustainable low external input agriculture, such as organic agriculture. Where the path of non-certified organic agriculture is chosen instead of high external input options, there is, in addition, an option for later entering into the organic market by certifying some of the organic production practices.

Box 3.4. Organics and vulnerability: the case of Uganda.
Michael Hauser*

The concept of organic agriculture receives particular attention in low-income countries where it is hoped to sustainably improve poor people's livelihoods. Given the increasingly globalizing nature of organic agriculture and organic businesses, a core question is to what extent these developments impact on poor people's vulnerability. Uganda, one of the sub-Saharan African countries with the most rapidly expanding organic sectors, is used to illustrate this. Given the risky environment poor people live in, this case outline explores the linkages between different organic strategies (non-certified and certified organics) and their outcomes in terms of vulnerability.

Organic agriculture in Uganda
Despite recent economic growth and an average rise of per capita income of about 6% per annum, Uganda is among the poorest countries in the world. Between 80 and 90% of the population live in rural areas and seek their livelihood in agriculture. Uganda is home to 3 million farm households with an average land size of 2 ha.

Non-certified, but IFOAM compliant organic agriculture has its roots in sustainable agriculture. Its formal promotion started in the year 1987, after years of political unrest. Certified organic agriculture is rooted in private sector initiatives. It is important to note that non-certified and certified organic initiatives have distinct characteristics (Table 1).

---

* BOKU – University of Natural Resources and Applied Life Sciences, Vienna, Dept. of Sustainable Agricultural Systems, Division of Organic Farming, Austria, E-mail: Michael.Hauser@boku.ac.at

This case outline is based on a paper by Michael Hauser: What 'rich organics' might mean for 'poor organics' – research and trade, presented at the international workshop 'Organic farming in a global perspective – globalisation, sustainable development and ecological justice', 22–23 April 2004, Palace Hotel, Copenhagen.

In 1995 about 40,000 farms were certified and inspected by internationally accredited certifiers and grow organic food and fibre for export. Arabica coffee, cotton, sesame, fresh fruits and dry fruits are important cash crops. By the end of 2003, 16 projects supported certified organic agricultural production. Estimating the number of non-certified, but IFOAM compliant organic farms is difficult, but probably exceeds three to four times the one for certified organic farm households.

Many of Uganda's organic advocates are organized in the National Organic Agriculture Movement of Uganda (NOGAMU). As a membership organization, NOGAMU supports information sharing, lobbying and advocacy as well as market development. In some ways, NOGAMU also protects the integrity of organic principles and values. UGOCERT, a local organic certification and inspection company, has been registered and has published national organic standards. Once accredited, this will greatly reduce certification and inspection cost in Uganda. As a result of NOGAMU's lobbying and advocacy work, pro-organic government policies are underway.

Table 1. Comparing (extreme ends of) non-certified and certified organic agriculture.

	Non-certified organics	Certified organics
Principal thrust and orientation	Increasing well-being through sustainable natural resource and community development	Increasing income through commodity development and niche marketing
Legal basis	Neither certified, nor inspected, no contracts between 'buyers' of organics and farmers	Certified and inspected, contracts between 'owner' of the certificate and farmers
Technology change	Complex and comprehensive change in farm management, including the introduction of new agronomic practices for ensuring ecological sustainability	Complex and comprehensive change in organizational and logistical arrangements, introduction of new agronomic practices for ensuring product quality
Type of extension services	Participatory extension and technology adaptation by farmers	TandV like extension and technology adoption/rejection by farmers
Potential for scaling-out	Spontaneous diffusion of technologies and concepts possible, socio-economic barriers	Some diffusion of technologies, but not of the entire concept, market access regulates participation
Public and policy perception	Low-input system for increasing food security	Compliance with liberal agendas and plans for the modernization of agriculture
Research support	Indirectly through research on sustainable agriculture	No research devoted to certified organics
Incentives and likely benefits	Non-monetary incentives, little re-investable income, direct livelihood benefits (through deliberate interventions)	Monetary incentives, re-investable income and thus indirect livelihood benefits

Organics, assets, and vulnerability
Given the expanding organic sector and the enormous 'hype' about organics especially in Uganda, the following observations can be made with respect to the likely vulnerability-reducing effects of organic agriculture. These effects are different for certified and non-certified organics.

Both certified and non-certified farm households are exposed to similar or the same threats (trends and shocks), such as increasing population pressure, natural resource degradation, pest and disease outbreaks, health threats, market dynamics or political instability. At the same time, there are threats (trends and shocks) that are only relevant for one of the two groups.

Threats that are specific to non-certified organics include falling farm gate prices for cash crops, late payment of the buyers or information cut-offs. Threats that are more specific to certified organics include unforeseeable market breakdown (due to new fraud cases) as well as resource degradation(!). The latter is a risk where organic agriculture becomes 'conventionalized' (i.e. narrowest possible interpretation of organic standards).

Through organic initiatives (and their interventions) farmers are able to build assets that help to buffer non-specific threats. However, certified organic initiatives tend to build monetary buffers (through the organic premiums) and non-certified initiatives tend to build non-monetary buffers (due to triggered community development processes).

From an agro-ecological point of view, susceptibility (i.e. defencelessness of the system) is lower and resilience (i.e. the ability of the system to return to its initial state) is higher in some of the non-certified initiatives. There, risk management is more one of 'ex ante' (i.e. a kind of precautionary principle). Non-certified organic initiatives may have a higher ability for risk management that is 'ex post'. Measured in terms of 'functional integrity', the extreme end of non-certified organics may come off better than the narrowly interpreted end of certified organics. It is not clear if higher income (through premiums) provides sufficient input into farmers' livelihood systems to 'purchase' assets that have a buffering effect.

Following the planned start of UGOCERT and the growing domestic markets for organic produce, the following may happen. First, increasing number of organic producers. Second, falling prices for organic products (especially falling organic premiums). It is important to note that organic standards do not encompass compulsory premiums (as it is in fair trade). Rising demand for organic food in the north may change the configuration of farming and livelihood systems in the south. Overdependence on single commodities and experts, unresolved conflicts over trade-offs (more of the one may mean less of the other), falling commodity prices when organics are being mainstreamed are all potential threats.

Understanding livelihood dynamics before and after 'conversion' to non-certified or certified organic strategies is essential to fully assess the benefits of organics to sustainable livelihoods and thus sustainable development. Our findings in Uganda indicate that organics can reduce vulnerability, but it also exposes farmers to new vulnerabilities.

Challenges
The growing organic movement faces all sorts of challenges. One of the prime issues is to endogenously develop and build an organic identity that is distinct from those

overseas. Some of these can and must be addressed in alliance with partners around the world. The following list is an overview:

- Strengthening human and social capital – to reduce the overdependence on external ('international') experts and expertise.
- Improving organic technologies – to reduce bio-physical risks and increase the buffering capacity in connection with natural shocks and trends (such as weather or climate change).
- Localizing food systems – to decrease the overdependence on export markets overseas through local and domestic organic markets.
- Maximizing benefits from organic trade – to keep the value added in-country and make it available for re-investments into the sector.
- Developing pro-organic research systems – to actively respond to 'burning' issues in the area of production, processing, transport as well as broader livelihoods aspects.
- Creating enabling policy environments. Examples include explicit organic policies and organic standards that are relevant to the local agro-ecologies of low-income countries.
- Ensuring ownership – to develop to reduce the dependency on external players, rebalance power relations and stakes.

Conclusion
- In Uganda and elsewhere, certified organics receives most of the attention. However, appraised in connection with vulnerability reducing aspects of organic agriculture this is not always justified. There is some scope for learning from both approaches.
- It is undisputed that the Ugandan organic sector has benefited from globalizing markets. Without the pull effects of growing organic markets in developed countries, certified organic production systems would be inexistent in developing countries.
- There is danger of 'conventionalizing' certified organics that may lead to a loss in 'functional integrity' of these organic systems.

Organic agriculture as an alternative example for agriculture, research and society

As indicated in previous sections, there are other, more political ways to promote ecological justice in agriculture. The organic movement may seek to influence governmental regulations of markets and the development of supranational institutions to consider the issues of ecological limits and ecological justice, but these developments depend on the general political understanding and motivation in different societies. Organic agriculture cannot decide the implementation of ecological justice at the national and supranational level. But if and when organic agriculture has more fully incorporated the principles of ecological justice, these efforts may serve as an alternative example for mainstream agriculture, for research, and for the broader implementation of ecological justice in other areas of society. This has been expressed in a distinctive way by Laura Raynolds:

> The fact that the international organic and fair trade movements have successfully created new niche markets for alternative products is no small feat. Yet I suggest that their true significance lies not in their market share (which will presumably always be relatively small), but in the challenge they raise to the abstract capitalist relations that fuel exploitation in the agrofood system. Both initiatives critique the subordination of agriculture and food to capitalist market principles that devalue, and thus encourage the degradation of, environmental and human resources, particularly in countries of the South.
>
> (Raynolds, 2000: 298)

The existence of alternative practices and products is not only important from the consumer point of view, but also for agriculture, research and society. Many of the solutions to environmental problems that are offered by the organic practices have been picked up by mainstream agriculture so that it can meet societal demands. The same process of adaptation may work with regard to the broader issues of ecological justice. From the perspective of research, organic agriculture offers established alternative practices and networks that can be utilized to gain a better understanding of agricultural systems and alternative forms of trade. Organic agriculture also poses new problems and issues for inquiry that are not noticeable in conventional agriculture. With regard to ecological justice new research and new measures are needed. Studies of food mileages, energy costs and nutrient flows will have a role to play as will more elaborate notions and methods such as ecological footprints (or rucksacks) and life cycle analyses. But these measures and calculations must be developed in interaction with the comprehensive, integrated approaches of organic agriculture.

Existing sustainability indicator frameworks may be modified or supplemented to capture the issues of ecological justice. If the goals of sustainability and ecological justice can be realized in the organic practices, and if this can be shown with widely accepted indicators, this can inflict on the market preferences being exerted for organic products and thus lead to the promotion of these goals. The challenges to the realization of these promises are, however, many and varied. First of all, indicators are measures and therefore do not include areas of ignorance. With regard to ecological justice there are many such areas, connected to e.g. long-term impacts that are not known at present and impacts that are different to different communities, ecologies and geographical areas. Targeted and participatory research efforts can go some way towards augmenting sustainability indicators as tool for awareness. But in general, the use of indicators will have to be supplemented with more general means of raising the awareness of ecological justice issues, in line with reflexivity, precaution, moral consideration, responsibility and participation.

More generally, the implementation and institutionalization of ecological justice in an alternative food system such as organic agriculture may function as an example for the broader implementation of ecological justice in other areas of society. The existence of such alternatives may work to broaden the discourse of sustainability and raise the general awareness of the issues of ecological justice. Moreover, it may contribute to the education of responsible citizens and function as a model for political visions.

Conclusions

In the present chapter we have investigated the role that ecological justice may have in relation to the present challenges for the global development of organic agriculture, starting from four questions: What is the meaning and context of ecological justice? How can these ideas help resist the pressures of globalization? How can ecological justice be implemented in relation to organic production and trade? And how can organic agriculture contribute to ecological justice in a global perspective? We analysed three key challenges: the commodification of hitherto commons, the externalization of environmental and social costs, and the growing distances of trade and ownership due to globalization. Finally, three ways of putting the idea of ecological justice into organic practice were identified: certified organic agriculture, non-certified organic agriculture, and organic agriculture as an alternative example for agriculture, research and society.

Broadly, we conclude with the following points:

- Ecological justice is a more comprehensive form of the well-known liberal idea of justice – extended to incorporate, first of all, the ideas that human communities and individuals have claims on their environments and that we share environments; and, second, the idea that justice and fairness concern not only humans, but animals and other living organisms as well.
- Certified organic agriculture is a proven form of institution to implement environmental ideals (and thereby elements of ecological justice) in globalised food systems, but the current standards have yet to fully meet the challenges of commodification, externalities and distant trade.
- Incorporating a measure of 'nearness' into the system, based on the ideas of transparency, substitutability, regional rules based on common principles, comprehensive tools to assess external costs, and participation could help organic agriculture to counter the ill effects of globalization.
- An alliance of organic and fair trade certification can go some way towards meeting the aims of ecological justice by incorporating the broader context of production, processing and transport into the market, though a simple combination of the two will not be adequate for the development of a fair organic trade.

- But leaving the aims of ecological justice to alternatives such as organic and fair trade within the market put great demands on the awareness and responsibility of the consumers.
- In addition to implementing ecological justice more fully into the organic certification standards, we suggest an alternative path towards implementing ecological justice through the promotion of 'non-certified organic agriculture' to develop local sustainable communities and food security based on the principles of organic agriculture.
- If and when the aims of ecological justice are well implemented into organic practices, the role as an alternative example for agriculture, research and society may be more important than the actual benefits to ecological justice due to these practices in themselves.

Acknowledgements

We thank Helena Röcklinsberg, Brian Baker and Egon Noe for the valuable comments they have given on earlier versions of this chapter.

References

Alcântara, T. and Dom Pedro de Alcântara (2004). Development of local organic markets: The final letter, International workshop on alternative certification. Ecology and Farming No. 37. See further on http://www.ifoam.org/about_ifoam/standards/pgs.html

Alrøe, H.F. and Kristensen, E.S. (2003). Toward a Systemic Ethic: In search of an Ethical Basis for Sustainability and Precaution. Environmental Ethics 25(1), 59-78. Available online at http://orgprints.org/552

Alrøe, H.F. and Kristensen, E.S. (2004). Why have basic principles for organic agriculture? And what kind of principles should they be? Ecology and Farming No. 36: 27-30. Available online at http://ecowiki.org/IfoamPrinciples/EcologyAndFarming

Alrøe, H.F. and Kristensen, E.S. (2005). Organic agriculture in a global perspective. In: Wulfhorst, J.D. and Haugestad, A.K. (eds) Building Sustainable Communities: Ecological Justice and Global Citizenship, Rodopi (forthcoming). Available online at http://orgprints.org/3855

Alrøe, H.F., Vaarst, M. and Kristensen, E.S. (2001). Does organic farming face distinctive livestock welfare issues? A conceptual analysis. Journal of Agricultural and Environmental Ethics 14: 275-299.

Baxter, B. (1999). Ecologism: An Introduction, Edinburgh: Edinburgh University Press.

Baxter, B. (2005). A theory of ecological justice. Routledge, London and New York.

Bell, D.R. (2003). Political liberalism and ecological justice. Paper presented at ECPR General conference, Environmental politics section, 19 September 2003. Available online at http://www.essex.ac.uk/ECPR/events/generalconference/marburg/show_panel.asp?panelID=46

Bell, D. (2004). Environmental Justice and Rawls' Difference Principle. Environmental Ethics 26: 287-306.
Bond, P. (2002). Unsustainable South Africa: environment, development and social protest. University of Natal Press, Scottsville. Merlin Press, London.
Brown, L.R. (2002). Restructuring the economy for sustainable development. International Review for Environmental Strategies, 3(1): 73-80.
Buck, S.J. (1998). The Global Commons: An Introduction. Island Press, Washington, DC.
Byrne, J. and Glover, L. (2002). A common future or towards a future commons: Globalization and sustainable development since UNCED. International Review for Environmental Strategies 3(1): 5-25.
Byrne, J., Glover, L. and Martinez, C. (eds) (2002a). Environmental Justice: Discourses in International Political Economy. Transaction Publishers, New Brunswick, NJ and London.
Byrne, J., Glover, L. and Martinez, C. (2002b). The Production of Unequal Nature. In: Byrne, J., Glover, L. and Martinez, C. (eds), Environmental Justice: Discourses in International Political Economy. Transaction Publishers, New Brunswick, NJ and London, pp. 261-291.
Christoff, P. (1996). Ecological Modernisation, Ecological Modernities. Environmental Politics 5(3): 476-500.
Cone, C.A. and Myhre, A. (2001). Community-Supported Agriculture: A Sustainable Alternative to Industrial Agriculture? Human Organization 59: 187-97.
DARCOF (2000). Principles of organic farming: Discussion document prepared for the DARCOF Users Committee, Danish Research Centre for Organic Farming. Available online at http://www.darcof.dk/discuss/Princip.pdf
Davidson, L. and Agudelo, D. (2004). The globalization that went home: Changing world trade patterns among the G7 from 1980 to 1997. Regional Integration Project Working Paper, Center for International Business Education and Research, Indiana University. Available online at http://www.bus.indiana.edu/ciber/research.cfm (accessed 8. March 2005).
Douglass, G.K. (1984). The meanings of agricultural sustainability. In: Agricultural Sustainability in a Changing World Order. Westview Press, Boulder, Colorado, pp. 3-29.
Ferry, L. (1995). The new ecological order. University of Chicago Press, Chicago.
Fuchshofen, W.H. and Fuchshofen, S. (2000). Organic Trade Association's Export Study for U.S. Organic Products to Asia and Europe. Organic Trade Association, Greenfield, MA. Online at http://www.ota.com/organic/mt/export_form.html
Goodpaster, K.E. (1979). From Egoism to Environmentalism. In: Goodpaster, K.E. and Sayre, K.M. (eds) Ethics and Problems of the 21st Century. University of Notre Dame Press, Notre Dame, IN.
Haen, H. de (2000). Producing and marketing quality organic products: Opportunities and challenges. In: Quality and Communication for the Organic Market. Proceedings of the Sixth IFOAM Trade Conference. Tholey-Theley, Germany: International Federation of Organic Agriculture Movements. Available online at http://www.fao.org/organicag/doc/IFOAMf-e.htm
Hajer, M.A. (1995). The Politics of Environmental Discourse: Ecological Modernization and the Policy Process. Clarendon Press, Oxford.
Halle, M. (2002) Globalization and sustainable development. International Review for Environmental Strategies 3(1): 33-39.
Hardin, G. (1968). The tragedy of the commons. Science 162 (December): 1243-48.

Jonas, H. (1984). The Imperative of Responsibility: In: Search of an Ethics for the Technological Age. University of Chicago Press, Chicago.
Kaul, I., Conceicao, P., Le Goulven, K. and Mendoza, R.U. (eds) (2003). Providing Global Public Goods: Managing Globalization. Oxford University Press, New York. Excerpts available online at http://www.globalpublicgoods.org
Low, N. and Gleeson, B. (1998). Justice, Society and Nature: An Exploration of Political Ecology. Routledge, London and New York.
Lund, V. (2002). Ethics and animal welfare in organic animal husbandry: An interdisciplinary approach. Acta Universitatis Agriculturae Suecia, Veterinaria 137, Swedish University of Agricultural Sciences.
Lund, V., Anthony, R. and Röcklinsberg, H. (2004). The Ethical Contract as a Tool in Organic Animal Husbandry. Journal of Agricultural and Environmental Ethics 17(1): 23–49.
Luttikholt, L. (2004). Principles of Organic Agriculture: History and process of rewriting. Ecology and Farming 36: 22–33. Available online at http://ecowiki.org/IfoamPrinciples/EcologyAndFarming
O'Riordan, T. and Cameron, J. (eds) (1994). Interpreting the precautionary principle. Earthscan, London.
Raffensperger, C. and Tickner, J. (eds) (1999). Protecting public health and the environment: Implementing the precautionary principle. Island Press, Washington, DC.
Raynolds, L.T. (2000). Re-embedding global agriculture: The international organic and fair trade movements. Agriculture and Human Values 17: 297-309.
Rawls, J. (1971). A Theory of Justice. The Belknap Press of Harvard University Press. Cambridge, MA.
Rigby, D. and Bown, S. (2003). Organic Food and Global Trade: Is the Market Delivering Agricultural Sustainability? School of Economic Studies Discussion Paper Series No. 0326, University of Manchester.
Ritzer, G. (2003). Rethinking globalization: Glocalization/Grobalization and Something/Nothing. Sociological Theory 21(3): 193-209.
Schlosberg, D. (2003). The justice of environmental justice: Reconciling equity, recognition, and participation in a political movement. In: Light, A. and deShalit, A. (eds) Moral and political reasoning in environmental practice. MA: MIT Press ,Cambridge.
Scialabba, N.E.-H. and Hattam, C. (eds) (2002). Organic agriculture, environment and food security. Environment and Natural Resources Service, Sustainable Development Department, FAO, Rome.
Shrader-Frechette, K. (2002). Environmental justice: Creating equality, reclaiming democracy. Oxford University Press, Oxford and New York.
Singer, P. (1979). Not for humans only: The place of nonhumans in environmental issues. In: Goodpaster, K. E. and Sayre, K.M. (eds) Ethics and Problems of the 21st Century. University of Notre Dame Press, Notre Dame, IN.
Spangenberg, J.H. (2002). Environmental space and the prism of sustainability: frameworks for indicators measuring sustainable development. Ecological Indicators 2: 295-309.
The Ecologist (1993). Whose Common Future? Reclaiming the Commons. Earthscan Publications, London. A paper drawn from this book: Nicholas Hildyard, Larry Lohmann, Sarah Sexton and Simon Fairlie (1995), 'Reclaiming the Commons', is available online at http://www.thecornerhouse.org.uk/item.shtml?x=52004

Thompson, P.B. (1996). Sustainability as a norm, Techné: Journal of the Society for Philosophy and Technology 2(2): 75-94. Online at
http://scholar.lib.vt.edu/ejournals/SPT/v2n2/pdf/thompson.pdf
Tybirk, K., Alrøe, H.F. and Frederiksen, P. (2004). Nature quality in organic farming: A conceptual analysis of considerations and criteria in a European context. Journal of Agricultural and Environmental Ethics 17(3): 249-274.
Valentin, A. and Spangenberg, J.H. (2000). A guide to community sustainability indicators. Environmental Impact Assessment Review 20: 381-392.
Volger, J. (1995). The Global Commons: A Regime Analysis. John Wiley and Sons, Chichester, UK.
WCED (World Commission on Environment and Development) (1987). Our Common Future, Oxford University Press, New York.
Woodward, L., Flemming, D. and Vogtmann, H. (1996). Reflections on the past, outlook for the future. In: Østergaard, T.V. (ed.). Fundamentals of Organic Agriculture, Proc. Vol. I of the 11th IFOAM International Scientific Conference, August 11–15, 1996, Copenhagen, pp. 259-270.

4
Ecological economics and organic farming

Paul Rye Kledal,* Chris Kjeldsen, Karen Refsgaard and Peter Söderbaum

Introduction.. 114
Ecological Economics as a trans-disciplinary approach 114
    Interactions between ecological, economic and social systems................ 114
    Thermodynamics in ecological economy 116
    Ecological economics and strong sustainability 117
Political economics and the conception of time and scale.......................... 119
Farming, production time, nature's time and scale..................................... 121
Organic farming... 124
The ecological economic perspective and organic farming........................ 126
Frameworks for decision-making 127
Conclusions... 130

Summary

Ecological economics (EE) is proposed as an approach to decision making and planning in organic farming. It is argued that EE is better suited for this task than the conventional neoclassical economy approach. The contribution that EE can make to the organic farming movement is apparent on the ontological level, through its focus on socio-economic systems as nested subsystems of the ecosystem. In addition, EE's stance on the issues of allocation, distribution and scale seems to constitute a more appropriate conceptualization about the interaction between socio-economic systems and the environment, which is more closely aligned to the principal aims of the organic farming movement. The concepts of time and scale are used as examples of how EE, with input from political economy, can help highlight problematic issues regarding the interaction between farming systems and their biophysical environment, which are not addressed in the neoclassical approach. Material Flow Accounting and Analysis (MFA) and Multicriteria Analysis (MCA) are discussed as practical

---

* Corresponding author: Danish Research Institute of Food Economics, Rolighedsvej 25, DK-1958 Frederiksberg C, Denmark. E-mail: paul@foi.dk

examples of the framework that EE can provide for decision-making. It is concluded that, by reconceptualizing the way in which organic farming manages the complex interrelations between ecological and socio-economic systems, the EE paradigm and its frameworks for decision-making can be of considerable value to the organic farming movement.

Introduction

Traditional, neo-classical economic theory limits itself to monetary assessments of production efficiency and economic aspects of different production systems and their use of resources. This seems unsatisfactory when analysing the differences between organic and conventional farming, because the rationale behind organic farming includes non-economic aspects such as minimizing the use of non-renewable resources and pollution and improving animal welfare. Ecological economics (EE) has been proposed as a trans-disciplinary framework, which moves beyond the approaches employed in traditional economics in that it considers the natural environment as an integrated part of sustainable development (Costanza *et al.*, 1997). What does EE, as an analytical tool and decision-making framework, have to offer the organic farming movement, and where does EE differ from more traditional economic approaches? In this chapter we will first present how the economic system works from a political economy approach, and show how the functioning of a capitalist market economy has an inherently contradictory approach towards the larger natural ecosystems of which it is part. Secondly, we will present new theoretical insights on how organic farming, with its rules and regulations, can be regarded as a response trying to overcome the environmental consequences of these contradictions in agriculture. Thirdly, we will give some examples of how EE, as a trans-disciplinary approach, can be of theoretical and methodological support to the organic farming movement.

Ecological economics as a trans-disciplinary approach

Interactions between ecological, economic and social systems

Ecological economics primarily differs from traditional neoclassical economics by being a trans-disciplinary field of study, which examines the interactions between economic and ecological systems from a number of related viewpoints. Ecological economics focuses on the human economy both as a social system, and as one constrained by the biophysical world. Therefore EE often focuses on areas where economic activity comes into conflict with the well being of the ecological and the social systems. The first of these systems ultimately supports all activities, while the second is the system to which the benefits of economic activity should ultimately be directed (Edwards-Jones *et al.*, 2000).

EE is therefore automatically concerned with three analytic focus points:
- the ecological system;
- the economic system;
- the social system.

where the last two are considered open subsystems of the ecological system: a system that is finite, not growing and materially closed (though open to solar energy) (Figure 4.1).

EE emphasizes the relationships between these systems at a number of levels and scales, from the local to the global. It treats human beings as integral components of, and active participants in, the ecological systems that support them, rather than as external to these systems. It searches for ways in which analyses of these different systems can complement and support each other.

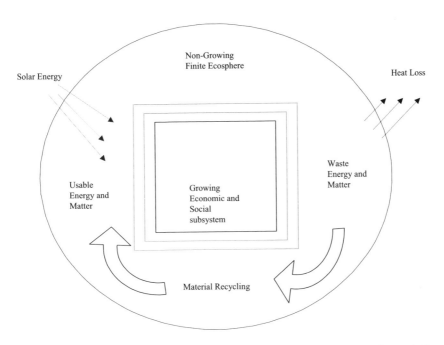

Figure 4.1. Ecological economics sees the economy as an open, growing, wholly dependent subsystem of a materially closed, non-growing, finite ecosphere (Rees, 2003).

The overall scale of the economic system, relative to the wider environment, is a key issue in EE (Daly, 1973). Daly argues for a 'steady-state economy' where the throughput flow of the economic system should be lowered to a minimum, because the throughput is the inevitable cost of maintaining the stocks of people and their wealth (Daly, 1991).

Thermodynamics in ecological economy

Due to the EE view of the economy as an integrated part of the biosphere – as an open subsystem of the environment it is essential to focus on the flows of matter and energy through the system, and the thermodynamic laws governing these processes. The concept of entropy and the laws of thermodynamics highlight how resource and energy scarcity, as well as the irreversibility of transformation processes, can constrain economic action (Georgescu-Roegen, 1971; Baumgärtner *et al.*, 1996).

The First Law of Thermodynamics says that in a closed system the amount of energy and matter is constant. This is the law that Boulding (1966) refers to when he describes the economic subsystem as a 'spaceman-economy'. There is a finite amount of energy and matter onboard Spaceship-Earth, and there is a limit, in time and scale, on how we can use it. The other thermodynamic law is the law of entropy. This describes how energy or matter is structured within a system. The higher the structure and organization is, the lower the level of entropy. The less structure and organization, the higher the entropy level.

Entropy can be interpreted as an indicator of the system's capacity to perform useful work. The higher the entropy value, the more energy already irreversibly transformed into heat, the lower the amount of free energy within the system and the lower the system's capacity to perform work. Most goods that we find useful have relatively low specific entropy per unit of mass (i.e. they 'wear-out' with use, becoming more and more 'mixed-up' with the environment (Bisson and Proops, 2002)). On the other hand a large part of our production is derived from raw materials that have rather high specific entropy (e.g. iron ore), but are extracted with the help of low specific entropy fuels. However, the production of a low specific entropy object, such as iron, generates other high entropy 'products', like solid slag, carbon dioxide and waste heat, thus 'all production is joint production' (Faber *et al.*, 1998). This is due to the Second Law of Thermodynamics which tells us that entropy increases throughout any production process.

So, every process of change moving us away from thermodynamic equilibrium requires low entropy energy. This is the case for natural ecosystems (e.g. a leaf growing on a tree) as well as for the human economy (e.g. the production of metal from metal ore) (Baumgärtner, 2002). However, there are at least two characteristic differences between natural and industrial metabolism (the material and energetic dimension of the economic process) (Ayres, 2001):

- The low entropy energy employed in modern industrial economies is typically not sunlight, as it is in ecosystems, but energy stored in materials, such as fossil or nuclear fuels.
- Material flows in our economic system are not bound into closed cycles, as they are in ecosystems but, to a large extent, are one-way throughputs. Materials are taken from reservoirs outside the economy and are ultimately disposed of in other reservoirs outside the economy. As a consequence, economies not only emit waste heat, as ecosystems do, but also generate vast quantities of material waste.

So, from a thermodynamic viewpoint, waste is an unavoidable and necessary joint product in the production of material goods. It is important to consider the (in)efficiency of the processes as well as the properties of the waste, and thus distinguish between high entropy waste, in the form of heat, and low entropy waste, in the form of waste materials. The former may be considered a minor problem since it can, in principle, be radiated into space, but can also cause harm when directly released into ecosystems or the ability to radiate heat may be impaired by the greenhouse effect (Baumgärtner, 2002). It is the latter, which accumulates in the biosphere, that causes major environmental problems. This is due to the available energy still contained in waste materials, i.e. the potential to initiate chemical reactions and perform work (Ayres, 1998).

Thermodynamic analysis can then be used to identify sustainable social modes of metabolism which, according to Baumgärtner (2002), conform with the following principles:

1. To not use material fuels as a source of available energy, but only sunlight.
2. To keep matter in closed cycles, i.e. let heat be the only true waste.
3. To carry out all transformations in a thermodynamically efficient way.

Thermodynamics is thereby a tool to identify feasible solutions and their physical efficiencies. However, before a choice can be made we need to know which criteria must be included in a valuation and how these criteria are going to be judged. This implies to include how the society perceives and values the different joint products, the processing of them and the waste or pollution they generate. This means that we need to link the material and energetic aspects of production with human perception and valuation of commodity products and waste joint products (Baumgärtner, 2002). See the example shown in Chapter 7 about different perceptions of waste.

Ecological economics and strong sustainability

The view of the economy as an open subsystem of a wider finite ecological system is in sharp contrast to that of neoclassical economics, where the economic system is viewed as an open system *independent* of the boundaries from the ecological system. The dependencies only become relevant for the economic system, when the ecological system constrains further growth, through natural resource scarcity or vulnerability to pollution. Environmental problems are seen as *externalities* that appear because of market failures, and should be solved through the market. This can either be achieved through higher market prices for scarce resources (reflecting laws of supply and demand) or through internalizing the costs of pollution.

Neoclassical economic theory assumes that, over time, the market can and will solve the constraints set by the ecological system in its interactions with the economic system, by generating new technologies, new ways of organizing production, or new substitutes for the depleted resources.

A common theme for both neoclassical and EE in relation to environmental concerns is the question of maintaining economic activity into the future, whether at the local or the global scale. These concerns have led to the ubiquitous concept of 'sustainable development', described as "development that meets the needs of the present without compromising the ability of future generations to meet their own needs' (WCED, 1987). In economic discourse, two competing positions, those of weak and strong sustainability, prevail today over the question of how to avoid compromising future generations' ability to meet their needs (Neumayer, 2003).

Weak sustainability argues for the need to maintain the *total capital stock* between generations. Total capital stock would be *natural capital*, like trees, fish, minerals, oil + *man-made capital*, such as machinery, houses, roads. In the weak sustainability approach it is acceptable to deplete certain natural capitals like oil resources if this leads to investment in man-made capital, such as universities generating new wealth, thereby securing the total stock of capital for the next generation.

The strong sustainability position focuses on natural capital, and argues for the need to maintain or increase the stock of this between generations. The wealth from using oil should therefore be directed to energy efficiency or renewable energy resources. The strong sustainability position therefore imposes some restrictions on the use of resources that imply stronger public interference in the market economy. The issues at stake here are those of *complementarity* and *substitutability* between natural resources turned into man-made resources. Ethical and philosophical values about nature influence the contrasting viewpoints about what should be handed on to future generations.

In general, neoclassical economists – including the larger part of environmental economists (e.g. Pearce and Turner, 1990) – favour the weak sustainability position, whereas ecological economists support the strong sustainability position. Table 4.1 compares the differing economic perspectives of EE and neoclassical economics.

Table 4.1. The differing economic perspectives of EE and neoclassical economics (Rees, 2003).

Neoclassical economics	Ecological economics
• Economic system is static, linear, deterministic • Economy separate from the environment • Models based on analytic mechanics • Substitutions are possible so there are: • No limits to GDP growth • Analysis preoccupied with growth • Efficiency oriented • Emphasis on production/consumption • Short-term frame • Favours monetary assessments	• Complex systems are dynamic, non-linear, self-producing • Economy as a subsystem of ecosphere • Models recognize thermodynamics • Complementarity dominates so there are: • Constraints on growth • Analysis focused on development • Equity oriented (intra- and intergenerational) • Emphasis on well-being (social capital) • Long-term horizon • Favours biophysical assessments

These contrasting perspectives between EE and neoclassical economics generate different precepts and implications on values, justice and policy prescriptions. According to Vatn (forthcoming) the systems perspective on nature demands a view of societal processes while the individualistic perspective of neoclassical economics adapts a more 'itemized' perspective of nature. In the institutional perspective for environmental management, the focus is first on the rights structures involved – i.e. who gets access to which resource, how different uses are allowed to affect other uses, and how the institutions involved treat such conflicts. A secondary question is that of how different regimes motivate actions and influence values (Vatn, forthcoming). This discussion parallels that about ecological justice (see Chapter 3).

In what ways are these different perspectives and values about the interrelations between economic activities and the environment of interest to the organic farm movement?

Political economics and the conception of time and scale

'Time is money' as the old saying goes, but it carries a central truth when it comes to understanding the depletion of many ecological systems caused by the workings of the economic system.

From a political economics point of view money (M) (or capital) is the starting point to understand the workings of a capitalist market economy. As illustrated in Figure 4.2 money (M) is used for buying commodities (C) such as natural resources, labour and technology. Through production these commodities are organized as efficiently as possible to produce a new commodity (C) sold at a market. The intention is that the money received is higher than that invested (M becomes M1), and production can be maintained by M1 being reinvested.

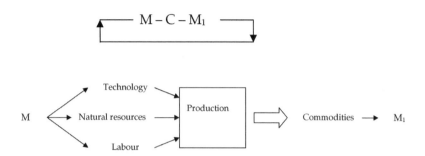

Figure 4.2. Resource flow from the political economic perspective.

In the political economy model money has a cycle (Marx, 1970, [1867]).

The circuit of money introduces time into the model, and 'time becomes money', influencing how commodities are produced and distributed to the market, in order to accelerate (i.e. shorten the return time) the return of M.

Secondly, the organization of a market economy, protected by institutions securing private property and a competitive market environment, enforces individual producers to be constantly alert for new technologies, new ways of organizing production or utilizing new resources to reduce costs, if they want to stay in business. This forced creativity, driven by the market's competitive downward pressure on prices, generally leads to producers following one, or more, of three logical paths:

1. Expanding production by using economies of scale and/or of scope.
2. Extracting or exploiting the input factors more efficiently.
3. Shortening production time by reorganizing labour, take advantages of the division of labour, apply new technologies, make better use of resources etc. Shortening production time reduces the time needed to reproduce capital, hence the cost of M invested becomes less.

This insight from political economy, on the pressures within the economic system to constantly grow in scale, and shorten production and distribution times, takes us to the heart of understanding some of the major contradictions inherent in the capitalist market economy and its relations with the ecosystem.

As we recall, from the perspective of EE, the economic system is viewed as a subsystem of the ecological system. This connectedness becomes clearly evident when we focus on time and scale.

Behind every effort in the economic system to reduce time and expand scale and the resources used, there exists another time and finite level of scale in the ecological system. There is the bio-spherical time, which created the mineral resources long before man was born. There is the time needed for nature to break down waste and to reproduce renewable resources. There is time needed for humans to reproduce themselves (physically and socially), as well as the workings of the more general time of various natural, social and cultural processes taking place in the world around us.

In a world made up by natural systems (living systems, ecosystems, climatic systems, socio-cultural systems, farm systems etc.) the ongoing pressure within the economic system for shortening production time and growing in scale, inherently potentially collides with the various times and scales required by the larger ecosystem to produce, or reproduce, itself.

How, where and when such collisions will occur is a complicated dialectic process that depends on the type of resource extraction, technology used, cultural knowledge and social morality of man, as well as the scale of intervention by the economic system into the ecosystem. This is one of the main reasons why advocates of EE emphasize a trans-disciplinary approach to better understand the changes in the environmental system in relation to the impacts of growth in the economic (market) system.

Farming, production time, nature's time and scale

In agriculture the borderline and contradictions between the economic system and the environmental system become evident, when we examine the production time of agricultural commodities and the scale of output.

In contrast to industrial production, using non-living raw materials, commodities in agriculture are living species that tend to slow down the reproduction (turnover) of capital. Since firms extract profits from each cycle of capital, they can only use these profits to replenish and expand their production when the production cycle is over and the product sold.

Figure 4.3 illustrates how production time consists of both labour time and nature's time. Production time can be prolonged due to drought, diseases or other more uncontrollable natural causes, *so unsteady nature time* has been added to the total production time. The arrows show the deliberate attempts (mainly by research and other efforts) to reduce production time either by shortening labour time or the time it takes for nature to produce a certain agro-commodity. Such attempts will include innovations from farmers, agro-corporations and researchers as well as governmental schemes all designed to help agro-capital achieve a better, and less risky, profit. These attempts can also be driven by *indirect* pressure via retailers and food processors squeezing farmers on price margins or imposing specific requirements on production size and time of delivery (Kledal, 2003: 19).

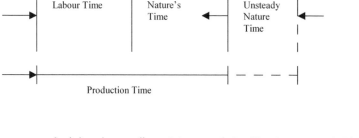

→ *Symbolizes human efforts aiming at reducing Time i.e. costs and risks.*

Figure 4.3. Production time in agriculture.

As organic farming relies on the utilization of natural resources, and focuses on sustainability, through (among other things) recycling resources and reducing pollution, it is an endeavour that shares many of the values and perspectives of ecological economics. However, the ecological economics literature has paid little attention to exploring how the principles of organic farming combine economic with ecological benefits for society as a whole.

Attempts at reducing labour time could typically include specialization, division and enlargement of agro-production so the farmer, or farm workers, only have one or few work processes, so as to better utilize economies of scale. For example, one farm takes care of only farrowing, another produces only hogs, but they can both produce more per labour unit.

Examples of attempts to shortening nature's time could be new genetics, or better management and feed systems that speed up growth. Reducing unsteady nature time could involve the implementation of technologies like pesticides, GMO, precision farming (GPS: Global Positioning System) etc.

As well as the noted differences that exist between agriculture and industry in relation to the cycle of capital and the relation to turnover time, there are considerable differences between different agro-commodities in regard to both production and labour time.

Figure 4.4 shows the production time of fattening hogs and wheat. The turnover frequency of hogs can be almost four times per year, whereas for wheat it is only one (in the northern hemisphere at least).

These two examples, one a plant the other an animal, show that, in general it has been easier for humans to shorten production time for animals, whereas for plants the push from capital has been to raise output (through higher yields). In the southern hemisphere, though, it has been possible to expand the cycles of plants, such as maize and rice, and thereby shortening the return time on capital invested in plant production.

In Table 4.2 a few examples are used to illustrate increases in farm productivity through shortening production time for animals (speeding up the production cycle) and raising yields in plants in conventional farming in Denmark.

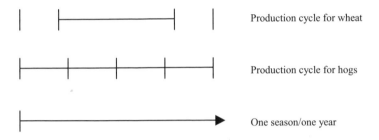

Figure 4.4. The number and length of production cycles for wheat and hogs during a one-year season.

Table 4.2. Rise in productivity of different agro commodities in Denmark between 1980 and 2004 (Pedersen *et al.*, 2001; Landskontoret for Svin, 1980-81; Jultved, 2004; Danish Agricultural Advisory Service, 2004 (www.lr.dk/budgetkalkuler2004)).

	1980	2004
Broilers	33 g/day	50 g/day
Fatteners	600 g/day	833 g/day
Winter wheat	6700 kg/ha	8300 kg/ha

The ability to raise productivity in animals and plants has given rise to various environmental and animal welfare problems. For example, the increased growth rate in broilers has led to serious leg problems because of weak bones, and the higher yields in cereals have led to increased leaching of nitrogen and problems of pesticides in ground and drinking water. The development of organic farming is closely related to these environmental and animal welfare problems, and the principles and standards of organic farming implicitly place constraints on the returns to capital investments in agriculture.

Organic farming:
a response to ecological damage caused by growth in scale and shortening of production time

Agriculture is unique in the sense that economies of scale and production time in the economic system are very closely connected to the various time requirements and scales of the ecological system that it relies upon. Therefore, conflicts and constraints between agriculture, as an economic subsystem that works with nature and living species, and the ecological system of which it forms part, are more evident than they are in industrial production.

Capital's ongoing push for maximizing profits (or minimizing costs) will, at certain points, encounter different types of constraints. This is shown in Figure 4.5, where constraints are encountered when trying to raise labour productivity, shorten the biological time on animal reproduction. Ecosystem constraints, such as polluting the environment, can be encountered, from trying to raise output. The black arrows in Figure 4.5 illustrate this. The more the market economy pushes for shortened production times and increased output, the more constraints it will encounter. These constraints can, at some point, lead to various types of societal conflicts (or externalities): alienation from how food is produced, environmental degradation, inadequate food safety and animal welfare, as well as concerns about the marginalization of farmers and rural areas.

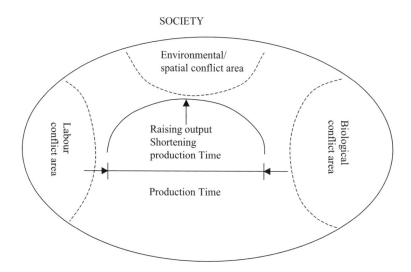

Figure 4.5. The connection between farming and areas of potential environmental and societal conflicts.

Organic farming can be viewed as a response to these conflicts. The rules and regulations set up by the organic farmers and consumers are, in many respects, counter rules that either *extend* nature's time, labour time and thereby total production time, or constrain scale and resource use, by limiting certain technologies or inputs. These include rules about animal welfare with regard to space and access to the open air, bans on the use of pesticides, limits to fertilizer use and basing nutrient supply on improved soil fertility, principles of self-sufficiency etc. (Kledal, 2003: 23).

Table 4.3 shows differences in productivity between conventional and organic farming for certain agro-commodities, illustrating how the rules and regulations of organic farming impose lower productivity and higher feed consumption (in the case of fatteners) as a trade off in addressing certain environmental and animal welfare problems. At first glance it might seem odd that it takes more feed to produce organically. It should be kept in mind though, that the overall fossil energy input of organic farming often is lower than that of conventional farming. However, the overall economic performance of organic farms is not necessarily lower than in conventional farming (FØI, 2004). This shows that the apparent trade-off between productivity and solving environmental and animal welfare related issues is not as clear-cut as it might seem.

It is these self-imposed constraints on input use and cost–minimizing efforts that reduce productivity and make organic products more expensive within today's institutional market regime. This is because organic farmers and consumers have voluntarily internalized the value of the ecological system as a reservoir of wealth creation for future generations. The challenge is how to translate these so called *private costs* internalized by organic farmers and consumers into *social benefits* for the whole of society.

Table 4.3. Productivity differences between Danish organic and conventional production, year 2004. (Pedersen *et al.*, 2001; Landskontoret for Svin, 1980-81; Jultved, 2004; Danish Agricultural Advisory Service, 2004 (www.lr.dk/budgetkalkuler 2004); Ørum and Christensen, 2001).

	Organic	**Conventional**
Broilers	24 g/day	50 g/day
Fatteners (Energy consumption)	3.16 FE/kg	2.86 FE/kg
Winter wheat (output)	5500 kg/ha	8300 kg/ha
Winter carrots (output)	35 t/ha	50 t/ha

In this regard the framework of EE can provide a structure for addressing some very important research questions within and about organic farming. One of these is assessing the sustainability position brought about by organic farmers' self-imposed lower productivity at the farm level (higher energy consumption in feed, lower yields, more land needed for the same amount of food production etc.), and the societal 'trade off' generated by maintaining a steadier state economy, with lower waste from using less energy, no fertilizers or pesticides and a minimal use of veterinary medicine. Another is to address the sustainability position of organic farming in a global perspective, in relation to world food consumption (see Chapter 11), increasing global food trade as well as the need for a more just distribution of resources and access to them.

The ecological economic perspective and organic farming

EE is mainly about scale, distribution and efficiency and addresses these questions from the perspective of its vision how the economic system is nested within the social system, which in turn is nested within the ecosystem (Vatn forthcoming).

The issue of 'scale', which refers to the physical size of the economy relative to the containing ecosystem (see Figure 4.1), is not recognized in standard economics. EE claims that *sustainable scale* and *fair distribution* are both problems that logically demand solution prior to *determining efficient allocation*. Scale determines which natural resources are scarce from an ecosphere point of view (Figure 4.1) and what is free or unlimited. Distribution determines who owns scarce goods or services. Only after these issues have been determined is the market able to effect exchanges, determine prices and allocate resources efficiently (Daly, 2003).

The role of entropy and the finite nature of the ecological system should also lead us to reconsider our conceptions of evolution, progress and the production of material things. Thermodynamics and biology will force us, over time, towards a state of minimum production of entropy and conservation of resources. To maintain the energy flow at a low level, slowing down the entropic process, we must look towards a more decentralized, small-scale organization that uses renewable resources (Tiezzi, 2002).

The principles of organic farming share very similar lines of thought. By setting up its own democratic counter rules, regulations and values on farm production methods and distribution, organic consumers and farmers have created a social setting trying to implement:

- sustainable limits on output (e.g. max 1.4 LU per ha);
- letting the resource flow on the farm depend as much as possible on the farm systems' own reproductive abilities;

- connecting social justice and farm production with environmental issues. Examples of the latter could be Community Supported Agriculture (CSA), Ecological Villages, or various types of closer links between producers and consumers sharing economic and environmental responsibilities, both locally and globally.

These policies and actions within the organic movement, designed to address natural resource and environmental constraints, constitute a complex system with many components and where many actors are interacting to produce self-organized systems, which can only be adequately evaluated by analysing and examining the ecological and the economic systems simultaneously. An analytic framework encompassing both properties is therefore an obvious choice of tool (Costanza et al., 1991).

Frameworks for decision-making

Economists have attempted to help decision-makers by finding ways to measure the wide range of effects of environmental changes on a single monetary scale. The derivation of a monetary value for goods that do not have a market value – which is basically the case for many environmental goods and services – is an attempt to extend the utilitarian and democratic principle of the free market into environmental decision-making (Edwards-Jones et al., 2000). Thus traditional environmental economics has constructed a set of techniques in order to apply this utilitarian approach and thereby derive a market value for certain environmental goods or services. Three types of technique for such valuation can be discerned:

- conventional market approaches;
- implicit market;
- constructed market.

These techniques use various methods to try to measure either actual behaviour that occurs in the market or *potential* behaviour.

As described in the introduction, scholars within EE have raised many philosophical and ethical objections to the underlying assumptions behind this utilitarian, individual, free-market approach that underlies neo-classical economics and the shortcomings of such approaches which seek to value environmental goods and services in strictly monetary terms.

In this section we present some methods from EE which we argue are more appropriate for evaluating organic farming systems and informing decision making for these systems. We focus on three analytical tools from EE, Material Flow Analysis, multi criteria analysis and deliberative institutions, which can be used for valuing environmental "goods' and informing decision making. They do however, have different characteristics, as they belong to different value articulating institutions (Jacobs, 1997). As Vatn (2004: 9) writes:

> A value articulating institution is a constructed set of rules or typifications. It defines **who** shall participate and on the basis of which capacity – i.e. in which **role**… A value articulating institution also defines **what** is considered relevant data and **how** data is to be handled.

Thus different value articulating institutions tend to generate different outcomes. This implies that, if for example animal welfare issues and pollution issues are seen to be important for evaluation of a farming system, it may be proper to use value articulating institutions that consider such. These issues are about ethical values and not so much about individual values (Holland, 1995), i.e. they may better be handled through dialogue than through monetary assessment. Therefore multicriteria analysis may be a more appropriate value articulating institution than a contingent valuation study, because the first allows for discussion and incommensurable values while the last one is based on commensurability and financial capability.

The principal concept in EE of the economy as a subsystem of the environment dependent on a constant throughput of materials and energy underlies the *Material Flow Accounting and Analysis (MFA)*. MFA is a dynamic systems perspective and theory that draws on the central concepts of stocks, flows, feedbacks and delays. These concepts are well known to, and applied in, many disciplines within the social and natural sciences. In MFA, raw materials, such as water and air, are extracted from the natural system as inputs, transformed into products and finally transferred back to the natural system as outputs (waste and emissions). MFA offers the foundation for setting up a 'theory of waste' connected to the economic and social activities of society.

The main purpose of an economy-wide MFA is to provide aggregate background information on the composition and the changes of the physical structure of socio-economic systems. MFA represents a very useful methodological framework for analysing economy–environment relationships and deriving environmental and integrated environmental/socio-economic indicators. Material flow-based indicators can be aggregated from the micro to the macro level. They allow comparisons with aggregated economic or social indicators such as GDP and unemployment rates, thus providing policy–makers with information they are familiar with handling and helping to shift the policy focus from a purely monetary analysis to one which integrates biophysical aspects (Kleijn, 2001). MFA can also be used as a method to consider the scale and the environmental impact of the economy. But scale only determines what is scarce and what is free. Distribution is about ownership and equity.

Using scarce resources most efficiently is a major task in economics. Providing effective policy interventions concerning environmental protection are those that solve environmental problems at minimum cost while meeting social and cultural goals. Faced with limited budgets and with sets of conflicting uses for scarce natural resources, decision-makers seek guidance on how to trade-off between those possible uses so as to maximize welfare or utility overall. For an individual decision-maker this choice can be made with a direct knowledge of personal goals and preferences, whereas democratic governments must operate on behalf of all their citizens in determining how to achieve overall welfare.

To comply with a more democratic and ideological approach, methods like the *Multi Criteria Analysis (MCA)* have been applied in EE trying to encompass the benefits of environmental goods and services within the realm of a multidimensional social and economic sphere.

The MCA is designed to deal with complex decision-making for problems characterized by having many, often conflicting, objectives for the assessment of a diversity of possible alternatives and often involvement of several decision makers. There are two fundamental conflicts involved (Vatn, forthcoming). First, those between different interests, individuals or groups and secondly, we have conflicts between value dimensions or perspectives. The latter can be as relevant within a person as between persons. MCA is formulated so that it can handle values or criteria that are not easily transformed into one dimension like a monetary measure. This is actually the core of MCA as the name also indicates: criteria are *multidimensional*, and the method allows for handling criteria that are *incommensurable* (Martinez-Alier *et al.*, 1998). It can also handle the fact that weights may be considered *coefficients of importance*, not signalling trade-off capabilities (Munda, 1996).

There are many different MCA methods. Common for most of them is that they have a number of criteria for evaluation of multiple alternatives. Most MCA methods include to define and structure the problem, to generate the alternatives, choose a set of evaluation criteria, identify a preference system of the decision-maker, choice of an aggregation procedure and calculation of efficient solution and best "compromises' (Munda *et al.*, 1994; Lahdelma *et al.*, 2000).

MCA techniques have some clear advantages over more restricted decision-making techniques, such as cost-benefit analysis. Their popularity has increased very substantially with improvements in both methodologies and computer power. Furthermore, their suitability to environmental and natural resource planning is increasingly being recognized (Edwards-Jones *et al.*, 2000).

MCAs have also been designed and implemented to enhance public participation putting emphasis on the process – named participatory or deliberative processes (de Marchi and Ravetz, 2001). MCA offers a distinct response to the complex decision-making – for environmentally related challenges like organic farming – and often ill-defined problems. From this perspective MCA can be described as a structured search process where the analyst supports the decision-maker or the stakeholders in defining the problem, articulates their values and objectives, looking for alternatives, assessing their consequences, ranking the alternatives in relation to the objectives, maybe going back and formulating new alternatives etc. (Vatn, forthcoming). These processes generally aim to be exploratory or consultative with focus on participation in the decision-making. The currently most used and reported forms of participation include focus groups, in-depth groups, citizens' juries, consensus conferences and forums. In some way also multi-criteria methods are viewed as participatory approaches. The approaches are advocated on grounds of justice and democracy in procedure and an appreciation that complex, multi-attribute issues cannot be effectively evaluated by a one-dimensional numeraire based on simple consumer choices (de Marchi and Ravetz, 2001). During the 1990s the momentum for such processes has developed, and the initiatives under 'Local Agenda 21' are an example that encourages local participation in decision-making. For further examples see de Marchi and Ravetz (2001).

Multicriteria decision-making methods are designed to deal with complex problems such as how to deal with scarce resources, different notions of values concerning welfare and make use of opportunities now or for future generations etc. The challenge is to choose or to form a value articulating institution that fits the character of the problem or good at hand. Shortly we can say this is about how to solve questions related to who to be involved, how to involve them and what to be involved about (Refsgaard, forthcoming).

These non-monetary approaches have a better potential of valuating the societal benefits from organic farming systems. Organic farming systems need to be valued not only through their contribution with pure food products, but also by their contribution to the environment like for example reduced use of fossil fuel, contributions to biodiversity, nearness in the consumption–production cycle etc. On these matters use of a single monetary measure will be highly misleading, which, again, is where EE may contribute with a broader perspective. In addition, the multiplicity of users (and perspectives) also makes a unique ordering of values or prioritization difficult or impossible. In the valuation of organic farming systems we have both the different contributions, the different users of them and their different interests implying that a process for evaluation and articulation is needed.

Conclusions

This chapter shows how ecological economics explores the interrelations between the ecological, the economic and the social systems. The EE paradigm and its frameworks for decision-making could be an important tool for the organic farming movement, in conceptualizing the way in which it manages these interrelations and could constitute the intellectual underpinning on which to base the construction of future policy tools. The current worldwide growth of organic farming raises new challenges about how organic production relates with, and depends upon, our environment. Researchers and farmers involved in organic farming and food consumption need to be able to identify how new policies can be formulated, that help and promote organic farmers and consumers, and make these interrelations more harmonious and sustainable. So far very little has been done in this regard.

Ecological economics itself is a new and dynamic field as well as a pluralistic one. Its foundations, based on economy, ethics and ecology, offer a theoretically and methodologically wider perspective drawing on a more multidisciplinary approach which has the potential to generate a better understanding and evaluation of organic farming and its complex relations with the social, economic and biophysical spheres.

References

Ayres, R.U. (1998). Eco-thermodynamics: economics and the second law. Ecological Economics 26: 189-209.

Ayres, R.U. (2001). Industrial ecology: wealth, depreciation and waste. In: Folmer, H., Landis Gabel, H., Gerking, S. and Rose, A. (eds) Frontiers of environmental economics. Edward Elgar, Cheltenham, pp. 214-249.

Baumgärtner, S. (2002). Thermodynamics of Waste Generation. In: Bisson, K. and Proops, J. (eds) Waste in Ecological Economics, Edward Elgar, Cheltenham, UK, pp. 13-37.

Baumgärtner, S., Faber, M. and Proops, J. (1996). The use of the entropy concept in ecological econmics. In: Faber, M., Manstetten, R., and Proops, J. Ecological Economics: Concepts and Methods, Edward Elgar, Cheltenham, Chap. 7.

Bisson, K. and Proops, J. (2002). An Introduction to Waste. In: Bisson, K. and Proops, J. (eds) Waste in Ecological Economics, Edward Elgar, Cheltenham, UK, pp. 1-12.

Boulding, K.E. (1966). The economics of the coming spaceship Earth. In: Jarret, H. (ed.) Environmental Quality in a growing Economy. Resources for the future/Johns Hopkins University Press, Baltimore, pp. 3-14.

Costanza, R., Daly, H. and Bartholomew, J.A. (1991). Goals, Agenda and Policy recommmendations for Ecological Economics. In: Cosanza, R. (ed.) Ecological Economics. The Science and Management of Sustainability. Columbia University Press, NY.

Costanza, R., Cumberland, J., Daly, H., Goodland, R. and Norgaard, R. (1997). An introduction to ecological economics. Sct. Lucie Press, Boca Raton, Florida.

Daly, H.E. (1973). The Economics of the Steady State. In: Daly, H.E. (ed.) Toward a Steady State Economy. W.H. Freeman, San Francisco, CA.

Daly, H.E. (1991). Elements of environmental macroeconomics. In: Constanza, R. (ed.) Ecological Economics, the Science and Management of Sustainability. Columbia Press, NY.

Daly, H.E. (2003). Ecological economics: The concept of scale and its relation to allocation, distribution and uneconomic growth. Presentation at CANSEE, October 16–19 2003. Jasper, Alberta, Canada.

Danish Agricultural Advisory Service (2004). www.lr.dk/budgetkalkuler 2004

de Marchi, B. and Ravetz, J.R. (2001). Participatory approaches to environmental policy. Environmental Valuation in Europe. Policy Brief no. 10. C. L. Spash and C. Carter. Cambridge Research for the Environment, Cambridge.

Edwards-Jones, G., Davies, B. and Hussain, S. (2000). Ecological economics – an introduction. Blackwell Science.

Faber, M., Proops, J. and Baumgärtner, S. (1998). All production is joint production – a thermodynamic analysis. In: Faucheux, S., Gowdy, J. and Nicolaï, I. (eds). Sustainability and firms. Technological change and the changing regulatory environment. Edward Elgar, Cheltenham, pp. 131-158.

FØI (2004). Økonomien i landbrugets driftsgrene 2003. Serie B, nr. 88. København, Fødevareøkonomisk Institut.

Georgescu-Roegen, N. (1971). The entropy law and the economic process. Harvard University Press, Cambridge, MA.

Holland, A. (1995). The assumptions of cost-benefit analysis: a philosopher's views. In: Willis, K.G., Corkindale, J.T. (eds) Environmental Valuation. New Perspectives. CAB International, Wallingford, pp. 21-38.

Jacobs, M. (1997). Environmental valuation, deliberative democracy and public decision-making institutions. In: Foster, J. (ed.) Valuing Nature: Economics, Ethics and Environment. Routledge, London, pp. 211-231.

Jultved, C.R. (2004). Rapport over P-rapporternes resultater oktober 2003. Notat nr. 0405.

Kledal, P.R. (2003). Analysis of Organic Supply Chains – A theoretical framework. København, Fødevareøkonomisk Institut. Working paper 15/2003. http://www.foi.dk/Publikationer/wp/2003-wp/Wp15-03.pdf

Kleijn, R. (2001). Adding It All Up. The Sense and Non-Sense of Bulk-MFA. Journal of Industrial Ecology 4(2): 7-8.

Lahdelma, R., Hokkanen, J. and Salminen, P. (2000). Using multicriteria methods in environmental planning and management. Environmental Management 26(6): 595-605.

Landskontoret for Svin (1980-81). Håndbog for Driftsplanlægning, Landbrugets Informationskontor 1980-81.

Martinez-Alier, J., Munda, G. and O'Neill, J. (1998). Weak comparability of values as a foundation of ecological economics. Ecological Economics 26(3): 277-286.

Marx, K. (1970) [1867]. Kapitalen: Kritik af den politiske økonomi. 1. bog 1. Rhodos, København.

Munda, G. (1996). Cost-benefit analysis in integrated environmental assessment: some methodological issues. Ecological Economics 19(2): 157-168.

Munda, G., Nijkamp, P. and Rietveld, P. (1994). Qualitative multicriteria evaluation for environmental management. Ecological Economics 10: 97-112.

Neumayer, E. (2003). Weak Versus Strong Sustainability – Exploring the limits of Two Opposing Paradigms. Edgar Elgar publishing.

Ørum, J.E. and Christensen, J. (2001). Produktionsøkonomiske analyser af muligherderne for en reduceret pesticidanvendelse i dansk gartneri. Fødevareøkonomisk Institut, rapport nr. 128.

Pearce, W.D. and Turner, R.K. (1990). Economics of Natural Resources and the Environment. Harvester Wheatsheaf, NY.

Pedersen, N. m.fl. (2001). Slagtefjerkræ, Landbrugets Rådgivningscenter, Landskontoret for uddannelse, Landbrugsforlaget 2001.

Rees, W.E. (2003). Understanding Urban Ecosystems. An Ecological Economics Perspective. In: 'Understanding Urban Ecosystems – A new frontier for Science and Education'. Springer-Verlag.

Refsgaard, K. (forthcoming). Process-guided multicriteria analysis in wastewater planning. accepted for publication in Government and Planning C: Government and Policy.

Tiezzi, E. (2002). The Essence of Time. WIT. Ashurst, UK.

Vatn, A. (2004). Environmental valuation and rationality. Land Economics 80(1): 1-18.

Vatn, A. (forthcoming). Institutions and the Environment. Edward Elgar.

WCED (1987). World Commission on Environment and Development in: Our Common Future. Oxford University Press, Oxford.

5
Organic farming in a world of free trade

*Christian Friis Bach**

Introduction	136
The agricultural agenda	136
Export subsidies	137
Market access	137
Domestic support	138
Trade and the environment	139
Non-discrimination	140
Labelling	141
Preferential treatment	143
Conflicts with MEAs	145
Subsidies and environmental goods and services	148
Conclusions	149

Summary

This chapter attempts to place organic farming within the current WTO negotiations. It looks at the WTO negotiations on agriculture and on trade/environment linkages and points to possible consequences for organic farmers.

The trend toward agricultural liberalization may pose a threat to organic farmers, but may also strengthen agricultural development in a number of countries and lead to positive spill-over effects, where the food standards and organic farm principles gain a broader international recognition. Moreover there may be new opportunities for providing non-production distorting subsidies based on environmental and non-trade concerns; clearer rules for labelling, and improved market access for organic products. This could expand the options for organic farmers.

---

* Corresponding author: Vejgaard, Svinemosevej 7, DK-3670 Veksø, Denmark. E-mail: Christian@FriisBach.dk.

©CAB International 2006. *Global Development of Organic Agriculture: Challenges and Prospects* (eds N. Halberg, H.F. Alrøe, M.T. Knudsen and E.S. Kristensen)

To avoid conflict with the core WTO principles, policies and negotiation partners it is important to develop non-discriminatory rules and regulations for organic farming; to strengthen the capacity of developing countries to produce, certify and sell organic farm products and to have an internationally recognized definition and description of the organic farming practices and principles.

In the end, international efforts to carve out an important niche for organic agriculture will depend on the political will – which again depends on whether sufficient understanding and backing has been developed in both the North and the South. This is a huge challenge for the organic farming community. The organic farming community must seek international solutions and agreements to cope with the increasingly difficult challenges posed by the international agenda.

Introduction

It is increasingly evident that the conditions for organic farmers are framed by both a more competitive global market economy and a stronger international political cooperation. On one hand, markets are becoming more and more integrated and on the other hand international rules and regulations are being developed to cope with the global marketplace. Both tendencies will affect organic farmers.

Most importantly perhaps, the production of and trade in organic products may be affected by the current negotiations, the so-called Doha Development Round, in the World Trade Organization. The most important areas are the agricultural negotiations and the negotiations on trade and the environment. In both areas there are potential threats but also opportunities.

The agricultural agenda

The most important changes may arise from the agricultural negotiations in the Doha Development Round. Agriculture only entered the global trade negotiations during the Uruguay Round which ended in 1994 and has not been subject to the same discipline as industrial goods. Moreover, agricultural protectionism and agricultural subsidies are significant in most rich countries. The result has been restricted market access for developing countries and overproduction in many rich countries coupled with export subsidies, which has lead to lower and more unstable world market prices on most temperate agricultural commodities. Thus, in terms of potential impact and political tension agriculture is by far the most important topic in the current negotiations.

The negotiations focus on three major areas: export subsidies, market access and domestic support.

Export subsidies

The most contentious part of the negotiations so far has without doubt been the use of export subsidies primarily in the EU and the USA. Subsidization of exports is the most detrimental part of the agricultural policy in rich countries because the result is lower and more unstable world market prices. Moreover, export subsidies have been a critical part of the agricultural policy especially in the EU. It was simply impossible to get rid of surplus production without export subsidies, and they have thus been the safety valve in the Common Agricultural Policy of the European Union.

The EU has fiercely resisted a full elimination of export subsidies and this was part of the reason why the negotiations broke down at the Seattle and Cancun ministerial meetings. However, with the framework agreement of 31 July 2004 it is finally agreed to eliminate 'all forms of export subsidies and disciplines on all export measures with equivalent effect by a credible end date' (WTO, 2004a).

The implications of this for organic production may be positive. Export subsidies are not important for organic farm products as they are typically sold on niche markets, and the removal of export subsidies will lead to higher world market prices on a number of agricultural products, especially on beef, dairy products and sugar. This can spill-over into higher world market prices on organic farm products and stimulate production especially in lower and middle-income countries. However, there is still no end date agreed and the French agricultural minister has stated that the full elimination will not be in place before 2017 or perhaps 2015.

Market access

The next area is market access. There is still considerable protectionism in most rich country markets. In the EU the average tariff on agricultural products is 10 or 16.5% (depending on the product classification) and some agricultural tariffs are above 200%. Moreover, there are tariff rate quotas, import licences and even safeguard clauses that together frame a very effective but also very non-transparent shield against outside competitors (WTO, 2004b).

The negotiations in the WTO are still in their infancy and no concrete tariff reduction formulas have been agreed upon. However, it has been agreed that substantial improvements in market access will be achieved for all products and that progressivity in tariff reductions will be achieved through deeper cuts in higher tariffs. It is however also agreed that Members may designate an appropriate number, to be negotiated, of tariff lines to be treated as sensitive. This can become a loophole for further high protection rates on specific products as for instance sugar and dairy products in the EU and rice in countries like Japan and South Korea (WTO, 2004b).

It is likely, not to say unavoidable, that market access will become the most difficult and heated topic within the agricultural negotiations in the Doha Development Round.

In general however, the negotiations will lead to larger market access for outside competitors and this can of course put pressure on the organic farmers in Europe and in Denmark. All farmers will increasingly become part of a global agricultural market without the price security that has been provided by the Common Agricultural Policy in the EU.

Domestic support

Finally, domestic support has to be reduced and restricted. Here the current support measures are divided into three different boxes.

The amber box contains all domestic support measures considered to distort production and trade (with some exceptions). These include measures to support prices, or subsidies directly related to production quantities.

The blue box is the 'amber box with conditions' – conditions designed to reduce distortion. Any support that would normally be in the amber box is placed in the blue box if the support also requires farmers to limit production. The European hectare support is the best example. At present there are no limits on spending on blue box subsidies. But in the current negotiations, there will be demands for both lower amber and blue box support.

Finally, there is the green box. Here subsidies must not distort trade, or at most cause minimal distortion. They have to be government-funded (not by charging consumers higher prices) and must not involve price support. They tend to be programmes that are not targeted at particular products, and include direct income supports for farmers that are not related to (are 'decoupled' from) current production levels or prices. They also include environmental protection and regional development programmes. 'Green box' subsidies are therefore allowed without limits.

In the current negotiations the Cairns Group of agricultural exporters (led by Australia, but including South Africa and most of the South American countries) would like a tighter definition of the green box. They argue that measures to encourage farmers to avoid environmental damaging practices disregard the 'polluter pays' principle. Others, especially the EU, want the criteria to be even more flexible to take better account of non-trade concerns such as environmental protection and animal welfare. This viewpoint is partly reflected in the text of 31 July which states that the review and clarification of the green box should take due account of non-trade concerns (WTO, 2004a).

The most important aspect for organic farming is whether payments received through the agri-environmental programmes for environmental benefits remain in the green box. It is highly likely that they will. And if the concept of non-trade concerns is taken seriously it could benefit organic farming due to its focus on animal welfare and an ecologically sustainable way of farming.

Options may also be explored for providing WTO compatible support to the agriculture sector, particularly for research and development and quality assurance especially for reducing costs of certification of organic producers in developing countries by setting up local certification systems, promoting small-holder certification, and reducing the costs of international accreditation for certification in developing countries.

In this context, there is extreme urgency to develop international mechanisms to develop channels to provide market information and analysis about these products and strengthen capacity-building initiatives (Chaturvedi and Nagpal, 2003). Again international cooperation is critical.

Trade and the environment

Secondly, organic agriculture may be affected by the negotiations – or lack of negotiations – on trade and the environment in the Doha Development Round (this section is partly based on Bach, 2004).

The links between trade and the environment have been highly discussed and disputed for the past 15 years. Various disputes have arisen – from shrimp-turtle, American beef treated with growth hormones to asbestos from Canada and swordfish in Chile. These disputes arose because of a number of fundamental inconsistencies within international treaties and organizations, and may have implications also for organic agriculture.

The general problem is that while the rules in the World Trade Organization focus on removing obstacles to the free movement of goods, many environmental agreements and concerns are focused on protecting local environmental resources. While the rules of the WTO focus on national treatment and non-discrimination many environmental concerns attempts to respect local differences in natural and social conditions. These issues may not be in conflict. But sometimes they are. And the case of organic farming may be and become one of these cases.

The WTO provisions include several references to the environment, such as the Preamble to the Marrakech Agreement, which notes the importance of 'allowing for the optimal use of the world's resources in accordance with the objective of sustainable development, seeking both to protect and preserve the environment and to enhance the means for doing so in a manner consistent with their *[the WTO members, ed.]* respective needs and concerns at different levels of economic development'. But in practice this goal may be hard to square with the wish to promote free and unrestricted trade.

Non-discrimination

The core principle in international trade law is non-discrimination. This principle has three different components.

The first is *most-favoured* nation treatment, implying that all countries must be treated just as favourable as the most-favoured nation. It is therefore not allowed to discriminate between different countries. The two exemptions to this rule are that you can give more favourable treatment to: (i) other countries if you enter into a free trade agreement or customs union with them; and (ii) developing countries and the least developed countries.

The second non-discrimination principle is *national treatment*. It is allowed to have tariffs and border protection – although the ultimate goal of course is to remove these barriers. But whenever a foreign good has passed the border it is not allowed to discriminate between nationally produced and imported goods. This principle has given rise to problems for some organic labelling schemes, as they have required that the goods where processed at nationally certified processing plants. This made it impossible for foreigners to fulfil the requirements if they wanted to process the goods in other countries, and was in conflict with the national treatment principle.

Finally, the last non-discrimination principle is *'like products'*, which implies that products that are (physically) alike should be treated alike. This principle embodies a number of potential conflicts between trade rules and environmental concerns, as it can make it impossible to distinguish between goods according to how they are produced. It is allowed to discriminate between books and bananas but not between products that are 'like'. Importantly, like does not mean identical. Products that are not physically or chemically identical can still be considered 'like' products if, among other things, the products have the same end use, perform to the same standards and require nothing different for handling or disposal (UNEP/IISD, 2000).

This could have potentially important effects on organic farming and trade as the organic and non-organic products in a legal sense are physically 'like products', and mainly differ because of the production method. Thus, it is no surprise that the discussion on whether you should be allowed to distinguish products based on their production and processing method (PPM) has been one of the hottest issues in the trade and environment debate.

If you look at the WTO rules it is difficult to see exactly where in the rules discrimination based on PPMs is ruled out, and some argue that the same disciplines could be used for PPMs as those used for product-based distinctions. However, at the moment the political reality makes it difficult. It is particularly developing countries who are concerned about introducing PPM based criteria to the WTO. They fear that they could become a perverse tool in the broader WTO context and be used to undermine the market access or competitiveness of the weaker Members (ICTSD/IISD, 2003).

Labelling

The discussion about the possible use of PPMs has important implications for the potential use of eco-labels such as for instance organic labels. This issue has been bogged down in the WTO for years in a dispute between those who wish to develop guidelines with respect to environmental labelling schemes (mainly the EU and Switzerland) and those who do not (many developing countries) (ICTSD/IISD, 2003).

The question is whether labelling schemes are consistent with the WTO rules. Implicitly, eco-labelling is often an attempt to distinguish 'like' products based not on their physical characteristics or end use, but on the way they are produced or processed and based on their impact on the environment. As such, eco-labelling takes us to the very heart of the trade and environment dispute.

The Doha Development Agenda instructs the WTO Committee on Trade and Environment (CTE) 'to give particular attention to labelling for environmental purposes' (WTO, 2001a). However, progress has been disappointingly slow.

The proponents of eco-labelling systems argue that eco-labelling can be an efficient and trade-friendly way of achieving environmental objectives. The opponents argue that eco-labelling can become a disguised barrier to trade and thus constitutes a new set of non-tariff barriers. They stress that labelling requirements tend to deter importers from placing orders with developing countries' industries. This can become an obstacle to developing country companies – especially the small and medium sized companies.

The issue has led to tense and lengthy discussions in the WTO since 1995. However, the old frontiers between developing and developed countries are slowly being penetrated as more and more developing countries set up their own eco-labelling schemes. Examples of developing countries with an eco-labelling scheme based on a life-cycle approach and that are members of the Global Eco-labelling Network are Brazil, India, Korea, Taiwan and Thailand (see http://www.gen.gr.jp/). And others are in the process of setting up eco-labelling schemes. In Nicaragua the government has issued a mandatory technical regulation on the use of organic/ecological. Moreover, some developing countries see possibilities of carving out a market niche within existing schemes in rich countries. Finally, some international discipline is slowly being applied to the diverse eco-labelling schemes for instance through the development of standards for certain types of environmental labelling in the ISO 14200-series (see http://www.iso.ch). This can become important.

Most WTO members now argue that voluntary, participatory, market-based and transparent labelling schemes can potentially be efficient economic instruments to inform consumers about environmentally friendly products (WTO, 2003a), and the issue is coming closer to a possible agreement within the WTO.

However, this does not solve the problems for some of the poorest developing countries where lack of institutions and regulations leave them hopelessly behind in a global product race with more and more focus on non-content related issues. Higher health and environmental standards, labelling and demands for traceability can, as discussed earlier, effectively squeeze poor countries and especially the poor and small-scale producers out of the global marketing chains. This has been shown to happen for small-scale producers of fruits and vegetables in Kenya (Jensen, 2004).

For developing countries eco-labelling is simply seen as a cost-increasing measure. The cost of identifying and tracing the chain of custody – necessary to sell the product with an eco-label – has been estimated to be up to 1% of the border prices (Baharuddin and Simula, 1994). For Max Havelaar coffee the licence costs are 0.16 EURO/kg – which is less than 2% of the final retail price (information from Max Havelaar Denmark). This licence fee covers inspection as well as marketing efforts.

It should however, be recognized that labels – and the underlying standards – can also create advantages for developing country exporters as they add clarity to the product demands and requests on the export markets. Thereby, they can add to transparency and predictability compared to a situation with no labels and standards. This effect is often ignored in the discussions.

The discussions in the WTO have been made difficult by the multiplicity of labelling approaches. Environmental-labelling schemes can be classified according to their legal status (mandatory versus voluntary), according to the rule-setting body (governmental versus non-governmental), according to the review mechanism for criteria (static versus evolutionary), the geographic scope (national versus international) and according to whether they use criteria based on process and production methods (product-related PPMs versus non-product related PPMs).

The EC has proposed to focus the discussion on governmental and non-governmental voluntary eco-labelling schemes based on a life-cycle approach – that is using non-product related PPMs, and argues that they are legitimate within the rights and obligations of the WTO. Examples are the Nordic *Swan* and the EU's *Flower* label. Another example may be the joint European label for organic products.

The argument used by the EU is that these schemes are clearly environmentally focused, that they have been discussed at length in CTE, that they have received support at the World Summit for Sustainable Development in Johannesburg and that a 1999 ISO standard (no. 14024) has created internationally agreed criteria for such schemes (WTO, 2003b). The role of the WTO could be to receive notification of the schemes, ensure transparency and oversee that the schemes are applied in line with the principles in the WTO agreement on Technical Barriers to Trade (TBT). According to the EC there is no need to modify the existing rules to accommodate voluntary eco-labelling schemes.

A number of other countries have generally welcomed the focus on voluntary eco-labelling schemes. However, developing countries, and some developed countries, insist that the issue should be dealt with in the Technical Barriers to Trade Committee and not in the Trade and Environment Committee – thereby sending a clear signal that they still see eco-labelling issues as a technical barrier to trade rather than a necessary measure to ensure coherence between trade and environmental goals.

Some developing countries fear that the 'life-cycle approach' to eco-labelling will implicitly introduce non-product related PPMs through the backdoor, and question the legality of this approach within the WTO – a point supported by the USA. Generally, developing countries stress that it is more important to assist them in designing and accommodating eco-labelling schemes than to regulate labelling schemes internationally. They fear that international regulation will eventually lead to mandatory schemes and trade restrictions (WTO, 2003c).

Obviously an expansion and international recognition of eco-labelling schemes will require a further development of international standards, a more equal access to both the use and development of the schemes and technical assistance to developing countries in using the existing schemes and in setting up their own schemes. Producers of organic products in developing countries face several potential constraints related to conversion, production, marketing and government support policies. Constraints on conversion to certified organic agriculture include uncertainty about markets and price premiums. Certification costs, technical requirements and sanitary and phytosanitary (SPS) measures might act as obstacles to exports of organic food products from developing countries (Chaturvedi and Nagpal, 2003). These problems must be addressed if an international consensus is to emerge.

The EU tries to encourage companies from developing countries to apply for the EU eco-label by charging a special and reduced fee, and supports the establishment of Sustainable Trade Innovation Centres (STIC) in an attempt to help developing-country producers integrate environmental factors into their export strategy (one of the private–public partnerships launched at the World Summit for Sustainable Development in Johannesburg). The Global Eco-labelling Network has started technical assistance programmes. Here more could be done.

There is an urgent need to assist developing countries and especially small-scale producers in meeting the increasing environmental demands, standards and labels in rich countries.

Preferential treatment

Despite the outcome of the debate on labelling, voluntary eco-labels may not be sufficient to provide adequate incentives for organic products that face higher cost production and processing methods. Thus, another issue that has been discussed is the use of preferential treatment for organic products.

Preferential treatment has been around since the 1970s, where an enabling clause in the WTO gave way for the use of preferential tariff rates as a tool to ease the access of developing countries to the markets of the rich. However, the use of preferential treatment in pursuit of environmental goals is relatively new and primarily a European phenomenon.

Within the current Generalized System of Preferences (GSP) the European Union has created a system of 'special incentive arrangements' according to which developing countries may apply for further tariff preferences if they can demonstrate compliance with specified environmental standards (EC, 2001). The special incentives for the environment have only applied to sustainable managed tropical forests and have been quite limited in scope, but an extension to organically produced farm products has been considered and proposed (EU, 2004). The most recent modification of the GSP scheme has been extended to certain conventions relating to environmental protection (e.g. conventions designed to combat trafficking in endangered species and to protect the ozone layer), but apparently without a special focus on organic production (EC, 2004).

It is already evident, however, that the preferential treatment has not been effective in creating incentives for the compliance with international environmental agreements.

This is primarily because the preferences are not of a size or value that is comparable to the actual costs of compliance. This is partly because trade liberalization has brought down tariff barriers on a number of products to a level where preferential treatment will be of minor importance. And partly because the additional 20% tariff reduction granted if a country fulfils the conditions does not match the costs of fulfilling the conditions. As such the economic incentive structure is weak.

Another reason is that beneficiary countries have preferred not to have the content and implementation of their social and/or environmental legislation subjected to the rigours of scrutiny needed in order to grant the preferential treatment. Also the length and relative complexity of the evaluation procedures have made the arrangements even less attractive (EC, 2004).

Moreover, the EU schemes have been challenged by a number of developing countries, including India, Indonesia, the Philippines and Thailand, both in the political discussions in the WTO, in formal consultations and in a panel dispute. In the panel decision the European Community's GSP drug arrangements (specific preferences given to countries that attempt to fight the production of narcotics) were found to be inconsistent with the WTO rules (WTO, 2002a, 2003d). The Appellate Body, however, partly reversed this ruling and established that trade preferences to developing countries can be given according to their particular situation and needs, provided it is done in an objective, non-discriminatory and transparent manner, However, the Appellate Body found that the drug arrangements in the EU's current GSP drug regime were not based on objective and transparent criteria for the selection of the beneficiary countries

(WTO, 2004c). This can end the attempts to solve specific environmental problems by means of preferential treatment.

If objective and transparent criteria are developed then preferential treatment of organic products could be an option, but to make it a viable option the current resistance of developing countries must be overcome as well. This is a more difficult task and would require a stronger international collaboration within the organic farm community.

Conflicts with MEAs

Another important question is whether the multilateral environmental agreements (MEAs) can provide a sanctuary for the organic farmers. Can the various principles spelled out in the MEAs give some kind of backing for the environmental concerns in organic farm production? This question is unanswered at best and the interaction between the WTO rules and the MEAs is part of the current negotiations. This was one of the hottest international topics 10 years ago where many observes feared that the WTO rules would undermine a number of MEAs.

Luckily this has not happened. In fact, there has not yet been a case where a WTO member has had a measure challenged that was adopted to fulfil an obligation in a MEA. Not even the three MEAs with specific trade restrictions (the Basel Convention, Montreal Protocol and CITES) have lead to disputes although it has been a close call on several occasions (e.g. ivory trade between Japan and Zimbabwe or the swordfish dispute between the EU and Chile that was solved unilaterally before a ruling in the dispute settlement system). However, with the current US challenge against the European rules on genetically modified organisms the peace-period may be over (WTO, 2003e).

For most MEAs, the conflict with WTO rules is of minor importance, either because the problems are of a nature that can fall within the exemptions of the GATT devised to protect natural resources (Article XX), or because the – more recently negotiated – MEAs include specific language on the issue and attempt to place international environmental agreements more explicitly into the context of international trade rules.

Left are a few agreements where genuine problems could occur. Most prominent amongst them are the Convention on Biological Diversity and the Cartagena Protocol. The problems are here both the possible trade measures and the relationship between the agreements and the agreement on Trade-Related Intellectual Property Rights (TRIPs) in the WTO.

For the Convention on Biological Diversity there could be problems for instance if the provisions on access and benefit sharing for genetic resources lead to trade measures that are implemented in a manner which discriminates between national and foreign companies/products. This would conflict with the WTO principle on national treatment.

The other source of conflict is the relationship between the Convention and the TRIPs agreement (ICTSD, 2002). Here a number of issues are being discussed: access to genetic resources, traditional knowledge, access to and transfer of technology, handling biotechnology and the fair distribution of benefits from genetic resources (WTO, 2002b). According to the TRIPs agreement, plants and animals can still be excluded from patentability, but WTO members are required to establish a patent regime for microorganisms and this could interfere with a country's ability to preserve genetic resources and traditional knowledge (Brack and Gray, 2003). Moreover, plant varieties must be protected either by patents or by an effective *sui generis* system (GATT, 1994).

The potential conflict zone is even worse in the Cartagena Protocol on Biosafety from 2000. Here the EU rules on Living Modified Organisms can run into problems. One potential problem is the rules on a labelling and traceability system for genetically modified foods, which has repeatedly come under criticism in the WTO from the USA, Canada, Argentina and others. The USA – supported by the food industry – has insisted that 'tracing of products' is not equivalent to 'traceability', arguing that 'product tracing' is limited to 'one step forward and one step back' whereas 'traceability' of products refers to the whole production chain of a product (ICTSD, 2003). Indeed LMOs, and the differences between the WTO rules and the Cartagena Protocol, may become one of the future battlefields in the WTO. Here there could be important spill-over effects to organic agriculture.

A general – and profound – problem is that the WTO rules, in general, are based on a scientifically based risk assessment, while some environmental agreements lean more towards the precautionary principle, where countries can prohibit imports if scientific certainty is lacking due to insufficient information and knowledge. This displays a general difference between the American tradition of risk assessments and the European tradition of the precautionary principle, which can result in conflicts in other areas as well (EEA, 2001). This can also lead to conflicts. However, there is no general agreement on whether the assessment of risks is a purely scientific process, or whether, and to what extent, other factors may be taken into account when assessing a risk. One panel ruling – the hormones case – in the WTO has left room for the use of the precautionary principle and have not 'exclude a priori, from the scope of a risk assessment, factors which are not susceptible of quantitative analysis by the empirical or experimental laboratory methods'. It also recognized that measures could be based on a minority scientific opinion. Other panel rulings have confirmed that there is no minimum level of risk that must be found for a measure to be justified. Some argue that this leaves significant scope for the application of the precautionary principle (Bach, 2004).

In general the doomsday notion that the WTO rules would undermine most of the multilateral environmental cooperation is over. This is not least because the rulings of the WTO dispute settlement system have inched closer towards the issues of sustainable development, the environment and towards the agenda in the MEAs. This is especially true for the WTO appellate body, which in the Shrimp-Turtle case (WTO, 1998) recognized that import measures, based on *production and processing methods* (PPMs), and on *extraterritorial measures* (measures taken to protect the environment outside the territory of the state taking the measure), could be justified under the GATT's General Exceptions (Article XX). The measures must still, however, obey two criteria. They must be proven to be qualified for the exemption and they must not be applied in an either arbitrary or unjustifiably discriminatory way. The latter criteria in the ruling meant that: the measures must be flexible (allowing for different solutions); the state enacting the measure must have made good faith efforts to negotiate multilateral solutions; and there must be a reasonable phase-in time. Importantly, the ruling used MEAs to help interpret the GATT obligations and to assess the appropriate scope of unilateral action in the absence of an MEA (Mann and Porter, 2003). This is a loophole where multilateral environmental agreements can sneak in.

The Shrimp-Turtle case even affirmed the principle that 'ongoing serious good faith efforts to reach a multilateral agreement', *even short of finalizing an agreement*, can be sufficient to justify a unilateral trade related environmental measure (WTO, 2001b). This gives some hope for those arguing that MEA rules should prevail over the WTO rules. The important condition is that member countries attempt to negotiate international agreements.

However, there are still unsettled issues and as shown a number of potential conflicts. As such the search for more formal solutions should continue. Here the most promising strategy when linking existing international agreements with the WTO framework is the method used for food standards and standards related to animal and plant health standards. This was done in the WTO Agreement on the Application of Sanitary and Phytosanitary Measures (SPS Agreement) and in the Agreement on Technical Barriers to Trade (TBT agreement).

For the SPS agreement the link to relevant international standards is simply established by Article 3(2) with the wording: 'Sanitary or phytosanitary measures which conform to international standards, guidelines or recommendations shall be deemed to be necessary to protect human, animal or plant life or health, and presumed to be consistent with the relevant provisions of this Agreement and of GATT 1994.'

Moreover, the relevant international standards are defined as being the Codex Alimentarius Commission for food safety, the International Office of Epizootics for animal health and zoonoses, and the International Plant Protection Convention for plant health. Finally, for matters not covered by the above organizations reference is made to: appropriate standards, guidelines and recommendations promulgated by other relevant international organizations open for membership to all Members, as identified by the Committee (WTO, 2005).

This approach, which is sometimes referred to as a 'savings clause' in the WTO, could be extended to cover MEAs and is both simple and elegant. First, because the existing international standards are recognized, secondly because it creates a strong incentive for international harmonization and thirdly because the future development of the standards is maintained in the relevant organizations and not moved into the WTO. The result could be to strengthen both the WTO and the existing MEAs. At first only a few MEAs could be referred to and more MEAs could be included if for instance three-fourths of the WTO members were in favour.

In this way the environmental standards and policies will, using a phrase by the former Danish Minister for the Environment, Svend Auken, 'enter the WTO while simultaneously getting out of the WTO'. Environmental standards and policies must be developed in the appropriate environmental organizations but recognized by the WTO as being necessary and consistent with the WTO rules.

In any case the panel rulings and future approaches to tackling MEAs in the WTO constitute a wake-up call for the organic farming community. International agreements are critical. The approval of the Codex Alimentarius Guidelines for Organic Food was an important step towards the international harmonization of government regulations. The Codex guidelines can provide an internationally agreed framework for organic food moving in international trade (FAO, 2001). They acknowledge that organic farming standards are a legitimate means of recognizing product quality rather than a technical barrier to trade. But if future rights are to be defended it may be critical to develop even stronger international agreements on organic agriculture with clear and objective criteria and objectives.

Subsidies and environmental goods and services

Two additional issues within the trade and environment discussions may affect organic agriculture: environmentally perverse subsidies and environmental goods and services.

When it comes to subsidies, there is general agreement that if the traditional WTO agenda of removing trade-distorting subsidies and distortions can be targeted towards environmentally, perverse subsidies it can create win-win situations, which leads to both an improved environment and increased economic growth and development. Not utilizing these win-win situations bears a daily cost to both the environment and the economy.

One of the win-win situations that has been discussed in the WTO is trade- and production-distorting agricultural subsidies, which, apart from the negative impact on growth and development, are claimed to have a negative environmental effect in the countries where they are practised, and a negative impact on the environment of other countries, because they cause low and unstable agricultural prices.

Thus, the discussions put additional pressure on some of the unfair and production distorting subsidy schemes which implicitly can improve the competitive position of organic agriculture.

The other area where there may be potential implications for organic agriculture is removing trade barriers on environmentally sound goods and services. This is primarily about the global environmental industry, for example air pollution control, waste management, water treatment, noise control and energy management – an industry estimated at close to US$ 500 billion in 1998 (OECD, 2000) and considered to be one of the fastest growing global sectors. Despite the focus, this sector is still facing quite considerable trade barriers, both tariff and non-tariff barriers.

Liberalization of trade in environmental goods and services can bring about a number of positive effects: environmental improvements, transfer of knowledge, making environmental technologies cheaper and more readily available, realization of economies of scale, improved competition and increased local innovation and adaptation.

However, progress has been relatively slow. The negotiations in the WTO have concentrated on the definition of environmental goods and services. Most countries favour a relatively narrow definition focused on goods used to clean the environment or to contain or prevent pollution (OECD, 2003), while the EU argues that the list should also include products made in an environmentally sound manner, including products from sustainable (organic) agriculture, fisheries and forestry. This is rejected especially by developing countries. They fear that definitions based on the production and processing methods (PPMs), could lead to new protectionism. The final outcome is difficult to predict, but again it seems likely that a firm and internationally recognized definition of organic agriculture would increase the chance that organic products could gain faster market access than traditionally produced agricultural products.

Conclusions

The current WTO negotiations may yield both threats and opportunities for organic farmers.

Agricultural liberalization can increase competition and further marginalize organic farming practices. On the other hand removing trade-distorting subsidies and providing additional market access for developing countries to rich countries markets can strengthen agricultural development in a number of countries, and lead to positive spill-over effects, where the food standards and organic farm principles gain a broader international recognition.

Cutting the production-distorting subsidies may pose problems for organic farmers but the opportunities for providing non-production-distorting subsidies based on environmental and non-trade concerns may improve and expand the options for organic farmers. Here it is critical to have an internationally recognized definition and description of the organic farming practices.

The non-discrimination principles within the WTO, and especially the concept of like-products, contradicts the basic idea in organic agriculture. Here it is critical to identify non-discriminatory rules and regulations for organic farming; to develop clear and objective criteria for organic agriculture based on international agreements; and to overcome the resistance found in developing countries by strengthening the capacity of developing countries to produce, certify and sell organic farm products. This can lead to a stronger international dialogue on organic farming principles.

It is likely that the current negotiations on trade and the environment will result in clearer rules for labelling. There may be new opportunities for preferential treatment of organic farm products. And liberalizing environmentally friendly goods could carve out a niche for organic products.

To avoid the threats and explore the opportunities within the WTO negotiations the main requirement is to develop an internationally coordinated approach and effort within the organic farming community.

References

Bach, C.F. (2004). International Trade, Development Aid and the Multilateral Environmental Agreements. Paper prepared for the Ministry of Foreign Affairs in Denmark, Copenhagen. http://www.friisbach.dk/fileadmin/cfb/Consult/TAE/T-A-E.pdf

Baharuddin, Hj.G. and Simula, M. (1994). Certification schemes for all timber and timber products. ITTO, Yokohama, Japan.

Brack, D. and Gray, K. (2003). Multilateral Environmental Agreements and the WTO. The Royal Institute of International Affairs and the International Institute for Sustainable Development. http://www.riia.org/pdf/research/ sdp/MEAs%20and%20WTO.pdf

Chaturvedi, S. and Nagpal, G. (2003). WTO and Product-Related Environmental Standards. Emerging Issues and Policy Options. EPW Special Article. http://www.epw.org.in/showArticles.php?root=2003andleaf=01andfilename=5341andfiletype=html

EC (2001). Council Regulation no. 2501/2001 of 10 December 2001 applying a scheme of generalized tariff preferences for the period from 1 January 2002 to 31 December 2004 OJ 2001 L 346/1 ('GSP Regulation').

EC (2004). Developing countries, international trade and sustainable development: the function of the Community's generalised system of preferences (GSP) for the ten-year period from 2006 to 2015. Communication from the Commission to the Council, the European Parliament and the European Economic and Social Committee. COM (2004) 461 final.

European Environment Agency (EEA) (2001). Late Lessons from Early Warnings: The Precautionary Principle, 1896-2000. Copenhagen.

EU (2004). The European Union's Generalised System of Preferences – GSP. http://trade-info.cec.eu.int/doclib/docs/2004/march/tradoc_116448.pdf

FAO (2001). Codex Alimentarius – Organically Produced Foods. http://www.fao.org/documents/show_cdr.asp?url_file=/DOCREP/005/Y2772E/Y2772E00.htm
GATT (1994). Uruguay Round Agreements. http://www.wto.org/english/docs_e/legal_e/27-trips_04c_e.htm#5
ICTSD (2002). BRIDGES Trade BioRes, 26 September 2002, http://www.ictsd.org/biores/02-09-26/story1.htm, BRIDGES Trade BioRes, 11 July 2002, http://www.ictsd.org/biores/02-07-11/story1.htm, BRIDGES Trade BioRes, 18 April 2002, http://www.ictsd.org/biores/02-04-18/story1.htm
ICTSD (2003). Bridges Trade BioRes Newsletter. In Brief. Vol. 3, Number 13, July 11, 2003. See http://www.ictsd.org/biores/03-07-11/inbrief.htm
ICTSD/IISD (2003). Trade and Environment: Doha Round Briefing Series, vol. 1, no. 9. http://www.ictsd.org/pubs/dohabriefings/doha9-trade-env.pdf
Jensen, M.F. (2004). Food Safety Requirements and Smallholders: A Case Study of Kenyan Fresh Produce Exports. Chapter IV. Ph.D. thesis. The Royal Veterinary and Agricultural University, Copenhagen, Denmark.
Mann, H. and Porter, S. (2003). The State of Trade and Environment Law 2003: Implications for Doha and Beyond. International Institute for Sustainable Development and Centre for International Environmental Law. Available at http://www.iisd.org/publications/publication.asp?pno=570
OECD (2000). Environmental Goods and Services: An Assessment of the Environmental Economic and Development Benefits of Further Global Trade Liberalisation, COM/TD/ENV(2000)86/FINAL.
OECD (2003). OECD Joint Working Party on Trade and Environmental Goods: A Comparison of the APEC and the OECD lists. Information Note by the OECD Secretariat. Available at http://docsonline.wto.org as WT/CTE/W/228, TN/TE/W/33
UNEP/IISD (2000). Environment and Trade: A Handbook. http://iisd1.iisd.ca/trade/handbook/3_3.htm
WTO (1998): United States – Import of Certain Shrimp and Shrimp Products. Report of the Appellate Body. WT/DS58/AB/R, October 12, 1998. Available at http://docsonline.wto.org
WTO (2001a). Doha WTO Ministerial – Ministerial declaration. http://www.wto.org/english/thewto_e/minist_e/min01_e/mindecl_e.htm
WTO (2001b). United States – Import Prohibition of Certain Shrimp and Shrimp Products. Recourse to Article 21.5 of the DSU by Malaysia. AB-2001-4. WT/DS58/AB/RW.
WTO (2002a). Trade Policy Review Body – Trade Policy Review – The European Union – Draft Minutes of Meeting, WT/TPR/M/102. Available at http://docsonline.wto.org
WTO (2002b). Review of the Provisions of Article 27.3(b), relationship between the TRIPs Agreement and the Convention on Biological Diversity and Protection of Traditional Knowledge and Folklore. Information from Intergovernmental Organizations. Addendum. Convention on Biological Diversity. IP/C/W/347/Add.1. Available at http://docsonline.wto.org
WTO (2003a). Committee on Trade and Environment – Report of the meeting held on 29 April 2003. Note by the Secretariat. WT/CTE/M/33. Available at http://docsonline.wto.org
WTO (2003b). Labelling for Environmental Purposes. Submission by the European Communities under Paragraph 32 (iii). Committee on Trade and Environment. WT/CTE/W/225. Available at http://docsonline.wto.org or as http://europa.eu.int/comm/trade/issues/global/environment/docs/contrib030303.pdf

WTO (2003c). Committee on Trade and Environment – Report of the meeting held on 29 April 2003. Note by the Secretariat. WT/CTE/M/33. Available at http://docsonline.wto.org

WTO (2003d). European Communities – Conditions for the Granting of Tariff Preferences to Developing Countries. Panel report. WT/DS246/R. Available at http://docsonline.wto.org

WTO (2003e). European Communities – Measures Affecting the Approval and Marketing of Biotech Products – Request for the Establishment of a Panel by the United States. WT/DS291/23. Available at http://docsonline.wto.org

WTO (2004a). Text of the 'July package' – the General Council's post-Cancún decision. http://www.wto.org/english/tratop_e/dda_e/draft_text_gc_dg_31july04_e.htm

WTO (2004b). Trade Policy Review: European Communities. http://www.wto.org/english/tratop_e/tpr_e/tp238_e.htm

WTO (2004c). European Communities – Conditions for the Granting of Tariff Preferences to Developing Countries. Appellate Body Report. WT/DS246/AB/R. Available at http://docsonline.wto.org

WTO (2005). The WTO Agreement on the Application of Sanitary and Phytosanitary Measures (SPS Agreement). http://www.wto.org/english/tratop_e/sps_e/spsagr_e.htm

6
Certified and non-certified organic farming in the developing world

Nicholas Parrott,* Jørgen E. Olesen and Henning Høgh-Jensen

Introduction	154
Certified organic farming in the developing world	155
Non-certified organic farming	159
Explicit organic approaches	159
Agro-ecological approaches	160
Unlearning from the Northern experience	164
Organic farming and yields	165
The economic returns to organic farming	167
Multifunctionality	169
Livelihoods	170
Actors, networks and motivations	171
Questions to guide future studies in organic farming	173
Conclusions	176

Summary

This chapter analyses the dynamics behind the growth of organic farming in the developing world. It identifies two organizational trajectories within this; a highly visible, and rapidly growing, formal certified sector and a less easily quantified, informal or agro-ecological, sector. The former is clearly oriented towards global commodity chains and is intended to bring benefits to producers by offering premia for ecological production and, as such, can be viewed as a form of ecological modernization. The latter approach implies a reconceptualization of Northern perceptions about organic farming – in so much as those adopting this approach often report higher yields, incomes and net returns, leading to enhanced food and economic security. This chapter examines

---

* Corresponding author: c/o Dept of Rural Sociology, Wageningen University, Hollandseweg 1, 6706 JW, Wageningen, The Netherlands. E-mail: nick.parrott@wur.nl

these reported benefits through the prisms of agronomy, economics, multi-functionally and livelihoods analysis.

Introduction

This chapter examines the different forms that organic farming takes in the developing world and the complex layers of meaning, perception, institutional involvement and farmers' interest associated with these different forms. It starts with a brief analysis of the most visible and quantified element of organic farming, formally certified organic production. This is a rapidly expanding sector which is proving to be highly successful in increasing returns to farmers through meeting growing consumer demand for organically (and often ethically) produced food. It is a sector that is currently attracting great institutional interest, but is one that is almost exclusively oriented to producing food (and other commodities) destined for export to the North.

A second less visible aspect of ecological farming is that of non-certified organic production. Here definitions of what constitutes organic farming can become quite blurred and the boundaries become quite ill-defined. While some of those practising and promoting non-certified organic farming explicitly align themselves with the organic movement, many others do not – and some even appear to distance themselves from such an association.

Analysis and comparison of these types of organic farming raises a number of questions, which this paper attempts to address. The first concerns the extent of organic farming in the developing world. We argue that this is larger, probably much larger, than certified figures suggest. Indeed, certified organic land in the developing world might be seen as the 'tip of an iceberg' of far more widespread and culturally embedded farming practices that rely upon ecological principles and knowledge. The challenge of estimating the size of the sector, in terms of practice and knowledge base (both formal and informal), is complicated by the fact that information about the informal organic sector is highly diffused and much of the information that is available overlaps with other approaches. In this analysis these will be grouped together and termed 'agro-ecological approaches'.

The second issue concerns the benefits of organic farming. Certified organic farming in the developing world largely conforms to Northern experiences in that farmers adopt a set of standards and in return normally receive higher prices for their produce. In a world where the terms of trade are often stacked against primary producers this is undoubtedly a benefit (see Chapter 5). Many other benefits are also often evident. Organic production practices may lead to higher yields, in particular when the starting point has been extensive and low-input conditions. It may, however, also lead to lower yields as reviewed in Chapter 10. However, it generally tends to reduce exposure to toxic chemicals, and risk of crop failure, reduces outgoings and exposure to debt, builds environmental resilience and can build on local knowledge and socio-cultural practices. The widespread existence of non-certified organic farming suggests that these benefits may be sufficient in themselves to attract farmers, even those without access to (or interest in) export production. This has clear implications for the dominant (largely Northern) perception about the benefits of organic farming. In this chapter we attempt to assess these benefits of organic farming from three different perspectives: those of classical agronomy (focusing on yields and soil fertility), of agricultural economics (focusing on net returns), and of human ecology, which incorporates both multi-functionality and livelihoods approaches. This leads us to formulate a series of recommendations for a broad-based research agenda on organic approaches in the developing world.

Certified organic farming in the developing world

Each year IFOAM publishes a statistical handbook about organic farming (OF). In recent years these have shown a meteoric rise in the number of certified organic farmers and amount of land under organic management (Willer and Yussefi, 2004 and previous years). Analysis of these figures provides a useful starting point for understanding the status and trends of certified organic farming in the developing world. In overall terms OF is highly developed in Latin America, with almost 6 million ha of certified organic land, 24% of global certified organic land (and slightly more than Europe). Three Latin American countries – Argentina, Brazil and Uruguay – are among the 'top ten' countries in terms of total certified land. By contrast Asia and Africa have relatively little certified organic land, 0.9 and 0.3 million ha respectively: 3.7% and 1.3% of the global total.

Figure 6.1 shows the very rapid growth in the amount of certified organic land in the South over the past 3 years.[1] Three years ago only one country (Argentina) in the South had more than 1% of its agricultural land under organic management. By 2004 seven countries had broken through this 1% threshold. In Uruguay 4% of cultivated land is now certified as organic and Costa Rica is close to this figure. In Africa, Uganda has passed the 1% threshold, largely due to the Export Promotion of Organic Products from Africa programme (EPOPA; http://www.grolink.se/epopa/), which has engaged more than 30,000 Ugandan smallholder farmers in export oriented organic production.

---

[1] While some of this growth may be due to improvements in data collection methods, the majority is likely to be due to new conversions and the rates are in line with the growth in organic market demand (Sahota, 2004).

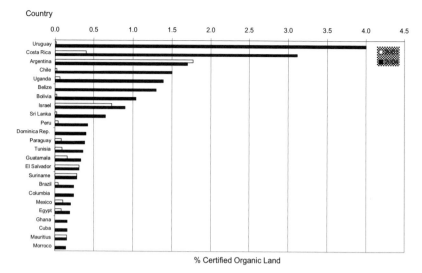

Figure 6.1. Growth in certified organic land by country: 2001–04 as proportion of total agricultural area (%) (adapted from Willer and Yussefi, 2001, 2004).

In Asia, Sri Lanka is a leading producer, with 0.7% of its cultivated land certified as organic; several other Asian countries (Indonesia, Kazakhstan and the Ukraine) have recently entered the organic market with a substantial production base, but because of their size the proportions remain relatively small. In most countries, however, the figures are still fractions of a percentage point, suggesting that organic farming is undeveloped. Yet at the same time there is no evidence of organic agriculture existing in many countries: this is particularly true of Africa, where about half of the countries have no recorded organic production (Parrott and van Elzakkar, 2003).

These statistics also reveal staggering differences between average farm sizes within selected countries (Figure 6.2). The mean size of organic farms ranges from more than 1000 ha (in Argentina and Uruguay) to less than 1 ha (in Indonesia, Senegal and Benin). While analysis of mean size can hide quite substantial deviations within individual countries,[2] and can be quite a crude tool, it does provide a basis for exploring the diversity, patterns and styles of organic farming that exists in different countries.

This suggests that, far from being uniform, the certified organic farming sector is highly heterogeneous, taking many forms, involving very different practices (and degrees of change in practice), and 'meaning' different things under different circumstances. Organic farming on extensive grasslands may not require very many changes in management practice. For instance, in Argentina the Association for the Promotion of Rotational Grazing was set up in 1965 to

---

[2] For instance, in Argentina 74% of the organic land is managed by 5% of farmers, who raise sheep in Patagonia.

develop ecological management systems for herds and pastureland. Few changes to long established practice were required when they 'became' organic (see Harriett-Walsh, 1998). By contrast, the organic production of plantation crops, which are likely to have been intensively grown in monocultures, may require significant changes in agricultural practice to maintain soil fertility, prevent infestations and achieve acceptable yields and returns. Here premiums may be necessary to offset initial declines in yields.

From a development perspective it is these smaller units that are of the greatest interest. The changes involved in smaller mixed farming will vary according to the regime employed prior to certification. They may involve a change of management techniques and/or a greater orientation towards markets in terms of crops and varieties grown, timing, continuity and quality control. They almost certainly imply significant organizational changes in order to link farmer groups with (international) markets and enabling them to comply with market requirements and organic standards (IFAD, 2003).

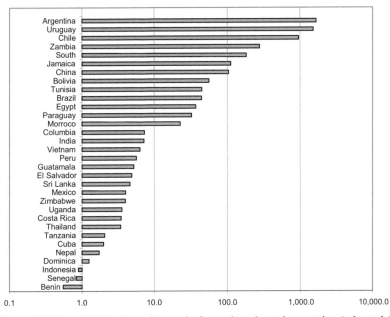

Figure 6.2. Size distribution of organic farms in selected countries (adapted from Willer and Yussefi, 2004). Average area in ha per organic farm, logarithmic scale.

Thus the orientation and motivation of farmers in these different groups will vary substantially. While farms in the larger categories are likely to be primarily focused on the market, small-scale farmers are likely to have a more mixed set of goals and a significant proportion of their farming will be subsistence oriented. Such farmers are most often the target of 'development' programmes, designed to improve living standards and reduce poverty. In this sense the involvement of the organic movement with small-scale farmers, through programmes such as EPOPA, represents a new turn on traditional development agendas by seeking to engage them with markets in a 'trade not aid' approach. Similar approaches can be found in moves for fair trade, local value adding, higher value crops, even the development of agri-tourism (e.g. Pinheiro *et al.*, 2002; Brozena, 2004). There are often large overlaps between these approaches.

Despite the differences in farm size and styles, one thing that these farming styles share in common is that they are geared to the global organic market. Production is primarily geared to meet growing demand in the rich Northern countries. Even Argentina, with a long established organic movement and (the crash of 2001 notwithstanding) a large and affluent middle class, exports 90% of its organic produce (Lenoud, 2004). In other Latin American countries the figure is higher still (ibid). This focus on inter-continental trade may partially explain the substantial differences in certified land in Latin America, Africa and Asia: a far higher proportion of Latin America's agricultural output is sold on distant markets. In Africa and Asia a higher proportion goes towards meeting regional food needs (Millstone and Lang, 2003). In many parts of Africa these basic needs still remain unmet for large parts of the population.

Thus certified organic production occurs in a context in which formalized market relationships predominate. The global market sets the terms of reference, defining what is (and what is not) organic, provides consumers with guarantees about production methods, producers with premia (often substantial) and prevents free riders. But these figures are a social construction, reflecting participation in global markets that require the setting of standards, certification and inspection and which provide the resources to do this. This reliance on international trade is recognized as a limiting factor and there is a growing move to develop informal and alternative certification systems that can be used to develop national and local markets. These have gained particular importance in parts of Latin America and serve to link producers with local consumers either directly through farmers markets etc., or indirectly through supermarket chains (e.g. Kotschi *et al.*, 2003; Agapito *et al.*, 2004; Boor, 2004; Fernando Fonseca, 2004). These are very important developments in terms of generating better returns for farmers and raising the profile of organic production in domestic markets. Their emergence is an important means of developing organic farming as a commercial activity in many countries as they provide an important 'third way' in which organic farming can be made both commercially viable and achieve a distinct local or national profile. It is not the purpose of this chapter to assess the development of this approach but rather to step back and look at the non-certified organic sector.

Non-certified organic farming

The previous section showed how certified organic farming is rapidly growing in the South. Yet, there is evidence that ecological farming practices are far more widespread than these figures suggest. Indeed, it is useful to think of certified organic farming as the tip of a much larger iceberg. Unfortunately we do not have the tools, or the resources, to measure the underwater part. We can, however, identify four main areas of this 'hidden' world of ecological farming, some of which are more visible than others.

Explicit organic approaches

First, it is possible to identify a number of organizations that explicitly recognize themselves as organic – but whose main concern appears to be with local development priorities. The IFOAM Membership Directory provides one avenue for identifying such organizations. Table 6.1 provides a breakdown of IFOAM members in the developing world. It is interesting to compare these membership figures with the figures for certified organic land in the previous section. Latin America accounts for 9% of IFOAM's membership (compared to 24% of certified organic land); Asia accounts for 19% of IFOAM's membership (compared to approx. 4% of certified organic land) and Africa 7% of IFOAM's membership (compared to approx. 1% of certified organic land). Some countries have many IFOAM members even though their formal certified organic sector is small. These include Kenya, Senegal and Benin in Africa; India, the Philippines and Nepal in Asia. Equally there are countries with IFOAM members where there is no record of any certified production. This includes Togo, the Democratic Republic of Congo, and Réunion in Africa, Malaysia, Iran, Singapore, Syria, Taiwan and the UAE in Asia and Trinidad and Tobago in Latin America. If we look at these organizations it is evident that many are development–oriented NGOs, with relatively little export orientation. This supports the notion of a world of organic farming removed from the imperatives of the global economy.

One such example is the Kenyan Institute of Organic Farming (KIOF). Widely regarded as a pioneer training centre it has trained more than 16,000 farmers in organic techniques over the past 20 years and is visited by farmers and advisors from all over East Africa. A recent letter from KIOF's Director summarized their achievements and *raison d'être*.

Table 6.1. IFOAM members by country (adapted from IFOAM, 2002, for Latin America and Asia; 2003 for Africa).

No. of IFOAM members	Latin America N=61 (9.5%)	Africa N=45 (7%)	Asia N=125 (19%)
> 30			India (35)
20–30			China (25) Japan (21)
10–20	Argentina (17)		
5–9	Chile (8) Mexico (8) Brazil (6) Peru (5)	Egypt (9) Kenya, Togo (8) South Africa (7) Uganda (6) Benin, Senegal, South Africa (5)	Philippines (9) Turkey (8) Nepal (5)
> 5	Costa Rica, Ecuador (4) Bolivia, Guatemala, Nicaragua, Paraguay (2) Colombia, Cuba, Trinidad and Tobago (1)	Burkina Faso, Dem. Rep. of Congo, Ghana, Malawi (3) Cameroon, Mozambique, Zimbabwe (2) Ethiopia, Madagascar, Mali, Mozambique, Réunion, Tunisia, Zambia (1)	Malaysia, Pakistan, Sri Lanka (4) Israel (3) Bangladesh, Indonesia, Thailand, Vietnam (2) Iran, Korea, Palestine, Singapore, Syria, Taiwan, United Arab Emirates (1)

The emphasis of the training programme is on organic farming for poverty alleviation and household self-sufficiency. Adopting organic farming techniques leads to substantial increases in productivity. It also leads to an increase in the diversity of food produced and generates surpluses for sale. Other benefits include savings in expenditure on chemical inputs and health benefits from not being exposed to these. Enhanced food security remains the most visible impact of organic farming among small scale low income groups.

John Njoroge (pers. comm.)

Agro-ecological approaches

Alongside organizations that formally align themselves to the organic movement there are a number of other organizations and movements that share very similar approaches to agriculture in their design. These include biodynamic agriculture, permaculture, nature farming, bio-intensive, ecoagriculture, and agro-ecology. Much has been written about the differences and similarities between these movements and the extent to which they are (or are not) compatible with organic farming. While farming according to these different styles may not always meet internationally recognized organic standards, the approach and philosophy that they employ is largely compatible with the organic world-view and rely on similar agronomic practices. Some of these movements have played a significant role in developing organic approaches in the developing countries. The biodynamic movement has for example played a key role in developing the organic movement in Egypt and the permaculture movement has launched a host of training initiatives in Zimbabwe (Parrott and van Elzakkar, 2003).

The activities of such organizations have led to significant levels of uptake of organic farming styles. For example in Argentina PRO-HUERTA has been supported by the Ministry of Social Development and trained almost two million families in backyard organic farming as a way of improving their nutritional status. As a result many of these families have been able to move from a food deficit to a surplus (Scialabba, 2000, cited in Parrott and Marsden, 2002). In southern and eastern Africa the Participatory Ecological Land Use Movement Association (PELUM) is a network of civil society organizations that share a commitment to fighting regional poverty, remedying social injustices, and capacity building. Ecological agriculture plays a central role in pursuit of these objectives. They currently have 138 member groups in nine countries. Experiences of some members of this network in Zimbabwe are compellingly told in Kotschi *et al.* (2003 pp. 121–130).

Low External Input Sustainable Agriculture (LEISA) has been promoted for around 20 years (Reijntjes *et al.*, 1992). The movement largely came about as a result of the failures or inappropriateness of 'Green Revolution' approaches to smallholder farming systems (see Shiva, 2001). LEISA stresses the importance of

> developing appropriate technologies and farming systems, through participatory approaches. They promote the optimal use of natural resources and local processes and, if necessary, the safe and efficient use of external inputs. (LEISA, 2003, p. 20)

Clearly this definition highlights a major difference in the visions of the LEISA and the organic movements. Despite this, LEISA often uses the same designs as the organic approach, but without necessarily sharing the same underlying philosophy. By way of example, ILEIA recently joined forces with Greenpeace, PAN Africa and the WWF to produce 'Farming Solutions': a website that catalogues the successful application of LEISA approaches (Greenpeace *et al.*, 2003). A high proportion of projects described on this site appear to be de facto organic; very few make any explicit reference to the safe and effective use of limited amounts of external inputs.

At the same time there are a number of international research institutes (IRIs) that have programmes, or elements of their programmes, that also appear to be consistent with the organic approach. One striking example is the Biological

Control Centre in Cotonou, Benin (part of the International Institute for Tropical Agriculture). This centre has developed several programmes that rely on organic principles (e.g. pest–predator relationships) or the development of locally produced myco-pesticides.[3] However, there is no explicit recognition of the 'organic' nature of their work and collaboration between the organic movement and BCC is almost non-existent (see also Chapter 12). Others like CIAT recognize agrobiodiversity as essential for fighting land degradation (CIAT, 2004).

Other IRIs (such as the World Agroforestry Centre) also have research programmes that broadly overlap with the organic agenda. At the national level institutes, programmes and individuals with an interest in organic approaches are likely to exist although tracking them down will be a major task. (The recent establishment of the International Society of Organic Research – ISOFAR – may help with this.) For example, in Burkina Faso much of the government-led programme to combat desertification and reclaim land relied on organic approaches (see articles in Djigma *et al.*, 1989). Cuba has a well-documented, locally based research and development programme for the biological control of pests and diseases (Rossett and Benjamin, 1996) and parts of Cuba's agricultural regime (particularly urban agriculture) incorporate many organic principles - including that of 'nearness' (Wright, 2004). With the exception of Cuba (whose move towards organic approaches has often been exaggerated (ibid.) there is rarely explicit recognition that such work is in fact organic.

Finally, we come to the most problematic area, that of resource-poor farmers' farming without the use of agrochemicals. This approach is often driven by poverty and lack of resources and characterized by low outputs and unsustainable practices, rather than by the conscious adoption of organic farming techniques or acceptable levels of productivity (FAO, 2003). Studying this group focuses attention on what constitutes organic farming and the extent to which traditional practices conform with organic designs. One recent report that examined the extent and potential of organic farming in sub-Saharan Africa identified that:

> isolated (organic) techniques are sometimes practised, (but) there is a general lack of an integrated approach to soil fertility and crop protection management and under-exploitation of the full range of techniques that would maximise the benefits of locally-available natural resources (Harris *et al.*, 1998)

It is exceptionally difficult to make a general assessment of the extent of organic farming practices in traditional farming systems. There are some shining examples of locally developed farming systems that make extensive use of available natural resources, maintain soil fertility and generate a wide range of products to both meet domestic needs and sell surpluses on local markets. The

---

[3] For details of this programme see http://www.lubilosa.org/INDEX.htm

Chagga tribe in the Kilimanjaro region of Tanzania is one example of this (see Parrott and Marsden, 2002, pp. 14–15). Miguel Altieri has catalogued many more such examples in Latin America (see Altieri, 2002; Altieri *et al.*, undated). Yet all too often traditional farming fails to produce sufficient food to meet domestic needs for poor rural families, let alone surpluses to meet other needs. The majority of the world's hungry live in rural areas and fail to generate sufficient food to meet their own requirements. In eastern and southern Africa, it is estimated that rural poverty accounts for as much as 90% of total poverty (Dixon *et al.*, 2001).

Sometimes improved and unimproved traditional farming systems can co-exist in the same area. Verkerk (1998) writes of visiting two farms within 3 km of each other in Zimbabwe. The first was organic by 'default' and gave the appearance of near-dereliction: aphids were rampant on mature plants, there was clear evidence of nitrogen deficiency, no evidence of organic matter having been applied into the soil or of attempts to control weeds, mulch the soil or provide shade. The vegetables were discoloured and malformed, and would have not been acceptable on local markets. By contrast, a nearby organic farm had healthy plants and a thriving nursery, showed evidence of extensive composting and mulching and of successful pest management.

This example highlights how difficult it is to generalize about the adoption of organic approaches in traditional farming systems and to identify the mechanism causing the system to perform better as it often is a combination of multiple factors, including knowledge. It is equally difficult to assess the potential for the uptake of organic approaches in addressing development issues. Harris *et al.* (1998) reported on organic farming in sub-Saharan Africa, noting that 'two thirds of farmers using organic methods said that they did so because they cannot afford fertilizers, pesticides or medicines for animals'. Yet the same report identified that 60% of the farmers whom they interviewed claimed that lack of knowledge prevented them from adopting organic methods: four times more than for any other single cause (ibid. p. 12). Thus many small traditional farmers may not use agrochemicals because they cannot access them. This is often termed organic by default, but is often far from sustainable, as it uses few, if any, organic methods and is not based on an organic philosophy. Lack of access to agrochemicals may constrain conventional intensification of these systems, but lack of knowledge seems to be as much of a constraint on their ecological improvement.

Some studies show how farmers reject the use of chemicals, hybrids and industrial farming techniques because they do not fit with their 'farming styles' or their cultural repertoires (see for example Hebinck and Mango, 2004). One could expend much effort in asking when a traditional farmer, employing local or imported knowledge of ecological farming strategies, becomes an organic farmer. A more relevant question is how farmers with detailed knowledge of ecological farming systems can be encouraged to share that knowledge with their less knowledgeable neighbours? Social connectedness has been shown to be important for knowledge adaptation (Hamilton and Fischer, 2003; Wu and Pretty, 2004). This has been found the major cause for the extensive conversion to organic farming in the southern part of Jutland in Denmark. A key-question is thus how horizontal information transfer can be promoted?

This review of agro-ecological approaches section raises three critical points. First, it shows that far more farmers use organic practices than figures for certified organic farming suggest, although it is difficult to ascertain how many because of the blurred boundaries with agro-ecological approaches and traditional farming. Second, it shows that a number of organizations are involved in doing research in, or disseminating, organic practices, although many do so without explicitly presenting their activities as organic. Indeed some appear to consciously distant themselves from being branded as organic. Third, there may be a potential for strengthening organic approaches if the appropriate dissemination strategies (and funds to implement these) can be put in place. In the following section we turn to assessing the benefits of ecological farming and to strengthening learning linkages between North and South.

Unlearning from the Northern experience: reconceptualizing the benefits of organic farming from a Southern perspective

The experience of organic farming in the North is specific to a unique set of circumstances. Spurred on by cheap and readily available supplies of inputs (including feed concentrates), agriculture in the North has, over the past 50 years, become highly industrialized. National (and regional) patterns of agriculture have, to varying degrees, become specialized, rationalized and increasingly geared to a global market economy. Under such circumstances most farmers who convert to organic farming experience a decline in yields, at least in the initial years, while the soil regains its fertility, while pest–predator relationships establish a new balance and while farmers learn how to adapt the practices to their conditions. In most Northern countries farmers are able to cushion themselves against this through a combination of premium prices and environmental subsidies.

This general experience has led to the perception that organic farming has little to offer third world farmers (FAO, 1998, p. 12). However, third world farmers' production systems range from intensive cropping exemplified by Bangladesh with 114 kg fertilizer nitrogen applied per hectare annually (FAO 2003) to subsistence farmers that do not use agrochemicals at all. Further, biophysical conditions range from very fertile soils with abundant precipitation to extremely depleted soils with erratic and insufficient precipitation. Consequently, there is a need for identifying the mechanisms inherent in the farming systems that under the given biophysical conditions can lead to predictions regarding yield developments after conversion to organic farming. This will be dealt with in detail in Chapter 10. Here we focus on the existing evidence from comparative studies from three different perspectives: the agronomic (in terms of yields and soil fertility), the economic (in terms of margins) and the human ecological in terms of multifunctionality and livelihoods. It should be noted that these effects will not be uniform as they will depend on where the farmer is starting from (in terms of intensity and type of production, market orientation etc.) and which organic technique(s) and strategies they employ.

Organic farming and yields

Much evidence has been published in recent years that contradicts the conventional wisdom that organic farming leads to a decrease in yields. Several of these studies have been criticized as being based on anecdotal evidence and not subject to peer review. It has also been argued that these improvements may result as much from periods of intensive input from extensionists, which would lead to increases of productivity whatever systems were being promulgated (see also Chapter 10). One of the most extensive studies of organic systems in developing countries has been carried out by Pretty *et al.* (2002) as part of a broader review of sustainable farming systems. Their results showed that conversion to organic farming has increased yields of different crops by between 30 and 500%. This led them to the conclusion that:

> in all cases where reliable data has been reported (from organic systems) increases in per hectare productivity for food crops and maintenance of existing yields for fibre have been shown. This is contrary to the existing myth that organic agriculture cannot increase productivity (Pretty, 2002, p. 142)

However, this analysis was carried out in the context of low input smallholder systems. Systems that were previously relatively input intensive are likely to experience an initial decline in yields, as discussed in Chapter 10. A review of organic farming systems in Latin America, commissioned by IFAD (2003), documented the experiences of 14 different farmers groups in six countries, covering more than 5000 farmers and 9000 ha. It found that some farmers (those using low or no input techniques prior to becoming organic) experienced increases of up to 50% in yields. Others maintained their yields at approximately the same level. Farmers who converted from relatively intensive use of agrochemicals experienced a drop in yields in the short-term.

These findings stress the point that the effects on yields of conversion to organic farming are highly dependent on the starting point of the farmers. Further, it suggests that farmers in low input systems with low productivity (often the target of extension programmes aimed at increasing food security) are those who are most likely to benefit in terms of productivity from adopting organic practices. Rather than regarding adoption of organic farming as a single trajectory, it is helpful to recognize that it can involve several distinct elements (or combinations thereof) whose effects will vary (see Box 6.1).

Yield comparisons form a central plank of classical agronomy. Such comparisons often provide the basis for extrapolations of different agricultural regimes to meet future national or global food requirements, and they can greatly influence the funding available for supporting and developing different approaches to farming. Yet farmers rarely base their decisions about what crops to plant or how to manage them solely on the basis of expected yields. They balance this information against many factors, including economic ones (the availability and cost of inputs and likely returns) and a range of broader human ecological considerations.

Box 6.1. Pathways to higher yields from ecological farming (adopted from Pretty, 2002).

Throughout this text (and the broader literature about organic farming) there is an unfortunate implication that becoming an ecological farmer involves one uniform process of transition. Previous discussions on the heterogeneity of certified organic farmers showed that this is far from the case. Farmers (certified or not) adopt different ecological approaches, singly or in combination. Pretty provides a typography of such changes. Whilst this typography may not be comprehensive it provides useful insights into the diversity of pathways that can be pursued and a possible basis for developing a more rigorous typography of transition.

1. Intensification of a single component of the farm system – such as home garden intensification with vegetables and trees.

2. Addition of new productive elements to a farm system – such as fish in paddy rice- that boosts total food production, but does not necessarily affect productivity of staples.

3. Better use of natural capital to increase total farm production, by water harvesting or irrigation scheduling enabling growth of additional dryland crops, increased supply of water for irrigated crops or both.

4. Improvements in per hectare yields of staples though the introduction of new regenerative elements into farm systems (e.g. integrated pest management) or locally appropriate crop varieties and animal breeds.

Trials comparing organic, chemical and mixed approaches to improving soil fertility often show that a mixed approach most often maximizes yields. Yet analysis of whether farmers adopt a mixed approach, a chemical one or an organic one would provide useful insights into the trade-offs that farmers make between what is technically optimal and what is socially and culturally optimal. Here organic research could well draw on that undertaken under LEISA approaches in order to gain greater understanding about the strengths and weaknesses of its own approach.

In the context of food security it is further important to remain aware of the key role that agricultural systems themselves play in determining entitlement to food. Systems that depend upon sustainable use of locally available natural resources and farmers' knowledge and labour are far more likely to meet the needs and aspirations of resource-poor farmers than those which require costly or scarce external inputs. In this respect organic farming is a technology that, to quote Gandhi, 'puts the last man first.'[4]

Thus, for the organic movement there are both advantages and disadvantages of becoming pre-occupied with debates about yields, yield potential and potentials for feeding the world. On the one hand such a debate (and research to inform that debate) is necessary to legitimate organic approaches as a strategy for meeting food security (particularly in the eyes of mainstream funding organizations). From this perspective, improving understanding of the effects of organic farming on yields (and the mechanisms employed) can be seen as a priority. Yet, on the other hand, food security is dependent on a far broader range of issues, and approaches that do not satisfy these other criteria are unlikely to be adapted by farmers.

The economic returns to organic farming

Improved economic returns to organic farming can be attained by a number of mechanisms, which include:

- Higher yields: leading to greater surpluses to sell at markets or increased food security throughout the year. This can reduce or eliminate the 'hungry period' when farm households have to buy in food stocks, often at high prices.

- Fewer financial outlays: farmers previously using artificial inputs save on the cost of these. Sometimes there is a trade-off between using fewer inputs and needing more labour, but often this can be achieved by utilizing underemployed family labour. This trade-off is explored in a little more detail in the section on livelihoods.

- Market access and premia: some farmers find that conversion to organic production opens up access to (foreign, certified) markets that previously didn't exist. Others find that they can command significantly higher prices through organic production. The levels of premia vary according to the crop

---

[4] And, as we have come to realize since Gandhi's day, the last man is more often than not a woman – see discussion under livelihoods.

in question. The IFAD survey of Latin American organic producers showed premia of between 22%, to banana producers in the Dominican Republic, and 150%, to cacao producers in Costa Rica (IFAD, 2003, p. 14).

- There are also often fewer fluctuations in the price of organic produce, which helps provide farmers with a more secure planning framework. Organic farmers also find it easier to tap into fair trade schemes and the longer term nature of relationships within organic supply chains (compared to the spot markets that dominate conventional production) also help generate better prices and greater price stability. These latter factors may not be related to organic production per se but still provide significant benefits to farmers.
- Finally, in areas where organic production involves a critical mass of producers, it can have a knock-on effect on the prices paid by conventional buyers, anxious not to lose their supply base (van Elzakkar and Tulip, 2000; cited in Parrott and Marsden, 2002, p. 90).

Farmers using organic growing methods but without certification or access to global markets can also benefit from enhanced prices. Zonin *et al.* (2000) examined an agro-ecological project in the Erixim region of Brazil, which sold organic produce locally and significantly contributed to farmers' incomes and countered the previous problem of rural out-migration. In other instances non-certified organic produce can command higher prices because of other superior qualities. One Indian farmer claims that he receives a 30% premia on his non certified organic rice as customers know that it tastes better (Faisal, pers. comm.; cited in Parrott and Marsden, 2002, p. 90). In other instances in India, wheat grown in rotation with certified organic cotton attracts higher prices for the same reason, and sugarcane, grown in the same system, attracts a premium from the mills because of its higher sugar content (ibid. p. 25).

Organic farming does not however offer a panacea for increasing the incomes of poor farmers. Many activities necessary for laying the foundation of a productive sustainable/organic farming system (e.g. double digging, tree-planting, rainwater harvesting etc.) require a substantial additional input of labour (Howard-Borjas and Jansen, 2000; IFAD, 2003). In some cases this can be generated from the farm household, in others it will involve hiring additional casual labour. Particularly in the latter case, the farmer will need convincing of the benefits before making such an investment. In many cases such investments, once they have proved their worth, have contributed to increases in rural employment opportunities, which also benefits the landless poor. In some instances they are reported to have offset or even reversed rural urban migration and created new markets for labour and land (see Hassane *et al.*, 2002).

Multifunctionality: organic farming and farmer's non-economic objectives

Organic farming also has other benefits besides economic ones. These relate to:

Risk aversion: organic approaches have been shown to be most resilient in bad years (of drought or infestation). This characteristic resonates strongly with the risk aversion strategies prevalent amongst smallholder farmers.

Health: Organics can significantly reduce exposure to toxic pesticides and herbicides – a benefit that accrues to both smallholder and plantation workers. This also permits pregnant women and those with young children to work in fields, free from the fear of exposure to such chemicals. Although widely disputed, organic food is widely viewed as having better nutritional qualities and this is particularly relevant in food insecure areas and areas where HIV/AIDS is prevalent.

Environmental resilience: organic farming promotes this in several ways: increased soil moisture retention capacity guards against effects of drought and helps reduce off-farm water flow; water and soil conservation techniques reduce erosion. One example of this phenomenon was found in a comparative study of the resilience of organic and non-organic farms in Honduras in weathering and recovering from Hurricane Mitch (Holt-Giménez, 2002).

Bio and Agro-biodiversity: at micro-level this helps increase productivity of soil and reduce infestation/disease. Diversity of crops (and emphasis on locally evolved ones) can help guard against crop (or market) failure, provide a longer growing season and provide a more balanced and nutritious diet. One study of the characteristics of organic and non-organic farms in India (funded by I-GO) showed that ecological farms had on average 200 trees per ha, compared to less than 40 on non-ecological farms (der Werf, 1993).

In assessing these benefits and their inter-relationships it is important to evaluate them from a farmer's perspective rather than from given (often mono) disciplinary standpoints. The emergence of alley cropping with leguminous hedgerows in the 1970s illustrates this point. Whilst it was a technically sound practice, it did not prove to be socially accepted. This illustrates the shortcomings of a technology transfer approach and the need to involve farmers in design and implementation. The development and promotion of green manures like Mucuna during the 1990s illustrate the importance of seeing the farm system through the farmers' eyes. Researchers initially promoted use of Mucuna because they saw the benefits in terms of increasing biological nitrogen fixation. When used as a green manure Mucuna gave very good yield responses (Figure 6.3). Farmers were much more involved in the development and evaluation of this technology and it emerged that farmers valued Mucuna more because of its very high weed-suppressing ability (Vanlauwe et al., 2001). Similar experiences of farmers adopting ecologically sound technologies can be found in Giller (2003) and Bunch (2003). Both authors stress that farmers need to see multiple benefits from green manures before adopting the technology. Often this will be the provision of food or fodder for livestock.

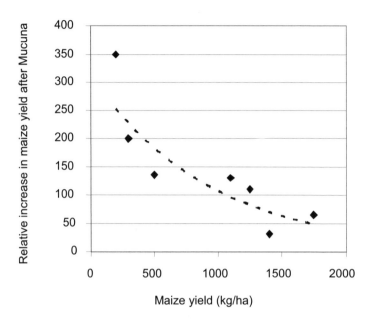

Figure 6.3. Maize grain yield after a Mucuna green manure crop relative to a control crop. Data summarized from trials in West Africa described in Vanlauwe et al. (2001).

Livelihoods

A final factor that needs to be considered is that farms do not operate in closed systems but as part of a larger social and economic environment, which presents unique sets of opportunities and constraints. These are often geographically, generation, and gender specific and so cannot readily be generalized. They do however play a significant role in determining decisions about allocation of labour and capital resources within the farm system. Their influence therefore needs to be recognized when evaluating the attractiveness of adopting organic farming approaches. Here we confine ourselves to briefly discussing three examples of how such issues can influence farm-based decisions about adopting organic approaches.

1. Insecure land tenure can provide a major constraint to making the initial investment required to improving a farm system. Farmers could be, quite logically, reluctant to invest in improving their land if they do not know whether they will have tenancy rights next year (especially if the improvements make the farm more attractive for repossession by the landowner). Landowners may also be opposed to farmers making improvements to the land, such as terracing, bunding or planting trees as these may grant usufruct rights in the future.

2. Several studies have noted that organic farming is a particularly appropriate option for female farmers. Women are more likely to be engaged with de facto organic farming, will have a greater affinity with organic approaches (Moali-Grine, 2000; Njai, pers. comm.) and may have pools of knowledge that can be drawn upon. Despite this, the role of women in organic agriculture is often overlooked, especially by extension services (Kinnon and Bayo, 1989; Kachru, 2000).

3. Competing labour opportunities: alternative, more remunerative and higher status, forms of employment may offer a significant pull factor, attracting particularly younger family members away from farm-based employment. Opportunities to work in e.g. the tourism industry are an obvious example of this (UNDP, 1992). Tourism can, however, also offer new opportunities to sell higher value organic produce, thus making running a farm more remunerative.

Actors, networks and motivations

The forms that organic farming takes at present are strongly influenced by the involvement of different actors and the creation of networks. The same factors influence the forms that organic farming will take in the future. In assessing the potential for scaling up organic approaches it is important to examine, albeit briefly, the actors currently involved and the potential for enrolling new participants.

The organic movement today in the developing world is dominated by a curious mix of NGOs and market forces (initially 'ecological entrepreneurs', but increasingly 'big players' in the global commodity markets). There is notably little involvement from governments (unless they see the potential for tapping into premia export markets) and, with relatively few exceptions, little in the way of explicit and concerted organic research.

Moreover the structure and dynamics of the actor network for organic farming in the developing world is very different from the one that exists in the North. In the North the organic movement has, over the past 20 years, successfully positioned itself in the minds of consumers as being able to provide safer, healthier, quality food, which also safeguards the environment and animal welfare. This is no small achievement. This has been done through maintaining a strictly defined identity, based on agreed and legally defined standards. In the South this division is, as shown in this chapter, not so clear-cut. There are many grey areas. The boundaries between organic farming, sustainable agriculture, and rural development are not so clear-cut.

This poses a dilemma when thinking about the future development of the organic movement in the developing world. Part of this dilemma is structural: standards are essential for maintaining consumer confidence, justifying market premia for ecologically produced food and creating a legally defensible organic identity. On the other hand the emphasis of the organic movements on standards alienates many within the LEISA movement, concerned about increasing the productive capacity of small and poor farmers. They feel it is quite wrong

> to add more production constraints on already hampered farmers and (that) there is no need to convince resource poor people to refrain from using chemical inputs.
> (P. Rotach, Brod für die Welt; quoted in Kotschi et al., 2003)

The lack of any form of standards within the sustainable agriculture has led to it having become a catch–all phrase and becoming 'a broad church, which attracts a diverse congregation with a range of different 'core beliefs' (Parrott and Marsden, 2002).

Yet there are also other barriers that need addressing. Many within the LEISA school find organic methods are often useful but have reservations about adopting the organic philosophy. For example, the organic view that the use of artificial inputs, such as fertilizers, undermines the long-term productive capacities of the agro-ecosystem is not universally (or even, maybe, widely) accepted amongst LEISA practitioners who see organics is as dogmatic, rigid, unscientific and sometimes mystical in its outright rejection of the use of (even small amounts of) mineral fertilizers and artificial pesticides. Advocates of LEISA can be excused for viewing organics as prescriptive and not focused on farmers' needs – a charge that the organic movement needs to be sensitive to. One of the key questions for the development of the organic (and LEISA) movement is whether they can set aside these differences and learn from each others' experiences. Both movements have great strengths and share an opposition to common threats, such as unjustified dissemination of GM crops and increasing emphasis on high input technologies being advocated by powerful corporations and research interests.

On the basis of this analysis we can identify some of the main opportunities and constraints facing the development of the organic movement (Boxes 6.2 and 6.3).

Box 6.2. SWOT analysis for organic farming from a small farmer's perspective.

STRENGTHS	WEAKNESSES
Often close to existing practice	OF can involve substantial additional labour input, especially in initial phase.
Fits with risk-averse strategies of farmers	
Can be based on, valorize and develop local knowledge systems	Investment of time and other resources might not seem worthwhile if benefits are not apparent, if the farmer has more pressing problems or lacks secure tenure.
Can valorize local resources, creating new uses for them	
	Lack of knowledge and readily available information
Reduces risk of exposure to harmful pesticides (which are often improperly used)	Poor (or non existent) extension services, which often promote (d) conventional approaches.
Reduces exposure to the debt required to purchase inputs	
	Lack of local markets
OPPORTUNITIES	Cost and complications involved in certification
Possibility for greater food security and selling surpluses locally	Low literacy rates hamper record keeping
Can create new on-farm income-generating opportunities	
	THREATS
Possibility for premia prices if certified	Belief in modernization
Opportunities for greater social contacts through project meetings etc.	Peer group ridicule

Questions to guide future studies in organic farming

The following questions are designed as a guide to some of the issues that might be tackled as a part of future studies of the potential spread of OF.

1. What is the extent of organic farming (including informal systems) in the developing world? Evidence suggests that this is much larger than figures for certified organic land suggest – but what is its true significance? Whilst it would be impractical to conduct a global survey, a small number of country or regionally based studies could help provide insights into the significance of organic farming on a larger scale. Such a survey would not be merely an exercise in gathering numbers but would provide a basis for developing a broader knowledge base, covering the following questions.

Box 6.3. SWOT analysis for organic farming from an institutional perspective.

STRENGTHS	WEAKNESSES
Donor agencies attracted to OF projects – but generally only for certified produce	Perception that organic production will lead to decline in yields and is not relevant to food-insecure countries
Governments attracted to OF as an opportunity to tap into premia export niche markets and increase overseas revenues and balance of payments	Non export/trade benefits of OF are rarely appreciated by governments
OPPORTUNITIES	Lack of published (and peer reviewed) data to support organic movements, claims that OF increases food security and farmers' self-reliance in the absence of export markets
OF can enhance national, regional and local self reliance	
OF can make substantial contributions to sustainable environmental resource use	Lack of supportive policy framework in most countries
Many NGOs have established core groups of trained farmers and in some cases national advocacy lobbies	Lack of demand (and disposable income) for developing domestic premia markets
	Lack of local standards, inspection and certification capacity
High levels of, and potential for, women's participation in OF	**THREATS**
Opportunity for tying OF in with other goals of the international policy community – notably with regard to biodiversity, anti-desertification, gender inequalities and potentially global warming (carbon sequestration effect)	OF runs counter to professional training and belief/value system of scientists and policy makers
	OF poses threat to established agri-business interests
	Lack of adequate extension services and research capacity in many countries

2. What are the benefits and drawbacks of organic systems? This chapter has already identified some of the initial published work in terms of yields, net returns and multiple objectives. Other projects are ongoing (e.g. IFAD have completed an evaluation of organic farming systems in Asia and are planning a similar exercise in Africa). Data gathered from question 1 above would help us to develop a better understanding of the economic, environmental and social implications of organic farming. What mechanisms are the most successful?

What pathways of transition are involved? What flanking activities[5] are required? Do the outcomes match with initial expectations? What constraints exist? To what extent do initiatives draw upon (or draw out) local knowledge? Such an analysis would also need to examine failed initiatives. These can tell us as much as the successful ones (although there is always a natural reticence to talk about them). Equally it is important to avoid overly focusing on 'project' based initiatives (which are often better recorded and evaluated) as this approach tends to lead to the neglect of grass-roots and spontaneously adopted initiatives. Research that compares organic with mixed and agro-chemical approaches also needs to be included in such a review.

3. Catalogueing the Organic Knowledge Base: this paper has suggested that there is much hidden organic research capacity and experimentation. There is a desperate need to draw this information together, to identify active and sympathetic researchers, institutes and development agencies, the knowledge that they already gathered and their current research and extension agendas.

4. Socially Grounded Research: it is important for organic farming research to be socially grounded in the practices, systems, aspirations and constraints of farmers. While the organic movement is keen to promote holistic approaches it is not always appreciated that this involves multi-disciplinary cooperation. In this respect the organic farming movement can learn much from the LEISA and the participatory approaches of research and implementation that it has developed over the years.

5. The Importance of a Multiple Objectives Approach. This relates to the previous point. Organic farming is not just about increasing yields, farmers' access to markets or whatever. It has a multiple objective orientation. Socially grounded research helps understand these multiple objectives, which may not always totally coincide with those of funding/donor agencies or researchers. Organic farming has the potential to contribute to meeting many objectives at both the farmer and institutional level. Combating food insecurity, desertification, global warming and promoting diversity and local self reliance are some of the institutional initiatives that organic farming may contribute to. Demonstrating the existence and extent of such benefits may open opportunities for accessing funds from various global initiatives to support organic farming. Research along the lines suggested above could contribute to this process and form a useful basis for developing local research capacity and increasing support for OF initiatives in the South

6. Spill-over from organic to conventional agriculture. One final issue, worthy of exploration is the actual and potential significance of the transfer of organic approaches to conventional agriculture. In Egypt the use of pheromone traps

---

[5] These might include enhancing management of the environmental resource base (e.g. water harvesting) or developing social and economic capacity, such as farm planning, record keeping or literacy.

(developed by the organic cotton sector) led to a 90% decrease in pesticide use in the conventional sector. The value (both economic and environmental) of such adaptation provides a further justification for expanding organic research capacity.

Conclusions

In conclusion this paper wishes to emphasize three main points. First, that the practice of, and knowledge base about, organic farming is likely to be far wider than we currently think. Published work in organic and agro-ecological approaches transcends a huge number of disciplines: soil science, ethno-botany, entomology, anthropology and so on. It is often difficult to identify such research because it does not specifically identify itself as organic. Secondly, organic research capacity in developing countries is probably far more developed than we imagine. For example, when IFOAM held a conference in Ouagadougou in 1987 a significant proportion of contributions came from researchers and extensionists from Sahelian countries (particularly Mali, Burkina Faso and Senegal) (see Djigma et al., 1989). Both these factors suggest that, rather than embarking on identifying new research programmes, what is really needed at this time is a knowledge synthesis. The second point strongly points to the need for the active involvement of partners in the South in designing and implementing this. Such partner(s) would ideally have prior knowledge of local networks, practices and priorities and would better placed for identifying what is possibly the most pressing problem for organic research in the developing world – the question of how this information can be made available to those who need it most; resource-poor farmers and those who are in contact with and support them.

Thirdly, there is no doubt that there is a need for more robust research on the effect of organic farming in enhancing yields (and the mechanisms through which this occurs). Such evidence challenges conventional wisdom. Accepting this evidence leads us to a position of arguing that market premia can be seen as a bonus, rather than the principal reason for considering adopting OF. The organic movement is aware of, and publicizes, this message (see Rundgren, 2002). It is a message that needs frequently reiterating in order to penetrate the received wisdom that associates organic farming with elite consumption patterns and the 'luxury' of environmentally benign production methods. This poses a dilemma in terms of research priorities, for on the one hand such research may well be necessary to broadening the legitimacy of organic farming to policy makers and fund-holders, yet at the same time such an approach may not be the most effective way of getting the organic message across to those farming communities themselves.

References

Agapito, J., Luis Jorquiera, J., Wu, S. and Schreiber, F. (2004). An Example from Peru. Ecology and Farming 38: 9-10.
Altieri, M. (2002). Non certified organic agriculture. In: Scialabba, N. and Hattam, C. (eds) Organic Agriculture, Environment and Food Security. FAO, Roma, 107-138.
Altieri, M., Rosset, P. and Thrupp, L.A. (undated). The Potential of Agro-ecology to Combat Hunger in the Developing World. Agro-ecology in Action. http://www.cnr.berkely.edu/~agro-eco3/the_potential_of_agro-ecology.html Viewed 1/7/03.
Boor, A. (2004). Local Marketing: News from IFOAM's Programme I-GO. Ecology and Farming 38: 8.
Brozena, C. (2004). The 'big picture' in northern India. Ecology and Farming 36, 10-12.
Bunch, R. (2003). Adoption of Green manure and Cover Crops. LEISA 19(4): 16-18.
CIAT (2004). CIAT in Perspective 2003-2004: Cardinal Points–Charting the Direction of Our Work. (at http://www.ciat.cgiar.org/newsroom/report2004.htm).
Dixon, J., Gulliver, A. and Gibbon, D. (2001). Farming Systems and Poverty. Improving Farmers' Livelihoods in a Changing World. FAO and World Bank, Rome and Washington DC. 412.
Djigma, A., Nikiema, E., Lairon, D. and Ott, P. (eds) (1989). Agricultural Alternatives and Nutritional Self-sufficiency. Proceedings of 7th IFOAM Conference, Ouagadougou, Burkina Faso.
Faisal (personal communication) Hyderabad, India.
FAO (1998). Evaluating the potential of organic agriculture to sustainability goals. FAO's Technical Contribution to IFOAM's Scientific Conference, Mar del Plata, Argentina, November.
FAO (2003). FAOSTAT, FAO, Rome.
Fernanado Fonseca, M. (2004). Formal organic certification: Not the only answer. Ecology and Farming 38: 39-40.
Giller, K. (2003). Kick starting legumes. LEISA 19(4): 19.
Greenpeace, ILEIA, Oxfam and PAN-Africa (2003). Farming Solutions: Success Stories for the Future of Agriculture. www.farmingsolutions.com. Viewed 16/6/03.
Hamilton, S. and Fischer, E.F. (2003). Non-traditional agricultural exports in highland Guatemala: Understandings of risk and perceptions of change. Latin American Research Review 38(3): 82-110.
Harriet-Walsh, D. (1998). Organic agriculture in Argentina. Ecology and Farming, September 12-17.
Harris, P.J.C., Lloyd, H.D., Hofny-Collins, A.H. and Browne, A.R. (1998). Organic Agriculture in Sub-Saharan Africa: Farmer Demand and Potential for Development. HDRA, Coventry.
Hassane, A., Martin, P. and Reij, C. (2002). Water harvesting, land Rehabilitation ad Household Food Security in Niger. IFAD/Vrije Universteit Amsterdam.
Hebinck, P. and Mango, N. (2004). Cultural repertoires and Socio technological regimes: a case study of local and modern varieties of maize in Luo land, West Kenya. In: Wiskerke, H. and van der Ploeg, J.D. (eds) Seeds of Transition: Essays on Novelty Production, Niches and Regimes in Agriculture. Royal van Gorcum, Assen (Netherlands).
Holt-Giménez (2002). Measuring Farms' Agro-ecological Resistance to Hurricane Mitch. LEISA 17: 18-20.

Howard-Borjas, P. and Jansen, K. (2000). Ensuring the Future of Sustainable Agriculture: what could this mean for Jobs and Livelihoods in the 21st Century? Paper Presented to the Global Dialogue EXPO 2000 'The Role of the Village in the 21st century: Crops, Jobs and livelihood. Hannover, 15–17 August 2000.

IFAD (2003). The adoption of Organic Agriculture Amongst Small Farmers in Latin America and the Caribbean: Thematic Evaluation. IFAD Report no. 1337.

IFOAM (2002). Directory of Member Organisations and Associates. Tholey-Theley (Germany) IFOAM

IFOAM (2003). Directory of Member Organisations and Associates. Tholey-Theley (Germany) IFOAM

Kachru, A. (2000). Organic agriculture and women: A third world women's perspective. Ecology and Farming, March, pp. 20-1.

Kinnon, F. and Bayo, B. (1989). How can more appropriate farming systems be developed? In: Djigma *et al.* Agricultural Alternatives and Nutritional Self-sufficiency. Proceedings of 7th IFOAM Conference, Ouagadougou, Burkina Faso, pp. 58-67.

Kotschi, J., Bayer, W., Becker, T. and Schrimpf, B. (2003). AlterOrganic: Local Agendas for Organic Agriculture in Rural Development. Proceedings of an International Workshop held in Bonn-Konigswinter, Germany 21–24 October 2002. Agro-ecol, Marburg, Germany.

LEISA (2003). Reversing Degradation. 19 (4), Leusden, ILEIA.

Lenoud, P. (2004). Latin America. In: Willer, H. and Yussefi, M. (eds) The World of Organic Agriculture: Statistics and Emerging Trends. IFOAM, Bonn, pp. 123-148.

Millstone, E. and Lang, T. (2003). An Atlas of Food: Who Eats What, Where and Why. Earthscan, London.

Moali-Grine, N. (2000). Ecodevelopment of mountainous regions in Kabylie (Algeria): Utopia or reality? In: Alföldi, T., Lockeretz, W. and Niggli, U. (eds) The World Grows Organic, Proceedings of the 13th International IFOAM Conference, Basle (Switzerland), August. IFOAM. Tholey-Theley, Germany.

Njai (personal communication). Gambian Agricultural Extension Service. p. 171.

Njoroge, J.W. (1996). Organic farming lifts the status of women; an African case study. Ecology and Farming, May, pp. 27-28.

Parrott, N. and van Elzakkar, B. (2003). Organic and Like Minded Movements in Africa: Development and Status. IFOAM, Bonn.

Parrott, N. and Marsden, T. (2002). The Real Green Revolution: Organic and Agro-ecological farming in the South. Greenpeace Environmental Trust, London.

Pinheiro, S., Cardosos, A.M., Turnes, V., Schmidt, W., Brilo, R. and Guzzat, T. (2002). Sustainable Rural Life and Agro-ecology – Santa Catarina State, Brazil. In: Scialabba, N. and Hattam, C. (eds) Organic Agriculture, Environment and Food Security. FAO, Roma, pp. 227-234.

Pretty, J. (2002). Lessons from certified and non certified organic projects in developing countries. In: Scialabba, N. and Hattam, C. (eds) Organic Agriculture, Environment and Food Security. FAO, Roma, pp 139-162.

Pretty, J.N., Morrison, J.I.L. and Hine, R.E. (2002). Reducing Food Poverty by Increasing Agricultural Sustainability in Developing Countries. Agriculture, Ecosystems and the Environment 95: 217-234.

Reijntjes, C., Haverkort, B. and Waters-Bayer, A. (1992). Farming for the Future: An Introduction to Low External Input and Sustainable Agriculture. Macmillan/ILEIA, Leusden (NL).

Rossett, P.M. and Benjamin, M. (1996). The Greening of the Revolution Cuba's Experiences with Organic Agriculture. Ocean Press, San Francisco.

Rundgren, G. (2002). Organic Agriculture and Food Security. IFOAM Dossier no. 1 Bonn, Germany.

Sahota, A. (2004). Overview of the Global market for Organic Food and Drink. In: Willer, H. and Yussefi, M. (eds) The World of Organic Agriculture: Statistics and Emerging Trends. IFOAM, Bonn, pp. 21-26.

Shiva, V. (2001) Yoked to Death: Globalisation and the Corporate control of Agriculture. Research Foundation for Science, Technology and Ecology, New Delhi.

UNDP (1992). Benefits of Diversity: An incentive Towards Sustainable Agriculture. UNDP, New York.

Uphoff, N. (ed.) (2002). Agro-ecological Innovations. Earthscan, London.

Vanlauwe B., Wendt, J. and Diels, J. (2001). Combined application of organic matter and fertilizer. In: Tian, G., Ishida, F. and Keatinge, J.D.H. (eds) Sustaining Soil Fertility in West Africa, SSSA Special Publication No. 58, Madison, USA, pp. 247-280.

Verkerk, R. (1998). Pest management in tropical organic farming: a cry for on farm research. Ecology and Farming, September, pp. 26-7.

der Werf, E. (1993). Agronomic and economic potential of sustainable agriculture in South India. American Journal of Alternative Agriculture 8(4): 185-191.

Willer, H. and Yussefi, M. (eds) (2001). Organic Agriculture Worldwide: Statistics and Future Prospects. Stiftung Ökologie and Landbau, Bad Durkheim (Germany).

Willer, H. and Yussefi, M. (eds) (2004). The World of Organic Agriculture: Statistics and Emerging Trends. IFOAM, Bonn.

Wright, J. (2004). Agricultura Organica y Seguridad Alimenteria. Existe una dependencia entre Ambas? Paper to VIth Congress de la Sociedad Espanola de Agricultura Ecologica. Almeira. 27 September – 2 October 2004.

Wu, B. and Pretty, J. (2004). Social connectedness in marginal rural China: the case of farmer innovation circles in Zhidan, north Shaanzi. Agriculture and Human Values 21: 81-92.

Zonin, W.J., Gonçalves Jr, A.C., Pereira, V.H. and Zonin, V.P. (2000). Agro-ecology as an option for income improvement in the family farming associations at 'Erexim' region, Brazil. World Congress of Rural Sociology, p.206.

7
Possibilities for closing the urban–rural nutrient cycles

Karen Refsgaard,* Petter D. Jenssen and Jakob Magid

Introduction	182
Recycling nutrients in society – an ecological economics perspective	184
Basic economic, institutional and social aspects of waste handling	186
Quantities of nutrients and organic resources	188
Ecological handling systems for organic waste and wastewater	191
Blackwater and urine diverting systems	192
Greywater	194
The cost of the handling system	196
Moral and cultural aspects related to recycling urban waste	197
Health aspects related to recycling urban waste	198
Recycling nutrients from urban waste – global examples	200
China	202
India	202
Botswana	203
South Africa	203
Malaysia, Kuching	204
Australia	204
Sweden	205
Norway	205
Conclusions	208

Summary

This chapter discusses the potential of organic farming for contributing to sustainable development, mainly in low-income countries, by integrating urban settlements with rural communities, through the recycling of domestic and household waste.

---

* Corresponding author: Norwegian Agricultural Economics Research Institute, P.O. Box 8024 Dep., N-0030 Oslo, Norway. E-mail: Karen.Refsgaard@nilf.no

The chapter links the thermodynamic laws for the transport of matter with economics, institutional and technological structures and the views of the stakeholders. The quantities of nutrients and organic matter in the production and consumption cycle under different conditions around the world are discussed. The amounts of plant nutrients and organic matter present in sewage, household waste and waste from food processing industries, are almost sufficient to fertilize the crops needed to feed the world's population. Conventional wastewater systems have been developed to ensure high local hygienic standards and to address some problems in the aquatic environment. However, sewage sludge is an unattractive fertilizer source, containing quantities of xenobiotic compounds and heavy metals. In agriculture in developing countries there is often a lack of nutrients due to limited capital, and a lack of organic matter on weathered soils. Due to limited availability and transport, urban organic wastes are predominantly used in urban and peri-urban agriculture. In dry parts of the world where water is highly valued (and may even be desalinized), use of wastewater for agriculture has been/is being developed. Ecological sanitation systems, based on biological (and technical) treatment in terrestrial systems, are able to recycle nutrients and organic matter from urine, faeces, greywater and organic waste. This chapter describes and evaluates the costs and benefits of such systems. It also discusses the moral and cultural challenges raised through the different ways in which science and religion deal with human behaviour when recycling urban waste. Combining these solutions with organic agriculture can contribute to improved recycling and sustainability. They imply institutional and economic challenges, but can contribute to improved health, agricultural production and social benefits, especially in low-income developing countries. While there are advantages in integrating urban settlements with rural ones through recycling human waste, it is paramount that health aspects of forming barriers against diseases are considered. The chapter ends with a presentation of examples of integrated recycling systems from urban to rural communities.

Introduction

The aim of this chapter is to look at the challenges and opportunities for organic farming to help integrate urban and rural settlements through utilizing flows of human waste and domestic waste: from the kitchen and toilet and from washing.

A basic principle in organic farming is the recycling of nutrients in order to reduce the use of non-renewable resources and the exploitation of fragile resources on a local scale. A central current focus of organic farming is on the recycling of nutrient at the farm scale. It is often a challenge for organic farmers to maintain nutrient levels within their production, the more so in plant production systems lacking access to animal manure, as there are fewer possibilities for substitution with input factors like mineral fertilizer, soil conditioner etc.

In Western societies the use of mineral fertilizer and imports of fodder results in a surplus of nutrients in certain regions. There is a flow of resources such as food, water, energy and minerals from rural regions and farms to urban regions, which also function as a sink for waste and emissions. In Denmark (5.5 million people) the nutrient turnover corresponds to the secretion from 120 million people (Magid, 2002), while in Norway (4.3 million people) it corresponds to 25 million people. This is due to a surplus of imports of farm products and agricultural inputs over exports. As such the nutrients from urban areas in countries like Denmark may only be of minor importance for conventional agriculture, but they may still be important for organic agriculture, if acceptable ways of recycling them can be achieved in practice.

Urban waste creates problems on a global scale; more than half of the world's population is city-dwellers and the proportion is increasing, implying that a major and growing proportion of waste is produced in urban agglomerations. Substantial amounts of plant nutrients and organic matter are present in sewage, household waste and waste from food processing industries (Skjelhaugen, 1999). Theoretically, the nutrients in domestic wastewater and organic waste are almost sufficient to fertilize the crops needed to feed the world population (Wolgast, 1993). However, conventional wastewater management systems have historically been developed with a view to sanitation standards and with little concern for recycling. As a result nutrients and organic matter are emitted to rivers, precipitated in sludge, or even incinerated and used as road material. More recently, environmental concerns have been the driving forces behind the technological development of sewage treatments that can biologically remove N, P and organic matter. This technology addresses some immediate problems in the aquatic environment, but the sewage sludge from the treatment plants contains xenobiotic compounds and heavy metals, and only a fraction of the nutrients that entered the urban areas, thus making the sludge an unattractive fertilizer source. In recent years there has been concern about the sustainability of this approach to wastewater handling, as well as concern about the fate of the final waste deposits in the environment (Magid *et al.*, 2001).

Agriculture in developing countries often faces a lack of nutrients. This is due to a range of reasons, including limited capital resources, limited access to organic matter and, in some regions, also because of highly weathered soils. Soils in the South are low in organic matter, implying that compost may be an appropriate alternative to artificial fertilizers. However, due to limited availability and transport, urban organic wastes are predominantly only used in urban and peri-urban agriculture. Given the one way flow of nutrients and organic matter from soils in, already nutrient-depleted, rural areas to urban centres, the use of wastewater treatment solutions does not seem sustainable or sensible. In dry parts of the world where water is a scarce and highly valued resource, and may have to be desalinized, the use of wastewater for agriculture has already been developed (Cross and Strauss, 1985).

The principles of ecological engineering offer a wide range of solutions for recycling nutrients and organic matter from different waste fractions: urine, faeces, greywater and organic household waste (Mitsch and Jørgensen, 1989). The combination of these ecological solutions and organic agriculture can contribute to improved recycling and sustainability. Yet they also raise institutional and economic challenges, which if successfully met may bring about improved health, agricultural production and other social benefits all of which are important in developing countries.

The chapter starts with a presentation of basic institutional and technological issues as well as the importance of integrating the views of the stakeholders. Then follows an overview of quantities of nutrients and organic matter involved in production and consumption cycles under different conditions around the world. A section presents ecological sanitation concepts or systems for managing wastewater from households in urban, as well as more rural, areas. Followed by evaluation of the benefits and costs involved and a discussion of the health aspects related to recycling urban waste. The last section of the chapter presents a few selected examples of recycling systems based on ecological sanitation.

Recycling nutrients in society – an ecological economics perspective

The term 'waste' in a society is a relative concept. The way we perceive the environment and the interactions between the environment and the economy, influence the way we view it, treat it and which solutions we choose to implement. Ecological economics views the economy as an integrated part of the biosphere – as an open subsystem of the environment. The focus is on the flow of matter and energy through the system, and the thermodynamic laws governing these processes; questions about the regulation of pollution might focus on the input, as well as the emissions, side of the economy (Vatn, forthcoming). (See also Chapter 4.)

When considering the recycling of nutrients it is essential to look at the total production and consumption processes in society and the material flows of matter and energy. The challenge is to find a 'proper' level of recycling nutrients and organic matter. From the thermodynamic interpretation, wastes are undesired joint products of the manufacturing process, however, from the economic interpretation, manufacturing is directed to satisfy consumption. Because wastes are undesired joint products, and (according to neo-classical economic theory) humans want to minimize costs, then no rational individual will want to pay for the waste. This implies that 'they are left where they fall' and may, according to their nature and location, cause pollution. There is also the problem of a response of inaction because, although the damaging effect on nature or society as a whole may be considerable, the direct effect on any individual will be quite small. To sum up, then from a thermodynamic point of view the ideal is not to produce any material waste, while from an economic point of view it is necessary to evaluate the costs for containing the spread or preventing the creation of waste against the costs of so doing. (For further elaborations on how the thermodynamic laws govern the production and consumption processes and how this interferes with microeconomics, see Chapter 4.) It is necessary to include the social and behavioural aspects in these evaluations, because the costs are also related to production and consumption patterns. Questions of personal and social responsibility play a role. For example are people prepared to take individual responsibility to separate, sort or otherwise deal with waste? Alternatively what are the costs of controlling or otherwise regulating waste production and disposal? (Vatn and Bromley, 1997).

Today, organic waste is not easily recycled back into food production systems in Western societies. Within most institutional regimes, waste is an externality, meaning that those responsible for creating the activity behind the waste affect the utility of those suffering from the problems of waste emission without compensating this decrease in utility by, for example, lowering the quality of bathing water to unacceptable standards, or smell from a disposal site etc. (Baumgärtner, 2002). But this perception of waste differs, between individuals and, particularly, between different societies.

Bisson and Proops (2002: 42) illustrate this by comparing different perceptions of waste in Europe from about the 16th century, where it was seen as a problem due to the relative abundance of animal manure as opposed to Asia, where it was seen as a resource:

> Agriculture in Europe was dominated by a mixed farming regime, with arable, pasture and grazing animals that served not only as a source of milk, meat and wool, but also as a nutrient pump from grazing ranges to the arable fields. The dung they produced was as valuable a resource as the other products that could be extracted from them. In such circumstances, human excrement from the cities was not considered a prime resource for agriculture. Japanese and Chinese towns, in contrast, relied on the supply of human excrement. Therefore, collection and transportation of nutrients from the cities back to the agricultural areas is economically feasible. Due to the very limited supply of animal manure in their agricultural systems, the rice-growing Asian agriculturalists needed not only to collect human excrement, but also to recycle such materials as oil-cake residues and ashes to fill the nutrient gap. The European solution came at a cost, because the production of manure via animals is an expensive solution in energetic terms. In solar based societies, energy means area, so the area needed to feed one person in Europe was much higher than in Asia, due to the extra area needed to produce animal fodder ... Another benefit was that cities in Asia were far more hygienic places than most European ones and water pollution due to faecal matter, one of the recurrent European problems, was almost unknown ...

This illustrates how the utilization of resources differs according to the natural conditions, thereby creating different cultural practices, in this case about perceptions for use of human waste. These issues are further elaborated below. In the next section we take, as a starting point, a model that considers both the scientific as well as the social and cultural aspects in the evaluation of systems for handling human waste.

Basic economic, institutional and social aspects of waste handling

In this section we consider the basic economic, institutional and social constraints and challenges of waste handling. A waste handling system is defined as comprising three different sub-systems (Figure 7.1), the users, the organization managing the system and the technological structure in itself.

The social understanding of waste, which means how waste is defined in terms of behaviour and perceptions and the role of it in the 'lived life', depends on the interplay of cultural concepts and material objects. Decisions about disposal, sewage, incineration and recycling of waste in different social contexts cannot be understood without considering both the material and the cultural contexts of waste. Firstly, stakeholders directly involved in the waste stream, in handling the products, through transportation, treatment and end disposal or use, have a central role. Waste can only become a resource if use of that resource is socially acceptable. There may be moral or cultural barriers against the use of, for example, human faeces; however, these aspects are discussed in more detail later. Anthroposophy, for example, does not accept the use of human waste in agriculture, thus precluding its use in biodynamic systems. This is also the case for organic farming where the regulations prohibit use of human faeces and urine as fertilizer in agriculture. However, in a survey among organic farmers in Norway about 40% were positive to utilization of human urine and 24% were positive to utilization of human faeces, but it is not allowed (Lystad *et al.*, 2002). These attitudes are mainly due to the nutrient recycling effect. On the other hand these organic farmers are reluctant to use these materials because of a perceived bad quality with risk for environment and health. Prices have to be comparable to competing products, but here the State can legitimately play a role in subsidizing such reuse by offsetting the saved cost of environmental externalities, such as water pollution. Also the responsibility for quality control and liability needs to be clearly defined and here the State also can contribute facilitating for contracts and collective agreements among the actors and with information and general knowledge. Thus both the organizational structures and the technological structures must function.

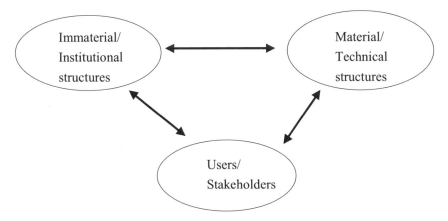

Figure 7.1. Institutional, technical and social components of waste systems. (Source: Söderberg and Kärrman, 2003.)

From a societal viewpoint it is important to ensure that controls are imposed on the recycling process, in order to eliminate the risk of disease vector transmission, and transmission of other substances that may compromise food safety. This principle of control is in accordance with existing practice in organic farming, whereby certification of products is often formalized in order to justify marginally higher economic returns. At present there are few examples of modern marketed organic agriculture based on urban fertilizers. The quality control aspect that would be necessary in developing such systems could be based on contract farming or long-term agreements on the management and use of urban fertilizers. This also applies to poor countries where urban agriculture plays an important role in sustaining food security.

As any input will cause emissions (see Chapter 4), the focus needs to be on both the input and the emission side of the economy. This means that our evaluations and decisions over waste handling systems need to look at how waste is created as well as how it is disposed (Vatn, forthcoming). Calculations of the marginal costs for treatment, compared to the marginal costs for emission, have to include the costs of the whole process from production to recycling back to agriculture. That means not only the treatment and handling costs but also the transaction and administrative costs involved in changing peoples' minds and behaviour. The costs are not only related to technical options for controlling and managing emissions, or whether to opt for fees or quotas, but are also related to consumption structures and behaviour. For example costs may increase greatly if households only make poor efforts at recycling. Yet, controlling households to improve the recycling rate may be very costly, due to their great number and limited size. From society's point of view it is essential to evaluate the costs related to the disposal or containment waste of material against the costs of prevention (Vatn and Bromley, 1997).

Nutrient recycling is not the only aspect of organic waste (water) handling. It is equally important to look at the total waste generation cycle from the consumption of food, creation, handling and emission of waste and attempt to reach a socially and economically optimal system. Considerable environmental benefits can be achieved by reducing nutrient emissions to water resources by removing urine from wastewater. This can be achieved by using source separating systems where the urine is collected separately and recycled to agriculture. Thus the benefits of implementing new waste management systems can counterbalance the costs incurred in so doing.

Quantities of nutrients and organic resources – from households to agricultural systems

In this section we give an overview of the production and consumption cycle for nutrients and organic matter. We show the composition of the nutrients and organic matter in the organic waste fractions from households and compare this with agricultural needs for nutrients and organic material.

The present day food production system is, to a great extent, a one-way nutrient flow powered by fossil energy input. Nutrients are purchased in the form of chemical fertilizer, applied to the (best) land, replacing those lost to the environment or removed in the crop which are sent to the city, passing through humans and are lost to wastewater and organic waste. However, there are several ways in which these cycles can be closed, at least to some extent (Hall *et al.*, 1992). First, the nutrients could be recycled from urban areas back to areas of agricultural production. Secondly, the nutrients could be more efficiently recycled between crop and animal production systems, which already occurs to some extent in organic (and mixed farming) systems. Thirdly, marginal lands could be brought into production through the design of integrated systems that promote solar-powered energy flows (i.e. photosynthesis) and nutrient cycles.

In Western societies the growing use of imported fodder and fertilizer in agriculture implies a build up of excessive[1] nutrients on the farm. Only a small proportion of imported nutrients leave the farm in agricultural products. Bøckman *et al.* (1991) provide several examples showing that only 10–30% of the total nitrogen input into agriculture is recovered in the products used for human consumption. From 18–30% on dairy farms and 30–40% on pig farms of the nitrogen from plant production is converted into consumable protein in the form of dairy or meat products, but the number seldom reaches 50% (Halberg *et*

---

[1] It is appropriate that there is some additional nutrient input to provide a safety net for recovery, as there is always a risk of some loss. However, within more intensive production systems these losses are excessive, to the point that in some countries they now have to be strictly controlled.

al., 1995; Kristensen *et al.*, 2005). The remaining nutrients are recycled back into plant production or, to some extent, lost in 'leaky' systems. The main input of nitrogen in conventional agricultural systems comes from atmospheric nitrogen, made available to farmers through industrial processes (Bøckman *et al.*, 1991). For phosphorus the situation is different as the major phosphorus loss in agricultural production is related to erosion. Secondly, phosphorus is a non-renewable resource and in the long run we are forced to find solutions for recycling it.

The energy required for the processing, transportation and use of mineral fertilizer is about 38 MJ or 10.5 kWh per kg nitrogen (Refsgaard *et al.*, 1998). This implies that agricultural production systems relying on mineral fertilizer use a great amount of low entropy fossil fuel to produce food. In organic agriculture the fossil fuel used for nitrogen production, producing about the same amount of food, is substituted with solar energy although this requires a larger hectarage. However, today's organic production of nitrogen is mainly generated by animal husbandry production which, as discussed above, has a low level of energy efficiency. Still, the process is changed from use of fossil fuels (social metabolism) to use of solar fuels (natural metabolism).

Table 7.1 shows where the nutrients in the household waste stream are concentrated.

Table 7.1. Resources from households in kg per person and year. (Sources: Wolgast, 1993; Jenssen and Skjelhaugen, 1994; Polprasert, 1995 and Mosevoll *et al.*, 1996.)

	Blackwater[1]		Greywater[2]			Organic household waste
	Urine	Faeces	Kitchen	Laundry	Shower/bathtub	
Nitrogen	3.57	0.49	0.18	0.15	0.11	0.73
Phosphorus	0.32	0.16	0.07	0.03	0.01	0.12
Potassium	0.26	0.11				
BOD[3]			5.11	2.92	2.19	12.41
COD[4]			12.41	5.11	2.56	
Total quantity excreted from a person in a year	< 500 l	50–180 kg (wet weight)				35 kg

[1] Blackwater is wastewater from the toilet
[2] Greywater is wastewater from kitchen, laundry, shower and bathtub
[3] Refers to Biological Oxygen Demand
[4] Refers to Chemical Oxygen Demand

It is relevant to consider how much grain can be grown with the nutrients present in human waste. Urine accounts for 88% of the nitrogen and 67% of the phosphorus produced by humans. In addition it is virtually sterile and easy to spread. This ease of handling and spreading combined with high nutrient content (which is higher than animal urine) make human urine the most favourable fraction of the waste stream for recycling. By contrast, the benefits of using human faeces in agricultural production are more related to its organic matter content. Use of blackwater, which contains a mixture of urine, faeces and some flush water, is another possibility for re-utilizing human nutrients and organic matter.

Using human excreta as fertilizer in organic production systems increases the possibility for plant production without animal husbandry. Assuming a need for about 150 kg nitrogen per ha per year, the excreta from one individual can theoretically fertilize 372 m^2, this not taking into account other nutrients from kitchen waste.

However, there are losses of nutrients during the recycling process back to agriculture. Wrisberg *et al.* (2001) report that composting etc. significantly reduces the nutrient content, especially for nitrogen (43–86%), diminishing the fertilizer value of the product. These losses depend on how they are treated and distributed. Given that the great majority of nutrients are found in urine, and bearing in mind the cost of composting, direct separation would seem a preferable and more cost effective approach from the viewpoint of nutrient recycling. However, where increasing soil organic content is an issue, as it is in many developing countries, composting, in order to maintain this resource, becomes relatively more attractive.

Nitrogen is, however, perhaps not the most critical nutrient to be considered. Nitrogen losses and shortages may be offset by judicious cropping systems including nitrogen fixing plants, as currently used in many organic systems. The return of phosphorus, potassium and micronutrients to agriculture is, however, essential for sustaining plant production.

There are vast differences between developed and developing countries in the levels of consumption and production of household waste. The volumes and mass increase in relation to the more industrialized the countries are. For example the production of waste in India is about 0.25 kg per person per day, while in the USA it is 1.25 kg per person per day. The composition of this waste also differs between developing and developed countries. Most of the waste stream in developing countries is organic, whereas in (for example) north-European cities, glass, metals and dust account for a much higher proportion (see Table 7.2), which have important implications for recycling.

Table 7.2. Percentage composition of waste. (Sources: Dalzell *et al.,* 1987; Deelstra, 1989.)

Waste type	Accra	Indian city	South America	Middle East	North European city
Organic	87.1	75.0	55.0	50.0	3.0 –16.0
Paper	5.7	2.0	15.0	20.0	2.7 – 4.3
Metals	2.6	0.1	6.0	10.0	7.0 –10.0
Glass	0.7	0.2	4.0	2.0	10.0 –11.0
Textiles	1.2	3.0	10.0	10.0	3.0 – 7.0
Synthetics	1.3	1.0	*	*	3.0
Various	1.4	7.0	10.0	0.0	1.0 – 3.0
Dust		12.0	0.0	8.0	13.0 –16.0

* Textiles and Synthetics are listed together under the same category.

Ecological handling systems for organic waste and wastewater

As shown in the last section, the composition of organic waste sources clearly indicates that the major part of the nutrients within household waste is contained in the urine and only a minor part in the faeces. These fractions constitute approximately 1% of total household waste volume, but contain around 82–87% of the nutrients (Magid, 2002).

This section presents sanitation systems designed to manage this waste in urban, as well as more rural, areas. Most of the systems are based on biological (and technical) treatment. To different degrees they offer opportunities for water saving, recycling nutrients and organic matter and, in some cases, for energy recovery. The systems have different specific effects, on, for example, the amount of phosphorus removed, or the emissions of nitrogen to water resources. The different systems also have different implications and requirements for people's behaviour, responsibility and control.

Experience from Norway shows that almost complete recycling and zero emissions can be achieved by separating the treatment of blackwater and greywater. Organic household (kitchen) waste can be treated jointly with the blackwater and, thus, increase the yield of the produced fertilizer, soil amendment and energy recovery (Jenssen *et al.*, 2003). Water consumption can be reduced by almost 50%, without any reduction in the standard of living. Compact and technically simple solutions for greywater treatment can allow decentralized treatment facilities, even in urban areas (Jenssen and Vråle, 2004). This further reduces the need for a secondary piping and pumping system for transporting untreated wastewater. The treated blackwater can be injected directly into the ground and fertilize the soil with little, or no, odour (Morken,

1998). This substantially contributes to reducing air pollution in comparison with traditional surface spreading.

Blackwater and urine diverting systems

As illustrated in Table 7.1, blackwater contains some 90% of the nitrogen, 80% of the phosphorus (when only phosphate-free detergents are used) and 30–75% of the organic matter associated with wastewater. New toilet technologies like urine separating, composting, or extreme water-saving toilets, facilitate nutrient collection and recycling (Jenssen, 1999). This concentrated toilet and organic household waste can be used for energy recovery through aerobic or anaerobic processes. At the same time source separation eliminates the major sources (industry and road runoff) of micro pollutants, thereby facilitating source control, which is crucial when using waste as a source for plant fertilizer.

Source separation and collection is possible with toilets based on vacuum and gravity, that use only 0.5–1.5 l per flush. This means that an average (Norwegian) family may reduce its volume of blackwater to 6–9 m^3 per year, whereas conventional toilets produce 6–15 times more. Such volumes can be handled and treated locally. However, even when the amount of flush water is reduced to only 1 l, the mix will still contain less than 1% dry matter. Organic household waste, animal manure or residues from food processing can all be used as additives to increase the dry matter content to a level required for successful composting (Jenssen and Skjelhaugen, 1994).

The blackwater can be treated aerobically in a liquid composting unit, which leaves only sanitized and odourless effluents (Jenssen and Skjelhaugen, 1994). By recovering heat generated by the composting process the unit delivers surplus, usable, energy. Anaerobic treatment is another attractive treatment possibility, due to the high methane content of the biogas produced and because this process requires only small amounts of energy. Efforts are being made to develop small-scale anaerobic digesters for use in cold climates.

In countries such as China and Malaysia both the capital and technology are available for development of such systems. They are also relevant in other developed countries, as some of the examples at the end of the chapter show.

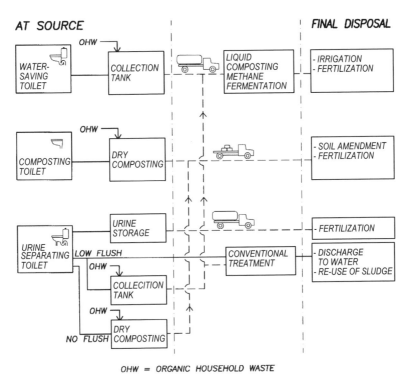

Figure 7.2. Infrastructure of blackwater/urine handling. (Source: Jenssen and Etnier, 1997.)

The processed blackwater can be applied and used in agriculture. A mobile direct ground injection system (DGI) has been designed for the purpose of injecting liquid organic fertilizers directly into the ground (Morken, 1998). One characteristic of this equipment is that penetration of the ground is not necessary, rather high-pressure injections shoot the fertilizer directly into the ground. This creates immediate contact with the soil, securing the absorption of ammonia and ensuring improved accessibility of the nitrogen content. The effect is to reduce ammonia losses by 15–20% compared to traditional surface spreading methods (where the losses typically amount to 70–80%). The equipment also makes it possible to combine sowing and fertilizing in one operation and can be used for any type of liquid organic fertilizer, including urine. The yields using the DGI method compare well to conventional methods using mineral fertilizer (Jenssen et al., 2003). However, this system is only appropriate in countries where agriculture is relatively highly mechanized and plots are sufficiently large to accommodate this kind of machinery.

Urine-diverting toilets come in two versions, single or dual flush. With a dual flush, urine diverting toilet the faecal fraction may either be collected separately or discharged with the greywater. The greywater may be either discharged to a secondary collecting sewer or treated on site. However, if it also contains the faecal fraction the treatment requirements increase, due to larger loads of nutrients, organic matter and, especially, pathogens. This increases the area of land required for nature based systems, like wetlands, sandfilters etc. and decreases the possibility of finding the available space to install such systems in urban settings.

While it is possible to collect the faecal part within a dual flush urine diverting toilet this gives excessive amounts of dilute blackwater. The flush for faecal matter uses between 2 and 4 l of water, reducing the dry matter content to $\ll$ 1%. This creates treatment problems for liquid composting and anaerobic digestion, making this option relatively expensive.

With a single flush, urine diverting toilet the faecal fraction is collected dry. This is normally stored in a removable chamber. Since no urine is present the collected faecal matter has much less odour than the combined urine/faecal mixture in e.g. a composting toilet. The experience with the present single flush urine diverting toilet systems is that the faecal fraction is too dry or desiccate when stored under the toilet so that a composting process does not start. In order to achieve composting the faecal matter should be removed from the collection point under the toilet and then composted.

Greywater

Greywater includes wastewater from bath, washing machine and kitchen, and includes that part of kitchen waste which is not collected in solid form. Greywater treatment constitutes an important aspect of ecological sanitation. Systems for greywater treatment have been successfully demonstrated with simple light-weight aggregate biofilter systems, in combination with man-made wetlands (Jenssen and Vråle, 2004). A source-separating-complete-recycling system is conceptually shown in Figure 7.3.

Greywater usually contains only minor amounts of nitrogen and phosphorus, but rather substantial amounts of organic matter (Rasmussen *et al.*, 1996). The extent of treatment depends on the final discharge standards required and use of the water. In Norway for example, discharges to the sea require only a simple (or no) treatment, while a more efficient treatment is recommended for discharges to lakes or rivers. It is necessary to improve the hygienic parameters (i.e. reduce bacteria levels) prior to discharge to small streams or for use in irrigation or groundwater recharge. This can be achieved by means of sand filters or by combining a biofilter and a subsurface flow constructed wetland using light-weight aggregates or similar porous media (Jenssen and Vråle, 2004).

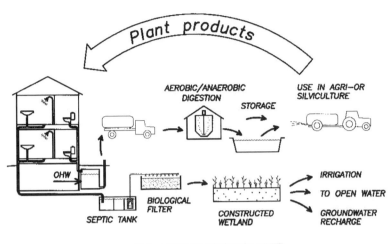

Figure 7.3. A complete recycling system based on separate treatment loops for blackwater and greywater. (Source: Jenssen *et al.*, 2003.)

A single-pass biofilter aerates the wastewater and reduces the biological organic degradation and the bacteria. Tests show that a biofilter of 1 m² surface area is capable of treating greywater from about ten persons (assuming a greywater production of 100 l per person per day). It does so with a 70–90% reduction in biochemical oxygen demand (BOD) and 2–5 log reduction of the indicator bacteria, depending on the loading rate (Jenssen and Vråle, 2004). This implies that very compact biofilters can be made. The key to their successful operation, however, is a uniform distribution of the liquid over the filter media and intermittent dosing with the greywater (Heistad *et al.*, 2001).

Such treatment facilities can be compact enough to be located in urban settings. With an integrated biofilter, as in Figure 7.4, the total surface area required is about 2 m²/person, with the wetland having a depth of a minimum of 1 m and the biofilter of 0.6 m. Typical effluent values from such a configuration are BOD < 10 mg/l, suspended solids (SS) < 5 mg/l, total nitrogen < 5 mg/l and faecal coliforms (FC) < 1000/100 ml. This last figure conforms to the European standard for bathing water quality, which requires FC values of < 1000. Thus, treated water can be discharged directly into local streams or water bodies, or used for irrigation or groundwater recharge, thereby eliminating the need for connections to the subsurface sewer system.

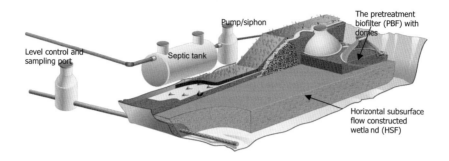

Figure 7.4. Constructed wetlands for cold climates with integrated pre-treatment biofilter in Norway. (Source: Jenssen and Vråle, 2004.)

The cost of the handling system

Technical aspects of nutrient recycling are not the only issues that need to be considered with respect to wastewater handling. Considerable environmental benefits can be had from reducing nutrient emissions to the aquatic environment, but the costs of installing the infrastructure of, for example, urine separating systems are considerable. In a social economic evaluation the benefits of implementing new waste management systems must balance the costs.

Sewerage systems are one of the most capital-intensive infrastructures in both developed and developing countries (Gupta *et al.*, 2001), demanding construction, operation, maintenance and rehabilitation. According to Otis (1996) 80–90% of the total capital cost of sewage treatment systems is due to these pipelines. In Norway numbers reported for investments in the sewage system due to pipeline construction are from 69% to 87% (Mork *et al.*, 2000; Finsrud, 2003). These figures do not include the operational and maintenance costs which are also high, as it difficult to identify and rectify problems in the performance of submerged (and therefore invisible and inaccessible) pipelines in efficiently transferring wastewater to the treatment facility (Gupta *et al.*, 2001; Tafuri and Selvakumar, 2002). Recent calculations by the sector itself in Norway show that the investments necessary to rehabilitate the existing conventional sewage systems in Norway were in the region of € 26 billion (Finsrud, 2003), corresponding to approximately € 1,330 per household. Further the report stressed the urgency of carrying out such work as the pipelines are in critical condition, with a daily loss of 225 l per person per day from water and wastewater pipelines (30 to 50% of the water supplied is leaking out either through the water or the wastewater pipelines). Finsrud (2003) estimates equivalent figures of 7% and 14% for Denmark and Sweden respectively. This shows the extent of capital investment required in existing systems, and the potential for investing in more sustainable solutions.

Refsgaard and Etnier (1998) have compared the economic and environmental implications of nature-based and decentralized wastewater treatment systems with conventional treatment systems in Norway. To secure a proper comparison, all stages of the handling process were considered where changes would occur if nature-based and decentralized systems were introduced. This includes collection at the household level, transport, treatment and disposal or spreading in agriculture. For investment in solutions for single households in new development areas the costs ranged from € 730 to € 2430. These calculations include costs for investment and operation of the total system, taking into considerations lower costs for organic waste handling and lower costs for drinking-water pipes. By using joint solutions for several households the costs decrease although the economy of scale differs according to the type of system employed. Variables in the calculations include the natural conditions for nature-based systems and, in conventional systems, the costs of expanding the sewer network vary in relation the length of the pipes. Today, in densely populated areas, these costs can be reduced through use of pressure-systems that use thinner pipes. The comparisons also include figures for cost-effectiveness, recycling and emission quantities for nutrients and organic matter that reflect the better performance of the source separation solutions in meeting environmental standards and helping offset other defensive expenditure.

On an institutional level, the involvement of the (organic) agricultural sector as a 'customer' of the end products of decentralized sewage solutions means that the sector would wish, or need to, take over part of the 'responsibility' for treating wastewater and organic waste, so as to gain more control over the quality of the organic fertilizer produced.

Moral and cultural aspects related to recycling urban waste

Despite the strength of scientific and economic arguments for linking the rural–urban nutrient cycle, the implementation of such systems can give rise to conflicts. This is partly due to more informal institutional aspects of how science and religion view human behaviour in relation to wastewater treatment. Just as science introduces new concepts and modifies behaviour, so religion generally preserves old beliefs and maintains traditions (Warner, 2000). Science emphasizes dispassionate reasoning while religion demands blind obedience to ritual.

The influence of religion on such patterns of behaviour varies between different cultures and places. Unlike Western societies, the Far East evolved cultures that accepted and indeed required the re-use of excreta. More than two-thirds of farmed fish come from Asia, where ponds are fertilized with excreta

(Mara and Cairncross, 1986). Necessity, and the pragmatic nature of Buddhism both probably played contributory roles (Cross and Strauss, 1985). Warner (2000) argues that there are differences in the way in which Judeo-Christian and Buddhist doctrines evolved and influence wastewater practices, compared to Islamic and Hindu edicts. Moslem doctrine prescribes strict procedures to limit contact with faecal material as it is considered impure. Moreover, scientific evidence may have far less influence in theocratic societies (notably Moslem ones) where 'religion is the law' than in secular societies, where law is 'the religion' and is much more influenced by scientific evidence. Therefore one especially must consider and work with the established religious doctrines when trying to modify behaviour and attitudes (Warner, 2000). The agronomic, social, environmental and economic arguments need to be couched within, and be compatible with, the prevailing religious orthodoxy, in order for farmers to benefit from adopting such practices and thereby create a demand for such products.

Health aspects related to recycling urban waste

The most important function of sanitation systems is that they form a barrier against the spread of diseases caused by pathogens in human excreta (Jenssen et al., 2004). In the short run, one of the greatest challenges to recycling human organic waste will be the awareness of health aspects for the consumers (and animals) that the managers of the system must become. In the long run the challenges will include monitoring for any unknown, and unexpected, negative effects on soil quality and on the integrity of agricultural production systems.

Ecological sanitation implies separate, often dry, handling of faecal matter with the objective of recycling the resources contained therein back to agriculture. Keeping human waste separate from the water cycle, helps avoid contamination of surface and ground water, which is important from a public health point. Sanitation systems also face specific challenges in counteracting pathogen transmission in the handling of material and its use on agricultural land. Farmers in Western societies today are reluctant to use sludge from conventional treatment systems, mainly because of the risk of contamination by organic pollutants, pathogens and industrial residues (Jenssen et al., 2003; Magid, 2004; Refsgaard et al., 2004). However, wastewater recycling to agricultural land (for example through irrigation) is widely practised in many developing regions and carries potential health threats (Ensink et al., 2002; IWMI, 2003).

In many ecological sanitation systems, the primary treatment is done at the household level instead of at professionally run, centralized, treatment plants. This implies the challenge of establishing simple systems, which are easy to handle and manage and, at the same time, do not increase the risk of disease transferral. Stenström (2001) reports that dry sanitation systems may be as, or more, effective as conventional systems in reducing the risk of exposure to pathogens. Based on current knowledge the WHO is preparing new guidelines for excreta and greywater re-use that will be available in 2006.

Larger systems with urine separation give a high level of protection prior to agricultural application on crops (Höglund, 2001). Treatment of the faecal fraction, using either dry, anaerobic or aerobic systems, can provide pathogen reduction that is in compliance with existing regulations for application to agricultural land (Jenssen *et al.*, 2004). Further, the biological processes that occur in the soil further serve to reduce pathogen levels.

The recent SARS and Avian Flu epidemics are dramatic examples of zoonotic (transferred from animals to humans and back) diseases that can arise in highly intensive animal husbandry in urban areas of the developing world. Another disease (neurocysticercosis) that is less dramatic but never the less causing substantial human and animal health impact is presently spreading inexorably across the African continent, due to the increasing production of pigs in urban areas. Cholera, diarrhoea and other faecal–oral diseases are closely related to poor sanitation. Diversion of wastewater from open sewers for use on leafy vegetables and other crops in urban agriculture is frequently observed in developing countries (Magid, 2004). Scavenging poultry and other small domestic animals that are allowed to roam outside and subsequently enter the living quarters can transfer parasites as well. Thus, disease vectors thrive due to inadequate management of waste and mismanaged urban agriculture.

Antibiotics, other medicinal residues and hormones (especially from industrialized animal production) entering the water bodies through sewage are known to modify the characteristics of aquatic micro-organisms, flora and fauna, although little is actually known about the nature of these changes (see Vaarst *et al.*, this volume). In the industrialized world we are increasingly facing ubiquitous multi-resistant *E. coli* as well as other bacteria previously susceptible to antibiotics. Fish and other seafood organisms take up resistant organisms, hormones and toxic substances and relay them back into the human food chain. Resistant bacteria and parasites pose an increasing challenge to the industrialized world in both human and animal health. Some medical doctors believe that human and animal natural immunization may be able to provide a solution, but if not, appropriate solutions for recycling human and animal excreta to peri-urban agriculture may provide the best form of prevention.

Recycling nutrients from urban waste – global examples

While in theory there are many advantages of recycling and many problems due to lack of proper sanitation, real life examples of such systems in developing countries are still few and far between.

It has been suggested that urban wastes are most readily utilized in agriculture where alternatives are not available or are too expensive. Farmers may be prepared to buy bulk compost but, due to availability and transport, urban organic wastes are predominantly used in urban and peri-urban agriculture.

'Getting rid of the shit' is what matters to most people – and they are even willing to pay for it, or given a certain level of community cohesion, spend some of their own time on waste management. A few people take pride in composting organic waste to use it in their home gardening, but this practice normally requires a considerable effort to work as a feasible community solution in continuously changing urban settlements. In many urban environments informal recycling is practised by scavengers who corner a market by picking through the dumping grounds and, in some places, this is the basis of a substantial economy. Provided that such solutions can be practised or developed while avoiding communication of disease vectors, they can be seen as environmentally beneficial. However, in the most rapidly developing urban environments (even in China and Vietnam these days) human and animal excreta often pose insurmountable management challenges. Sanitation systems are often either non-existent or inadequate. In so far as economics do allow new sanitation systems to be developed, the systems that are promoted collect rainwater, greywater and blackwater, ensuring that all urban water is mixed and thus contaminated with high loadings of nutrients, organic matter and disease vectors. In most cases the costs of sanitation, storage and transportation of waste for productive and appropriate use in agriculture are perceived as prohibitive, and even in capital intensive animal production systems situated in the urban fringe, waste management is minimal due to a lack of an appropriate physical and administrative infrastructure.

Costs of managing waste and wastewater should be balanced against the costs of mismanagement, in terms of the longer-term impacts on human health and the total environment. Many interesting cases can be found where community organizations or private enterprises play a crucial role in local waste management, financed by local dwellers. However, in most cases that we know of in developing countries, the local management schemes cannot carry more than the costs of transporting the waste away from the local area. Paying for further treatment and the additional costs that would be involved in recycling is not seen as an immediate necessity, and therefore not given priority. Strategies for financing such activities are critical for achieving sustainable urban waste management. The cases given below provide some indication of what can be achieved. Yet, it is difficult to provide detailed cost figures for ecological sanitary systems because the local conditions, which they rely on, vary greatly. In general, figures from UNEP (2004) show that the annual costs for ecological sanitation options are lower than most conventional options (see Figure 7.5).

Not all the impacts of a change to a new sanitary solution can be expressed in monetary terms. Aside from the costs of toilets and treatment facilities, etc., there are many other important effects. These include environmental benefits, such as of reduced pollution of nearby rivers, improved health and increased availability of drinking water for the local population. They also include social aspects, like local employment from handling organic waste and increased food production through increased access to organic fertilizer resources for local farmers.

In this chapter a few selected examples of recycling systems based on ecological sanitation are presented. These examples are either drawn from the authors' direct experience or systems that the authors have secondary knowledge about through their academic networks.

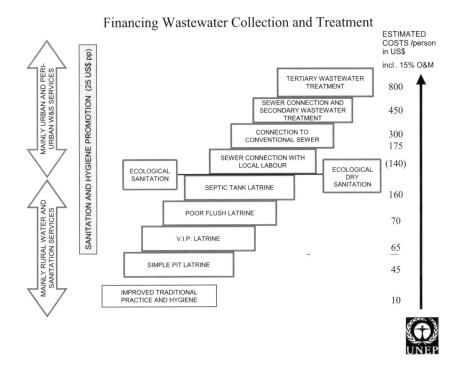

Figure 7.5. A ladder of sanitation options. (Source: UNEP 2004.)

China

China has a long tradition of effective management of natural resources. This includes reuse of garbage and human excreta in agriculture and aquaculture. The classical night soil system was reported to reuse as much as 90% in agriculture (Edwards, 1992). Tradition therefore facilitates implementation of modern ecological sanitation in China. In 1998, 70 households in the rural areas of Guangxi installed new urine-diverting toilets and by the end of 2002 more than 100,000 households had similar toilets (EcoSanRes, 2003). This has paved the way for urban implementation in China (Black, 2002). The reuse in aquaculture of wastewater from large cities started in 1951 in Wuhan, reaching about 20,000 ha by the 1980s (Edwards, 2000). The reuse of wastewater in aquaculture systems has been linked to traditional concepts of integrated farming and fish poly-cultures, which are seen as effective solutions for meeting a growing pollution problem in watercourses (Li, 1997). Irrigation with municipal wastewater reached about 1.5 million ha of land in 1995 covering around 1% of the total cultivated land of China (Ou and Sun, 1996). However, wastewater irrigation poses potential health problems that are not always properly dealt with.

India

A toilet centre provides sanitary facilities for 600–800 slum dwellers in Bangalore (Heeb, 2004). After storage, the urine is used as fertilizer and the faecal matter is composted with paper waste and garden waste and used for soil amendment. In addition to improving public health the toilet centre enhances the dignity of women through eliminating sexual harassment associated with the traditional practices of defecating in the open. The toilet centre, which generates 200 t of urine and 100 t of faeces per year, produces 50 t of compost, which in turns yields 50 t of bananas per year. The project has created eight new full-time jobs. The annual cost of the existing systems is approximately US$ 10 per user.

Wastewater aquaculture in Calcutta

The main sewers of Calcutta began functioning in 1875. In the 1930s sewage-fed fish farming started in the extensive pond system used for wastewater treatment. The fisheries developed into the largest single excreta-reuse aquaculture system in the world with around 7000 ha in the 1940s, supplying the city markets with 10-12 tons of fish per day (Ghosh, 1997; EcoSanRes, 2003). Today the Calcutta Wetlands, using wastewater both in agriculture and in aquaculture, covers an area of about 12,000 ha, known as the Waste Recycling Region (Ghosh, 1996). Wastewater-fed aquaculture systems like the Calcutta Wetlands represent controllable public health risks (Strauss, 1996). This is due to a combination of long retention times, high temperatures, high solar irradiance and natural microbiological activity, and adequate personal hygiene and food handling. Lessons learned from Calcutta are that a wastewater reuse system can meet modern criteria of sustainable development and hygiene, even for a mega-city. It does so through:

- providing low-cost wastewater treatment, storm-water drainage and a green area as a lung for the city;
- providing employment for about 17,000 poor people and production of about 20 t of fish per day for the urban poor (Edwards, 2000); and
- reducing environmental impacts of contamination from heavy metals from major industries, e.g. chromium from the tanneries in Calcutta (Biswas and Santra, 2000).

As such, the system serves as a model that could be replicated elsewhere in India and other countries.

Botswana

The villages of East and West Hanahai are located in Botswana's Kalahari Desert. On-site sanitation facilities allow families to produce their own soil conditioner and fertilizer for their vegetable gardens (Werner *et al.*, 2004). The toilet systems collect urine and faeces separately. After a period of awareness raising, information sharing and mobilization, which included meetings with the community chiefs and other events targeting all women and men in the villages, 20 families volunteered to pilot the concept of ecological sanitation. All of them selected urine diverting dry toilets, to provide privacy and comfort.

South Africa

After a successful pilot project involving 12 families, a new medium-income housing area for 3000 inhabitants in Kimberley will be equipped with ecological sanitation systems (SIPU International, 2002). The system will include the following features:

- Separation and collection of urine, which will be used by the forestry department as fertilizer for silviculture.
- Regular collection of faecal matter for composting.
- Treatment of greywater in soak pits and subsequent drainage to a wetland.

Malaysia, Kuching

The city of Kuching, capital of the state of Sarawak, Malaysian Borneo, has prepared a strategy for sewage management in the city, which combines conventional and ecological sanitation. A solution for the sewage is urgently needed. Currently the blackwater is discharged to the storm water drains through septic tanks, that are emptied, at most, every 4 years, and the greywater is discharged directly to the storm water drains. The result is a very high level of bacteria in drains and river tributaries (> 16,000 counts/ml) and high organic and nutrient load and a critical oxygen deficit. Outbreaks of cholera occur every year.

A proposal for a centralized sewage treatment system has been prepared. However, due to high costs and local physical conditions centralized sewage will only be suitable for the central business area of the town. The town is generally flat, with many low-lying areas, and only limited possibilities for gravity piping. In addition, a large part of the area is deep peat, which may decompose due to the draining effect of the sewers and thus lead to breaking sewers and rising-mains due to subsidence.

The city has therefore prepared a framework plan for integrated sewage management, implementing ecological sanitation for large parts of the city. The ecological sanitation will be based on local treatment of greywater and collection of the blackwater for centralized biogas and fertilizer production. Greywater pilot facilities were established in late 2003. The design of the biogas plant has commenced. Collection of blackwater has commenced for selected housing areas and institutions (pilot project) when septic tanks have been cut off from the storm drains and emptied on a regular basis. After hygienic treatment in the centralized biogas facility, the blackwater will be used as fertilizer in oil-palm plantations. A number of problems will have to be solved along the way, including mechanisms for cost recovery, traceability of waste products and a number of technical and biological problems that may arise when implementing this type of approach in a tropical environment.

According to the government plan, successful implementation of the pilot project will lead to an eventual extension of the ecological sanitation scheme to around 250,000 households.

Australia

In Melbourne, the Werribee wastewater system was opened in 1897. Half of the wastewater from the 4 million citizens is used for irrigating pastures for cattle and sheep. The public water company Melbourne Water manages 54% of its wastewater in 11,000 ha of ponds, wetlands and meadows, i.e. 500,000 m^3 of wastewater per day. At present livestock grazes on 3700 ha of pastures irrigated with raw or sedimented sewage and 3500 ha of non-irrigated pastures. The livestock yield a substantial return of about A$ 3 million per year, which significantly offsets the cost of sewage treatment (Melbourne Water, 2001).

Sweden

In the Swedish capital of Stockholm, urine diversion is used in several urban housing areas, e.g. Palsternackan (50 apartments), Understenshöjden (44 apartments), Gebers (30 apartments) and the newest Kullan (250 apartments). These are all family homes and show that people easily adapt to the new system (Johansson *et al.*, 2001). On the Swedish west coast Volvo has established a new conference centre (Bokenäs) for 500 people where blackwater and organic household waste are used for biogas production and the greywater is treated in a natural system. In several Swedish cities nitrogen-reducing wetlands have been shown to be cost efficient ways to meet increased water quality demands and some urine is utilized in agriculture.

Norway

Kaja – a complete recycling system at student dormitories in Norway

The Agricultural University of Norway is pioneering environmentally safe solutions to organic waste and wastewater treatment. In 1997, a first generation recycling system based on ecological engineering principles was built serving 48 students (Jenssen, 2005b). The system reduces water consumption by 30%, almost completely eliminates pollution, and produces a valuable plant fertilizer and soil amendment product from the waste material. The concept is based on:

- Separate treatment of toilet wastewater and water from kitchen and shower.
- Modern and reliable vacuum toilet technology with high comfort levels (see Figure 7.6a).
- Liquid composting of toilet waste and organic household waste for sanitation, stabilization, removal of odours and production of high quality liquid fertilizer (see Figure 7.6b). Liquid composting can be substituted for, or combined with, biogas production.
- Simple and reliable filtration of greywater for producing water of a suitable quality for irrigation, groundwater recharges or discharge to a nearby stream.
- A patented machine for fertilizer distribution that hydraulically 'shoots' liquid bio-fertilizer into the ground, resulting in higher yields and less pollution from run-off (see Figure 7.6c).
- Water-saving devices for showers, characterized by high comfort.

Liquid composting provides a sanitized mixture of organic household waste and blackwater. The liquid fertilizer is hydraulically injected into the soil and provides equivalent yields to mineral fertilizers. When the blackwater is removed, the remaining greywater meets drinking water standards with respect to nitrogen levels and bathing water quality standards with respect to bacteria. The greywater treatment systems are compact (1–2 m^2 per person) and can be landscaped.

Overall, this system:

- Recycles 80–90% of the nitrogen and phosphorus in the wastewater.
- Reduces nutrients and organic matter (BOD) by >95% hence, near zero emissions.
- Reduces the need for pipelines – the most expensive part of a traditional sewage network.
- Replaces expensive chemical fertilizer.
- Makes it possible to recycle nutrients locally, decreasing the need to transport fertilizer.
- Makes energy production from waste resources possible.
- Saves 30% of the domestic water consumption. Adding more water-saving devices makes it possible to save up to 50% or more.
- It is possible to use the separated greywater for irrigation or groundwater recharge after filtration, thus saving even more water.
- Greywater treatment facilities can easily be adapted to the terrain.
- Facilitates development of real estate in areas with no existing sewage network.

a   b   c

Figure 7.6. a) wall-mounted vacuum toilet, b) the liquid composting reactor, c) direct ground injection. (Source: Jenssen, 2005b.)

Decentralized urban greywater treatment at Klosterenga

At Klosterenga, in the capital of Norway, Oslo, the greywater is treated in an advanced nature-based greywater treatment system in the courtyard of the building (see Figure 7.7) (Jenssen, 2005a). The system consists of a septic tank, pumping to a vertical down-flow single pass aerobic biofilter followed by a subsurface horizontal-flow porous media filter. The Klosterenga system was built in 2000 and has consistently produced an effluent quality averaging to:

- COD 19 mg/l;
- Total nitrogen 2.5 mg/l;
- Total phosphorus 0.03 mg/l;
- Faecal coliforms 0.

For nitrogen the effluent has consistently been below the WHO drinking water requirement of 10 mg/l and for bacteria no faecal coliforms have been detected. The space required for this experimental system is about 1 m^2 per person, and part of the treatment area is also used as a playground (see Figure 7.7). Such high qualities of effluent water reduce the need for a secondary sewer collection system, because local streams or water bodies can safely be used for receiving the treated water, even in urban areas. The low area requirement of the system and the high effluent quality facilitates use in urban settings, discharge to small streams, open waterways or irrigation or groundwater recharge.

Figure 7.7. The Klosterenga greywater treatment system. Upper left; flowforms. Upper right; the wetland is in the foreground and the biofilter is underneath the playground behind the stone-wall. Lower left, the treated effluent is exposed in a shallow pond. Discharge to a local stream is possible as the stream was reopened. (Source: Jenssen, 2005a.)

Conclusions

There are both possibilities and challenges for recycling urban waste to agriculture. From an urban viewpoint the primary goal for waste and wastewater management is attaining a healthy local environment. Thus 'getting rid of the shit' is the main priority and people are willing to pay a certain amount of their income to attain this, or to organize themselves in ways that allow the disposal of waste from the surroundings. This implies the need to organize waste management schemes that are environmentally benign, and extend beyond the immediate local surrounding. This is particular problematic, yet also urgently required, especially in poor countries with weak institutions and weak environmental controls. Ecological sanitation systems can be combined with (organic) farming systems and contribute to sustainable development in such countries for the following reasons:

- There is a lack of established infrastructure for wastewater handling and scarce resources like water, capital, and fertilizers.
- Labour is cheap and available, whereas capital and water resources are often in short supply. Thus conditions in developing countries are more appropriate for developing ecological treatment systems rather than conventional ones which are both capital and water intensive.
- Ecological sanitation systems have advantages like low transport costs, lower requirements for water, and reuse of nutrients. Ecological sanitation saves at least 20–40% of domestic water consumption. After filtration, greywater can be used for irrigation, groundwater recharge or potable water. This is of key importance since water scarcity is a major limiting factor for development in many countries. Further, it is possible to recycle 80% to 90% of the nitrogen, phosphorus and potassium contained in excreta and wastewater from urban settlements into inexpensive local fertilizers for use in organic agriculture.

The thermodynamic principles analysis can be used to identify sustainable social modes of metabolism, figuring out the material and energetic efficiencies for different systems as for example use of urban organic waste in organic agriculture. However, these efficiencies also need to be evaluated and valued by the humans involved and the challenge is to find a 'proper' level of recycling nutrients and organic matter. From a thermodynamic point of view the ideal is not to produce any material waste, while from an economic point of view it is necessary to evaluate the costs for containing the spread or preventing the creation of waste against the costs of so doing. And questions of personal and social responsibility matter. For example, analysing if people are prepared to take individual responsibility to separate, sort or otherwise deal with waste, and considering what are the costs of informing, regulating and controlling waste production and use? Today, organic waste is not easily recycled back into food production systems in Western societies. Within most institutional regimes, waste is an externality, meaning that those responsible for creating the activity behind the waste affect the utility of those suffering from the problems of waste emission without compensating this decrease in utility, by for example lowering the quality of bathing water to unacceptable standards, or smell from a disposal site etc.

By sustaining peri-urban and urban agricultural production, ecological sanitation can play a multiple role in achieving development policy goals through enhanced food production thereby improving food security and reducing malnutrition, generating local business and job opportunities and thereby alleviating poverty. Organic agriculture is interesting in this perspective because of the focus on resource recycling in their aims. But there is also a real need for secure access to nutrients especially in systems with mainly crop production. However, there are great challenges for the farms by enlarging the nutrient cycle with the cities, related to dependency, soil health and human health.

Already proven technologies can be adapted and improved in new management systems, through collaborative schemes between stakeholders including CBOs, SMEs, municipalities and peri-urban farmers. Our case studies show that this can best be done by thinking big, but starting small. Research is needed to document, monitor and improve development of such systems and their environmental and health impacts (adverse as well as beneficial).

A more fundamental research need is that of addressing the question of risk management. Whether we choose to dispose of our waste (or resource) in the aquatic system (as we do for the time being in the industrialized world) or through the terrestrial system there are risks involved. We are currently not able to foresee the consequences of these in the longer term. There is much evidence that disposal through the terrestrial systems can make these risks more manageable and, in some cases, involves lower costs than conventional centralized sewage treatment.

References

Baumgärtner, S. (2002). Thermodynamics of Waste Generation. In: Bisson, K. and Proops, J. (eds) Waste in Ecological Economics, Edward Elgar, Cheltenham, UK, pp. 13-37.

Bisson, K. and Proops, J. (2002). An Introduction to Waste. In: Bisson, K. and Proops, J. (eds) Waste in Ecological Economics, Edward Elgar, Cheltenham, UK, pp. 1-12.

Biswas, J.K. and Santra, S.C. (2000). Heavy metal levels in marketable vegetables and fishes in Calcutta Metropolitan area, India. In: Jana, B.B., Banerjee, R.D., Guterstam, B. and Heeb, J. (eds) Waste recycling and resource management in the developing world, University of Kalyani, India and International Ecological Engineering Society, Switzerland, pp. 371-376.

Black, M. (2002). Official conference report. Report on first international conference on ecological sanitation, Nanning, China, 5–8 November 2001. http:// www.ecosanres.org/PDF%20files/Nanning%20Conf%20report%20%20final.pdf (23.10.2003).
Bøckman, O.C., Kaarstad, O.L. and Richards, I. (1991). Landbruk og gjødsling. Mineralgjødsel i perspektiv. Norsk Hydro as, Norway.
Cross P. and Strauss, M. (1985). Health aspects of nightsoil and sludge use in agriculture and aquaculture. Report no. 04/85, IRCWD, Ueberlandstrasse 133, CH-8600, Duebendorf, Switzerland.
Dalzell, H.W., Biddlestone, A.J., Gray, K.R. and Thurairajan, K. (1987). Soil management: Compost production and use in tropical and subtropical environments. FAO Soils Bulletin 56, Soil Resources, Management and conservation Service, FAO Land and Water Development Divison.
Deelstra, T. (1989). Can cities survive: solid waste management in urban environments. AT Source 18(2): 21-27.
EcoSanRes (2003). Guangxi Autonomous Region, China. EcoSanRes. http://www.ecosanres.org/asia.htm (07.10.2003).
Edwards, P. (1992). Reuse of human wastes in aquaculture, a technical review. UNDP-World Bank, Water and Sanitation Program, 350 pp.
Edwards, P. (2000). Wastewater-fed aquaculture: state of the art. In: Jana, B.B., Banerjee, R.D., Guterstam, B. and Heeb, J. (eds) Waste recycling and resource management in the developing world. University of Kalyani, India and International Ecological Engineering Society, Switzerland, pp. 37-49.
Ensink, J.H.J., van der Hoek, W., Matsuno, Y., Munir, S. and Aslam, M.R. (2002). The use of untreated wastewater in peri-urban agriculture in Pakistan: Risks and opportunities, International Water Management Institute, Research Report 64.
Finsrud, R. (2003). Gjenanskaffelskostnader for norske vann- og avløpsanlegg. NORVAR-rapport 130/2003.
Ghosh, D. (1996). Turning around for a community based technology, towards a wetland option for wastewater treatment and resource recovery that is less expensive, farmer centered and ecologically balanced. Calcutta Metropolitan Water and Sanitation Authority, 21 pp.
Ghosh, D. (1997). Ecosystems approach to low-cost sanitation in India: Where people know better. In: Etnier, C. and Guterstam, B. (eds) Ecological engineering for wastewater treatment. Proceedings of the International Conference at Stensund Folk College, Sweden, 24–28 March, 1991. 2nd Edition. CRC Press, Boca Raton, USA, pp. 51-65.
Gupta, B.S., Chandrasekaran, S. and Ibrahim, S. (2001). A survey of sewer rehabilitation in Malaysia: application of trenchless technologies. Urban Water 3: 309-315.
Halberg, N., Kristensen, E.S. and Kristensen, I.S. (1995). Nitrogen turnover on organic and conventional mixed farms. Journal of Agricultural and Environmental Ethics 8: 30-51.
Hall, C.A.S., Cleveland, C.J. and Kaufmann, R. (1992). Energy and Resource Quality. The Ecology of the Economic Process. University Press of Colorado, ISBN 0-87081-258-0.
Heeb, J. (2004). Source separation – new toilets for Indian Slums. In: Werner, C. et al. (eds) Ecosan – closing the loop. Proc. 2nd int. symp. ecological sanitation, Lübeck Apr. 7–11 2003, GTZ, Eschborn, Germany, pp. 155-162.

Heistad, A., Jenssen, P.D. and Frydenlund, A.S. (2001). A new combined distribution and pretreatment unit for wastewater soil infiltration systems. In: Mancl, K. (ed.) Onsite wastewater treatment. Proc. Ninth Int. Conf. On Individual and Small Community Sewage Systems, ASAE.

Höglund, C. (2001). Evaluation of microbial health risks associated with the reuse of source-separated human urine. PhD Dissertation, Stockholm.

IWMI (2003). Confronting the Realities of Wastewater Reuse in Agriculture, Water Policy Briefing, Issue 9, pp. 1-8, International Water Management Institute, Colombo, Sri Lanka, www.iwmi.org/health

Jenssen, P.D. (1999). An overview of source separating solutions for wastewater and organic waste treatment. In: Kløwe, B. *et al.* (eds). Management the wastewater resource. Proceedings of the fourth international conference on Ecological Engineering for Wastewater Treatment. Agr. Univ. Norway, Ås. 7–11 June 1999.

Jenssen, P.D. (2005a). Decentralised urban greywater treatment at Klosterenga, Oslo, Norway. In: van Bohemen, H.D. (ed) Ecological Engineering: Bridging between ecology and civil engineering. AEneas, Amsterdam.

Jenssen, P.D. (2005b). Kaja – a complete recycling system at student dormitories in Norway. In: van Bohemen, H.D. (ed) Ecological Engineering: Bridging between ecology and civil engineering. AEneas, Amsterdam.

Jenssen, P.D. and Etnier, C. (1997). Ecological engineering for wastewater and organic waste treatment in urban areas – an overview. In: Mellitzer *et al.* (eds) Water saving strategies in Urban renewal, Dietrich Reimer Verlag, Berlin, pp. 51-60.

Jenssen, P.D. and Skjelhaugen, O.J. (1994). Local ecological solutions for wastewater and organic waste treatment – a total concept for optimum reclamation and recycling. In: Eldridge Collins (ed) On-Site Wastewater Treatment: Proc. of the Seventh Int. Symp. on Individual and Small Community Sewage Systems, Atlanta, ASAE, pp. 379-387.

Jenssen, P.D. and Vråle, L. (2004). Greywater treatment in combined biofilter/constructed wetlands in cold climate In: Werner, C. *et al.* (eds) Ecosan – closing the loop. Proc. 2nd int. symp. ecological sanitation, Lübeck, 7-11 Apr. 2003, GTZ, Germany, pp. 875-881.

Jenssen, P.D., Heyerdahl, P.H., Warner, W.S. and Greatorex, J.M. (2003). Local recycling of wastewater and wet organic waste – a step towards the zero emission community. Paper presented at the 8th International Conference on Environmental Technology; Lemnos Greece 6–12. Sept. 2003. www.ecosan.no

Jenssen, P.D., Heeb, J., Huba-Mang, E., Gnanakan, K., Warner, W.S., Refsgaard, K., Stenström, T.-A., Guterstam, B. and Alsén, K.W. (2004). Ecological Sanitation and Reuse of Wastewater – ecosan – A Thinkpiece on ecological sanitation. The Agricultural University of Norway.

Johansson, M., Jönsson, H., Gruvberger, C., Dalemo, M. and Sonesson, U. (2001). Urine Separation – closing the nutrient cycle (English version of report originally published in Swedish). Stockholm Water Company. Stockholm, Sweden. Available at: http://stockholmvatten.se/pdf_arkiv/english/Urinsep_eng.pdf

Kristensen, I.S., Halberg, N., Nielsen, A.H. and Dalgaard, R. (2005). N-turnover on Danish mixed dairy farms. Part II. In: Bos, J., Pflimlin, A., Aarts, F. and Vertés, F. (eds) Nutrient management on farm scale. How to attain policy objectives in regions with intensive dairy farming. Report of the EGF workshop. Plant Research International 83, 91-109.
http://www.agrsci.dk/var/agrsci/storage/original/application/57b9a70804960c9f54ad2 55b11f22d.pdf

Li, S. (1997). Aquaculture and its role in ecological wastewater treatment. In: Etnier, C. and Guterstam, B. (eds) Ecological engineering for wastewater treatment. Proceedings of the International Conference at Stensund Folk College, Sweden, 24–28 March 1991. 2nd Edition. CRC Press, Boca Raton, USA, pp. 37-49.

Lystad, H., McKinnon, K. and Henriksen, T. (2002). Organisk avfall som gjødselvare i økologisk landbruk. Resultater fra spørreundersøkelser og identifisering av FoU-behov. Jordforsk-report 72-02.

Magid, J. (2004). Urban Ecology, Metabolism and Health – Recycling and agriculture: waste or resource? In: Design and Appraisal of Capacity Development Activities in Urban Environmental Management, Danida, pp. 49-53.

Magid, J. (2002). Byernes affaldshåndtering og næringsstofkredsløb. In: Jensen, E.S., Vejre, H., Bügel, S.H. and Emanuelsson, J. (eds) Vision for fremtidens jordbrug (s. 181-202). Gads Forlag 2002, Copenhagen.

Magid J., Dalsgaard, A. and Henze, M. (2001). Optimizing Nutrient Recycling and Urban Waste Management – New Concepts from Northern Europe. Chapter 3.6, In: Drechsel, P. and Kunze, D. (eds) Waste Composting for Urban and Peri-Urban Agriculture: Closing the Rural-Urban Nutrient Cycle in Subsaharan Africa. CABI Publishing, Wallingford, UK, pp. 137-139

Mara, D. and Cairncross, S. (1986). Guidelines for the safe use of wastewater and excreta in agriculture and aquaculture. WHO, Geneva, pp. 187.

Melbourne Water (2001). Infostream. Western Treatment Plant, PO Box 2251 Werribee, Victoria 3030, Australia. www.melbournewater.com.au

Mitsch, W.J. and Jørgensen, S.E. (1989). Ecological Engineering: An Introduction to Ecotechnology. Wiley Interscience, New York.

Mork, K., Smith, T. and Hass, J. (2000). Ressursinnsats, utslipp og rensing I den kommunale avløpssektoren. 1999. SSB-report 2000, 27.

Morken, J. (1998). Direct ground injection – a novel method of slurry injection. Landwards, winter 1998, 4-7.

Mosevoll, F., Andreassen, L., Gaarde, H. and Jacobsen, J. (1996). TA1374. SFT-report 96:19. National Pollution Control Authority.

Otis, R. (1996). Small diameter gravity sewers: experience in the United States. In: Mara, D. (ed.) Low-Cost Sewerage. John Wiley and Sons, Chichester, UK, pp. 123-133.

Ou, Z.Q. and Sun, T.H. (1996). From irrigation to ecological engineering treatment for wastewater in China. In: Staudenmann J., Schönborn, A. and Etnier, C. (eds) Recycling the Resource, Ecological Engineering for Wastewater Treatment. Environmental Research Forum Vols. 5–6 (1996): 25-34.

Polprasert, C. (1995). Organic Waste Recycling. John Wiley & Sons Ltd, London.

Rasmussen, G., Jenssen, P.D. and Westlie, L. (1996). Graywater treatment options. In: Staudenmann, J. *et al.* (eds) Recycling the resource: Proceedings of the second international conference on ecological engineering for wastewater treatment. Waedenswil, Switzerland, 18-22 Sept. 1995. Transtec. 215-220.

Refsgaard, K. and Etnier, C. (1998). Naturbaserte avløpsløsninger i spredt bebyggelse. Økonomiske og miljømessige vurderinger for kommune, husholdning og gårdsbruk. NILF-rapport 1998: 4.

Refsgaard, K., Asdal, Aa., Magnussen, K. and Veidal, A. (2004). Organisk avfall og slam anvendt i jordbruket, Egenskaper, kvalitet og potensial – holdninger blant bønder. NILF-rapport 2004: 5.

SIPU International (2002). Hull Street Integrated Housing Project, Kimberley, South Africa. SIDA, Stockholm.

Skjelhaugen, O.J. (1999). A Farmer-operated System for Recycling Organic Wastes. Journal of Agricultural Engineering Research 73: 372-382.

Söderberg, H. and Kärrman, E. (2003) (eds) MIKA. Methodologies for integration of knowledge areas. The case of sustainable urban water management. Chalmers University of Technology, Göteborg.

Stenström, T.A. (2001). Reduction efficiency of index pathogens in dry sanitation systems compared with traditional and alternative wastewater treatment systems. Proceedings First International Conference on Ecological Sanitation. www.ecosanres.org

Strauss, M. (1996). Health (Pathogen) considerations regarding the use of human waste in aquaculture. In: Staudenmann, J., Schönborn, A. and Etnier, C. (eds) Recycling the Resource, Ecological Engineering for Wastewater Treatment. Environmental Research Forum Vols. 5-6 (1996): 83-98. Transtec Publications, Switzerland.

Tafuri, A.N. and Selvakumar, A. (2002). Wastewater collection system infrastructure research needs in the USA. Urban Water 4: 21-29.

UNEP (2004). Financing wastewater collection and treatment in relation to the Millennium Development Goals and World Summit on Sustainable Development targets on water and sanitation. Eighth special session of the Governing Council/ Global Ministerial Environment Forum Jeju, Republic of Korea, 29–31 March 2004. UNEP/GCSS.VIII/ INF/4.

Vatn, A. (forthcoming). Institutions and the Environment. Edward Elgar.

Vatn, A. and Bromley, D.W. (1997). Externalities – a market model failure. Environmental and Resource Economics 9: 135-151.

Warner, W. (2000). The influence of religion on wastewater treatment: a consideration for sanitation experts. Water 21 Aug.: 11-136.

Werner, C., Mang, H.P. and Kesser, V. (2004). Key activities, services and current pilot projects of the international ecosan programme of GTZ. In: Werner, C., Avendano, V., Demsat, S., Eicher, I., Hernandez, L., Jung, C., Kraus, S., Lacayo, I., Neupane, K., Rabiega, A. and Wafler, M. (eds) Ecosan – closing the loop. Proc. 2nd int. symp. ecological sanitation, Lübeck, 2003, Apr. 7-11, GTZ, Eschborn, Germany, 75-82.

Wolgast, M. (1993). Rena vatten. Om tankar i kretslopp (Clean Waters, Thoughts about recirculation). Creamon, Uppsala, Sweden, 187 pp.

Wrisberg, S., Eilersen, A., Nielsen, S.B., Clemmesen, K., Henze, M. and Magid, J. (2001). Vurdering af muligheder og begrænsninger for recirkulering af næringsstoffer fra by til land. Miljøprojekt under Aktionsplanen for økologisk omstilling og spildevands-rensning. Miljøstyrelsen. 145 pp.

8
Soil fertility depletion in sub-Saharan Africa: what is the role of organic agriculture?

John Pender* and Ole Mertz

Introduction	216
Causes of soil fertility depletion in SSA	218
Approaches to restore soil fertility and improve productivity	220
Criteria for success	220
High external input agriculture (HEIA)	221
Low external input, sustainable agriculture (LEISA)	224
Organic agriculture	228
Integrated soil fertility management	231
Selecting the right approach	231
Conclusions	233

Summary

This chapter reviews the problem of soil fertility depletion in sub-Saharan Africa (SSA), its causes, and assesses the potential to address the problem using organic farming or other approaches, including high external input agriculture (HEIA), low external input sustainable agriculture (LEISA), or integrated soil fertility management (ISFM) approaches. We identify merits and drawbacks of each approach, and conclude that no single approach is likely to work in all of the diverse contexts of SSA, and that a pragmatic approach to the problem is needed. To help guide such efforts, more information about the potential recommendation domains for different approaches is needed, identifying where particular approaches are likely to be profitable, of acceptable risk, economically, environmentally and socially sustainable, and consistent with smallholders' resource constraints. We argue that HEIA and certified organic approaches have greatest potential in areas of favourable agro-climatic conditions and market access suit-

---

* Corresponding author: International Food Policy Research Institute, Environment and Production Technology, 2033 K Street, NW, Washington, DC 20006-1002, USA. E-mail: J.Pender@cgiar.org

©CAB International 2006. *Global Development of Organic Agriculture: Challenges and Prospects* (eds N. Halberg, H.F. Alrøe, M.T. Knudsen and E.S. Kristensen)

able to higher value products or higher use of inputs, while in areas of lower potential or poorer access, LEISA or ISFM approaches hold more promise.

Introduction

Soil fertility depletion is a major problem in many areas of sub-Saharan Africa (SSA) and contributes, along with other factors, to low and declining agricultural productivity and food insecurity in SSA. According to FAO statistics, food production per capita has declined 17% in SSA from an already low level since 1970, the most of any major region of the world (Figure 8.1). Cereal yields have remained stagnant in SSA since the mid-1970s while yields have doubled in other regions of the developing world, and now average only one-third of yields in other developing regions (Figure 8.2). However, these data conceal large differences between countries and regions within countries, and in some of the poorest countries of the Sahel there have been increases in both yields and per capita food production from the 1960s to the late 1990s (Club du Sahel, 1996; Toulmin and Guèye, 2003).

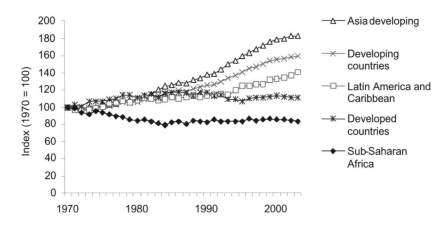

Figure 8.1. Food production per capita (FAOSTAT data 2004).

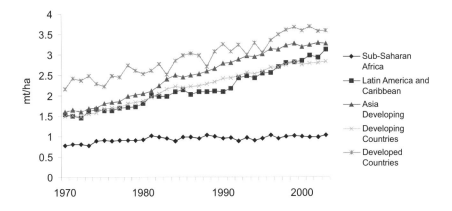

Figure 8.2. Cereal yields (FAOSTAT data 2004).

The dominance of agriculture in the livelihoods of most people in SSA means that poor agricultural performance likely translates directly into increasing poverty and food insecurity. Soil fertility depletion is regarded by some scientists as 'the fundamental biophysical root cause of declining per capita food production in sub-Saharan Africa' (Sanchez *et al.*, 1997). Although the problem of soil fertility depletion is widespread in SSA, it is not universal or inevitable. Numerous case studies have documented African farmers' ability to respond to population pressure, poverty, market changes and environmental stress by adapting land use strategies and adopting improved land management practices, leading in some cases to substantial environmental recovery and improved productivity (Tiffen *et al.*, 1994; Scoones *et al.*, 1996; Adams and Mortimore, 1997; Kaboré and Reij, 2004). In many areas, this has involved farmers' use of indigenous technologies, such as the zai (planting pits) used in West Africa to conserve soil moisture and improve fertility (Hassan, 1996; Kaboré and Reij, 2004); stone bunds and soil bunds in Ethiopia (Asrat *et al.*, 1996; Pender and Gebremedhin, 2004); raised bed (vinyungu) cultivation in valley bottoms in Tanzania (Lema, 1996); and many others. In other cases, land management technologies have been promoted by research or technical assistance programmes, such as agroforestry practices in Zambia and Kenya (Place *et al.*, 2002a); use of inorganic fertilizers as promoted by Sasakawa Global 2000; and integrated soil fertility management approaches such as promoted by the International Fertilizer Development Center (IFDC undated; Breman and Debrah, 2003), and others.

The debate about how to address the problem of soil fertility depletion in SSA is polarized. Advocates of high external input agriculture (HEIA) argue that only through greatly increasing use of fertilizer and other external inputs can African farmers increase productivity and hope to achieve food security, and that organic approaches would lead to environmental devastation as millions of hectares of land must be brought into production to feed growing populations (Borlaug, 1992, 1994; Avery 1995, 1998). Advocates of low external input sustainable agriculture (LEISA), agro-ecological farming, organic agriculture and similar approaches argue that the conventional HEIA approach has failed to reach the vast majority of African farmers, is environmentally destructive, that LEISA and organic approaches can increase yields dramatically and are more suited to the market and resource conditions of smallholders in Africa (Reijntjes *et al.*, 1992; Altieri, 1995; Pretty, 1999). Others argue for a more integrated soil fertility management approach, using complementarities between different types of inputs (Bationo *et al.*, 1996; Sanchez *et al.*, 1997; IFDC undated; Breman and Debrah, 2003). In this chapter, we review the causes of and possible solutions to the problem of soil fertility depletion in Africa, giving particular attention to the potential role of organic farming in addressing the problem.

Causes of soil fertility depletion in SSA

The direct causes of soil fertility depletion in SSA include declining use of fallow without substantially increasing the application of inorganic or organic sources of soil nutrients, cultivation on marginal and fragile lands with limited use of soil and water conservation measures, burning of crop residues, and other inappropriate land management practices. Use of inorganic fertilizer in SSA averages less than 10 kg per cultivated ha, less than 10% of the average intensity of fertilizer use in other developing regions of the world (Figure 8.3). Application of organic materials such as manure and compost, or use of nitrogen-fixing leguminous plants to restore soil fertility are also limited (Barrett *et al.*, 2002; Place *et al.*, 2003; Nkonya *et al.*, 2004;).

Underlying these proximate causes are many biophysical factors, including rugged terrain, intense and unreliable rainfall, marginal soils for agriculture, and pest and disease problems in many areas. The soils in much of SSA are very old and weathered and highly spatially variable in their deficiencies. Only 7% of the agricultural land in SSA is estimated to be free of major soil fertility constraints, less than in almost all other regions of the world (Wood *et al.*, 2000), and only 6% is estimated to be of high agricultural potential, considering both climate and soil constraints (Tegene and Wiebe, 2003).

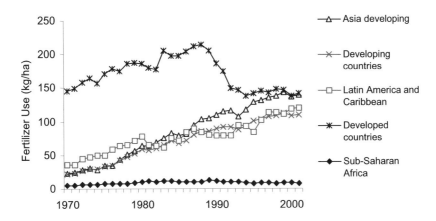

Figure 8.3. Use of inorganic fertilizer (FAOSTAT data 2004).

Probably even more important than biophysical constraints are socioeconomic factors such as: rapid population growth; limited access of African farmers to markets, infrastructure, technical assistance and credit; low economic returns and risks of many soil fertility management technologies; lack of farmer awareness of suitable and profitable technologies; insecure land tenure; poor management of common lands; poverty (Pender *et al.*, 1999); and increasing de-agrarianization in many communities (Bryceson, 2002). These socioeconomic factors are in turn affected by many government policies and programmes such as structural adjustment and market liberalization policies, investments in infrastructure and provision of public services, agricultural research and extension policies and programmes, land tenure policies, and others.

These biophysical, socioeconomic and policy factors influence land management and degradation in complex and context-specific ways. For example, population growth in some contexts contributes to increased land degradation (e.g. Grepperud, 1996; Pender *et al.*, 2001), but in other contexts may reduce it, by inducing investments in land improvement, infrastructure and other needs (Tiffen *et al.*, 1994; Templeton and Scherr, 1999). Poverty can cause households to degrade soils by forcing them to farm on marginal lands with little use of fallow, causing them to have a short-term perspective and limiting their ability to invest in land improvements (Cleaver and Schreiber, 1994; Pinstrup-Andersen and Pandya-Lorch, 1994; Reardon and Vosti, 1995; Pender, 1996). On the other hand, poor households may be more prone to invest in land as one of their most critical assets, especially where alternative opportunities for use of labour are limited (Pender and Kerr, 1998). Evidence on linkages between population growth or poverty and land degradation in SSA does not support a simple conclusion that either factor inevitably leads to land degradation (Tiffen *et al.*, 1994; Leach and Mearns, 1996; Scherr, 2000).

Other factors also have mixed and context-specific impacts. For example, land tenure insecurity is commonly cited as a cause of land degradation, since this may limit farmers' incentive to invest in land improvements. However, the extent to which African customary tenure systems are subject to tenure insecurity is very debatable (Atwood, 1990), and much of the available evidence suggests that land titling efforts in Africa have been of little benefit in promoting increased investment and agricultural productivity (Place and Hazell, 1993; Platteau, 1996). Land tenure insecurity may even promote land investment as a means of strengthening land claims (Besley, 1995; Otsuka and Place, 2001).

Approaches to restore soil fertility and improve productivity

Criteria for success

The complex and context-specific way in which such factors influence land management and degradation imply that no simple 'one-size-fits-all' strategy will suffice to address soil fertility depletion problems in the diverse circumstances of SSA. Technologies to restore soil fertility must be identified that are sufficiently profitable in the near term, of acceptable risk, and consistent with smallholders' constraints, to be widely adopted by smallholder farmers in SSA. Farmers in developing countries often are quite risk averse and have very high discount rates, due to poverty and limited insurance and credit options (Binswanger, 1980; Pender, 1996; Holden *et al.*, 1998; Hagos, 2003; Yesuf, 2004). As a result, farmers are unlikely to adopt technologies unless they are profitable in a relatively short period of time and of limited risk. Technologies must also be consistent with farmers' constraints. For example, lack of credit or cash may limit farmers' ability to use fertilizer or other purchased inputs, even if it is profitable and not too risky. Lack of land may prevent households from using land improvement practices such as improved fallow systems or even terraces. Labour constraints may prevent farmers from adopting very labour-intensive land management technologies; this can be an especially binding constraint for female farmers, who often have many competing demands on their time. Similarly, lack of access to oxen, equipment, or other forms of physical capital may prevent farmers from adopting capital-intensive technologies.

Beyond being sufficiently attractive and feasible to be adopted, technologies must also be economically, environmentally and socially sustainable over the long term if they are to provide a lasting solution to the problem of soil fertility depletion. To be economically sustainable, technologies must be profitable and of acceptable risk in the longer term. For example, reliance on subsidies to promote adoption of inorganic fertilizer has led in many cases to substantial adoption, but such subsidies in most cases could not be sustained economically, resulting in substantial disadoption after subsidies were removed. To be environmentally sustainable, technologies must not undermine future productivity by degrading the resource base or the supporting ecosystem. For example, excessive or unbalanced use of chemical fertilizers may lead to acidification of the soil (Scherr, 1999) or depletion of particular nutrients, undermining productivity in the longer term. Technologies and their impacts must also be consistent with cultural norms and broader social goals if they are to continue to be accepted by African farmers. For example, promotion of mechanical technologies such as tractors may favour wealthier households and contribute to land consolidation at the expense of poorer households, which could prove to be unsustainable in particular social and political contexts.

High external input agriculture (HEIA)

The Green Revolution, which led to dramatic increases in cereal yields in parts of Asia and Latin America beginning in the late 1960s, was successful in large part due to heavy use of inorganic fertilizer and irrigation, in addition to high yielding dwarf varieties. According to one study, use of inorganic fertilizer accounted for 50–75% of the increase in crop yields in Asia during the Green Revolution (Viyas, 1983). Unfortunately, conditions are less favourable in SSA for a replication of that experience. Only 4% of the arable area in SSA is irrigated, and the potential for rapid increase in irrigated area is likely to be limited by biophysical, social and policy factors. For example, Rosegrant *et al.* (2004) project that the irrigated cereals area in SSA will increase only from 4.5% to 4.8% of cereals area by 2025 under a 'business-as-usual' investment scenario. The soils in most of SSA are more marginal and less responsive to inputs than in Green Revolution areas of Asia (Voortman *et al.*, 2000), access to markets and infrastructure is more limited (Binswanger and Townsend, 2000), and the markets are much smaller due to lower population. The prevailing policy environment in SSA, which is generally opposed to efforts to stabilize commodity prices or subsidize fertilizer, other inputs, or credit, is also less favourable to adoption of Green Revolution technologies than the policies existing in countries where and when the Green Revolution occurred (Dorward *et al.*, 2004).

Despite these problems, there have been some notable success stories in promoting use of improved varieties, fertilizer and other complementary inputs in rainfed agriculture in SSA. Adoption of improved varieties of maize in combination with inorganic fertilizer in parts of eastern and southern Africa is an example of a 'qualified success' (Eicher and Byerlee, 1997). For example, the average maize yield in Zambia increased by nearly 5% per year in Zambia between 1970 and 1989, 2.2% per year in Zimbabwe (in 1980–89), 1.4% per year in Kenya (1965–80), and 1.2% per year in Malawi (1983–93) (Smale and Jayne, 2003).

However, these successes were driven mostly by adoption of improved varieties. Adoption of fertilizer lagged behind and is still low in many areas (Eicher and Byerlee, 1997). Such stepwise technology adoption can increase soil fertility depletion in the near term since high yielding varieties mine soil nutrients more rapidly than traditional varieties if adequate soil fertility replenishment does not occur. These successes also depended upon heavy government interventions in the form of state run seed systems, input and credit subsidies, and price stabilization measures, and yields have since declined in most of these countries as these supports have been withdrawn (Smale and Jayne, 2003). The sustainability of such successes on a broad scale in SSA has yet to be demonstrated.

Part of the reason for limited success of efforts to promote inorganic fertilizers is limited profitability and significant risk of this technology in many environments. Estimates of the incremental value–cost ratio (VCR) of fertilizer used on different crops in SSA vary widely, from over 20 to less than 1 (Yanggen *et al.*, 1998).[1] This variation depends on many factors, including the type of crop grown (higher VCR for more fertilizer-responsive crops such as maize, and for higher value cash crops), agro-climatic conditions (higher VCR where better soils and climate lead to better yield response), access to markets and roads (higher VCR resulting from higher output/input price ratio), crop and land management (higher VCR where management practices increase yield response to fertilizer), post-harvest management, storage and marketing technologies and institutions (higher where these lead to better farm-level prices of outputs), and government policies and programmes affecting these factors, as well as affecting commodity and input prices directly (via subsidies, taxes, exchange rates).

A critical factor in much of SSA is the high cost of transportation, leading to farm-level fertilizer costs commonly two to six times higher in SSA than other regions of the world (Millenium Project Task Force on Hunger, 2004). High transport costs, combined with agricultural subsidies and trade policies of many countries, also cause farm-level commodity prices to be much lower and more variable in SSA than in many other regions. These factors lead to wide variations in input–output price relationships for fertilizer across countries in SSA and over time (Figure 8.4). Assuming a typical yield response of improved maize (10–15 kg maize/kg N) (Yanggen *et al.*, 1998), the input/output price ratio would have to be less than 5–7 for the VCR to be greater than 2, a minimum level generally considered necessary for substantial adoption of fertilizer. As shown in Figure 8.4, this is not the usual case in SSA. The result is that fertilizer use is likely too unprofitable or risky for most SSA farmers outside of areas with favourable soils and climate and good access to roads and markets.

The profitability and sustainability of inorganic fertilizer use is also undermined by land degradation processes. The ability of crops to utilize fertilizer efficiently depends upon the capacity of the soil to store nutrients and water, which depends upon the organic matter content and texture of the soil. Where soil organic matter is being depleted, topsoil eroded, soils damaged by compaction, etc., the efficiency of nutrient uptake is likely to be declining. For example, in long-term experiments conducted in Kenya, maize yields declined by 50% over a 7-year period under continuous cropping with inorganic fertilizer inputs; while yields were higher and declined much less where manure was applied (Nandwa and Bekunda, 1998). Excessive use of inorganic fertilizers can contribute to some problems of soil degradation by accelerating the rate of decomposition of organic matter, acidification and depletion of soil nutrients not included in the fertilizer formulation (Reijntjes *et al.*, 1992). Such problems are exacerbated by the lack of crop and location specific fertilizer recommendations in SSA (Gruhn *et al.*, 2000).

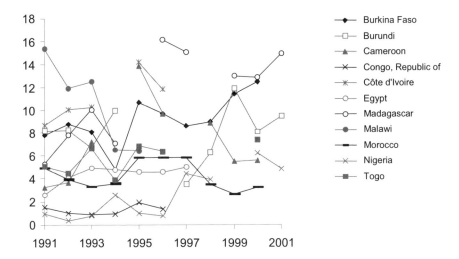

Figure 8.4. Ratio of average nitrogen fertilizer to producer maize price in selected countries (FAOSTAT data 2004).

Even where fertilizers are profitable, their use is often limited by farmers' constraints, especially access to cash or credit, and limited market development. Programmes such as Sasakawa Global 2000 have shown good success in stimulating farmer adoption and increasing yields in high potential areas by providing a package of improved seeds, fertilizer and credit (Quiñones *et al.*, 1997). In Ethiopia, the government incorporated the Sasakawa Global 2000 approach into its extension and credit programmes in the late 1990s and was successful in stimulating widespread adoption of fertilizer. This approach was successful in substantially improving cereal yields among farmers in higher potential areas of Ethiopia (Benin, 2003), but the long term profitability for farmers was undermined by a collapse in maize prices that occurred in 2001 and 2002 as a result of the production boom and limited market development (Gabre-Madhin and Amha, 2003). In drought-prone areas, by contrast, promotion of fertilizer use in Ethiopia has not been very profitable (Benin, 2003; Pender and Gebremedhin, 2004), and has contributed in some cases to worsening poverty as farmers were forced to sell assets to repay fertilizer loans (Demeke and Egziabher, 2003).

Low external input, sustainable agriculture (LEISA)

Problems with the HEIA approach in SSA have contributed to the search for effective alternatives such as low-external input, sustainable agricultural (LEISA) technologies (Reijntjes *et al.*, 1992). This approach involves limited or no use of external inputs (fertilizers, hybrid seeds and pesticides), and can include a wide range of technologies, such as soil and water conservation measures, minimum tillage, application of manure and compost, incorporation of crop residues, transfer of biomass and mulching, use of leguminous cover crops, shrubs or trees in improved fallows or intercropping systems, use of crop rotation to manage soil fertility and pests, etc. The ecological principles underlying these technologies involve: providing favourable soil conditions for plant growth by managing soil organic matter and enhancing soil biological activity; optimizing plant nutrient availability through biological nitrogen fixation, nutrient recycling and limited complementary use of inorganic fertilizers; minimizing losses by managing microclimates and water and preventing erosion; minimizing pest and disease problems through integrated management; and exploiting complementarities in use of genetic resources by combining these in farming systems with a high degree of genetic diversity (ibid.).

There is evidence that LEISA technologies can increase yields and farmers' incomes in parts of SSA. In a review of sustainable agriculture projects in 17 countries in SSA, Pretty (1999) found yield increases of 50–100% in 95% of projects that focused on increasing yields. Agroforestry approaches, including improved fallows using leguminous trees and transfer of high quality biomass (such as *Tithonia diversifolia*), in some cases combined with use of rock phosphate, are being used by more than 150,000 farm families in eastern and southern

Africa (Millenium Project Task Force on Hunger, 2004). These have been shown to increase maize yields two- to fourfold (Sanchez, 2002), and can increase farmers' returns to both labour and land substantially, especially when used on high value crops (Place *et al.*, 2002b). In the northern Ethiopian highlands, about two of every five farmers have invested in stone terraces to conserve soil and water since 1991 (Pender *et al.*, 2001), and these are increasing yields substantially, resulting in an average rate of return of about 50% on farmers' investment (Gebremedhin and Swinton, 2002; Pender and Gebremedhin, 2004). Reduced burning, reduced tillage, and use of manure and compost are also increasing crop productivity in northern Ethiopia (Pender and Gebremedhin, 2004).

In the semi-arid Sahelian zone of West Africa, tens of thousands of farmers are using planting pits to conserve soil moisture, often with manure or compost to increase fertility (Kaboré and Reij, 2004). These technologies can more than quadruple sorghum yields and provide a return to farmers' labour 33% higher than the rural wage rate (ibid.). Similar planting basin systems have emerged in Zambia, Cameroon, Nigeria, Uganda and Tanzania. In Zambia, tens of thousands of farmers have adopted conservation farming methods, including use of planting pits, minimum tillage, crop residue retention, leguminous crop rotations, and improvements in seeding and input application (Haggblade and Tembo, 2003). These technologies have been found to increase cotton yields by more than 40% and maize yields by more than 30%, and increase returns to peak season labour by 150% for cotton and 90% for maize (ibid.). The system of rice intensification in Madagascar, involving improvements in plant spacing, reduced use of flood irrigation, hand weeding and compost use, has been shown to increase returns well above the costs of additional labour (Uphoff, 2002), resulting in an increase of net revenue of more than 100% (Joelibarison, 2001).

These technologies are believed to reduce production risk and increase the sustainability of production, by helping to conserve soil and water and increasing soil organic matter levels (Reijntjes *et al.*, 1992; Hassan, 1996; Haggblade and Tembo, 2003; Kaboré and Reij, 2004). There is, however, little empirical evidence from tropical regions that soil organic matter can actually be increased, but benefits of preventing drastic reductions are well known from temperate regions and include reduced sensitivity to flooding and temperature extremes, erosion, and acidification (Alföldi *et al.*, 2002; Mäder *et al.*, 2002; Scialabba and Hattam, 2002). Crop rotation and other practices using leguminous plants such as improved fallows can help to restore soil nitrogen (Giller *et al.*, 1997). Furthermore, increasing crop biodiversity through crop rotation and other practices may provide opportunities to better utilize nutrients in the soil as a result of differential abilities of different crops and varieties to use nutrients of limited solubility (Høgh-Jensen, personal communication). Reduced use of agricultural chemicals helps to reduce environmental contamination, health risks and loss of biodiversity, and can reduce pest outbreaks resulting from destruction of beneficial insects that prey on pests (Altieri, 1995).

Not all impacts of LEISA technologies are necessarily favourable, however. For example, nitrate leaching can be as great from organic production as from conventional systems where similar levels of nitrogen are being input into the system (Kirchmann and Bergström, 2001). Many LEISA technologies do not replace soil nutrients that are exported from the farming system as harvest or residues, but simply recycle nutrients from one part of the system to another (e.g. applied manure may recycle nutrients originally taken from pasture land to cropland; biomass transfer and mulching move nutrients from one field to another). Such approaches may enrich soils in one part of the farming system, but cannot ensure sustainability of the entire system over the long term unless additional sources of nutrients are brought into the system to offset losses (whether through imports of fertilizer, animal feed, or other sources). Use of nitrogen-fixing leguminous plants through improved fallows, leguminous cover crops, etc. can be a very effective strategy for restoring soil nitrogen (Giller *et al.*, 1997; Sanchez *et al.*, 1997). However, replenishing other nutrients, especially phosphorus, is much more difficult without using mineral fertilizers, because the concentration of phosphorus in organic materials is generally low (Buresh *et al.*, 1997).

Cash costs of these technologies are low, making financial constraints and risks less of a concern than for purchased inputs (except where hired labour is needed to implement the technology). Many of these technologies also build upon traditional land management practices (e.g. crop rotation, fallowing, use of manure), facilitating farmers' ability to understand and adopt or adapt them to local circumstances.

Despite their advantages, LEISA technologies are often not adopted for many reasons. Such technologies do not always increase productivity or farmers' incomes. For example, recent research in Uganda has found limited yield impact and low profitability of several LEISA technologies, such as use of crop rotation, manure and compost (Woelcke *et al.*, 2002; Nkonya *et al.*, 2004; Pender *et al.*, 2004). Organic practices also were found to have insignificant or negative impacts on crop production in a higher rainfall area of northern Ethiopia (Benin, 2003). Evidence from Ethiopia also shows that soil conservation measures such as terraces can have negative yield impacts, especially in higher rainfall areas, by causing problems of waterlogging and increasing pest problems, as well as reducing cropped area (Herweg, 1993). The returns to biological nitrogen fixation methods such as leguminous crops or improved fallows may be limited by rainfall or soil quality in marginal areas (Kaizzi, 2002). For example, the benefit/cost ratio of using velvet bean (mucuna) as a leguminous cover crop has declined in Benin, possibly due to shortages of soil nutrients other than nitrogen (Pretty, 1999). Generally, the profitability and adoption of such technologies is greater when there is a market for the seeds, or when they serve other purposes in addition to restoring soil fertility, such as providing food, fodder, or weed control benefits (Place *et al.*, 2003).

Even when such technologies are profitable, they may not be adopted because of the constraints facing small farmers. Labour constraints are a major consideration for most of these technologies, especially for female-headed households and HIV/AIDS-affected households. For example, to provide 100 kg of nitrogen to a hectare of maize, farmers would need to apply more than 20 t of leaf biomass or manure, compared to only 217 kg of urea (Sanchez et al., 1997). The additional labour required for such large transfers could be much more than 1000 hours (Place et al., 2002a). Given the high costs of transporting such bulky inputs, African farmers find it difficult to use them on fields that are distant from the household compound. This is particularly problematic in areas such as the East African highlands where many farmers own numerous fragmented plots in different locations.

Access to organic inputs is also limited by access to land, livestock and biomass. In densely populated areas where most households own less than a few hectares of land, improved fallow and leguminous cover crop technologies are unlikely to be adopted on more than a very small area. For example, the average size of improved fallow among farmers adopting this technology in the densely populated highlands of western Kenya is only about 0.04 ha (Place et al., 2004). Lack of access to livestock often limits poor farmers' ability to use manure and compost (Clay et al., 2002; Mekuria and Waddington, 2002; Place et al., 2003; Pender and Gebremedhin, 2004). Scarcity of biomass and competing demands for other valuable uses (such as for construction materials, fuelwood, and fodder), also can limit farmers' ability and incentive to recycle biomass materials such as manure and crop residues to the soil, especially in densely populated lower rainfall areas, such as the northern Ethiopian highlands and parts of the Sahel.

Land tenure systems can also inhibit adoption of LEISA technologies. As mentioned above, fragmented landholdings reduce farmers' ability to transport and use bulky organic inputs. Land fragmentation also can inhibit investments in valuable trees or penning of animals on distant plots, since these may be subject to theft. Short-term tenants are unlikely to invest in land investments or management practices that may be seen as reducing the security of the landowner or yield returns beyond the period of the lease. The impacts of these factors have been shown in numerous empirical studies in SSA (e.g. Gavian and Fafchamps, 1996; Clay et al., 2002; Gebremedhin and Swinton, 2002; Benin, 2003; Deininger et al., 2003; Nkonya et al., 2004; Pender and Gebremedhin, 2004), even though, as mentioned earlier, lack of land title is not a major factor causing tenure insecurity in SSA. The land rights of women are an important issue that has been less well studied than tenure insecurity, though potentially even more important. In many parts of SSA, women often lack rights to make decisions about land use and investments or to inherit land, even though they are often the primary food producers (Abbas, 1997; Gladwin et al., 1997). Free grazing of livestock, which is common in many parts of SSA, also may undermine farmers' incentives to invest in some LEISA technologies, since these may be damaged or eaten by livestock.

Knowledge constraints also can inhibit adoption of newer LEISA technologies or their effectiveness. Many of these technologies are knowledge-intensive and need to be adapted to local circumstances to be effective. Traditional technical assistance approaches that focus on promoting specific technologies are thus less likely to be effective than approaches that teach farmers the agronomic principles behind successful LEISA technologies and develop their capacity to experiment and innovate.

Organic agriculture

Organic agriculture is a specific type of low–external input agriculture that adheres to certain principles in the production and transformation of agricultural commodities. All of the discussion in the preceding section on LEISA approaches therefore applies to organic agriculture. In this section, we consider the special characteristics of organic agriculture and what additional implications these have for the ability to address soil fertility depletion in SSA through this approach.

The International Federation of Organic Agriculture Movements (IFOAM) standards enumerate a list of general principles of organic agriculture (IFOAM, 2002). Of particular relevance for this discussion is the principle that organic agriculture seeks to 'maintain and increase long-term fertility and biological activity of soils using locally adapted cultural, biological and mechanical methods as opposed to reliance on inputs' (ibid.). The IFOAM standards specifically require the use of soil and water conservation methods and recycling of microbial plant or animal material to the soil as the basis for soil fertility management. The IFOAM standards do not completely prohibit the use of mineral fertilizers, but specify that such fertilizers shall be used only as part of a programme to manage long-term soil fertility together with use of organic approaches, and that the mineral fertilizers must be in their natural form and not rendered soluble by chemical treatment (IFOAM, 2002, Sections 4.4.5 and 4.4.6). This provision appears to allow the use of natural phosphate rock as a source of fertility, for example, but not the use of other processed phosphate fertilizers.[1] The implications of this for addressing soil fertility in SSA are discussed below.

Organic agriculture may be either certified or non-certified, as discussed in Chapter 6. With certification, price premiums of 10 to 50% are common for developing country exports of organic products (Scialabba and Hattam, 2002). Producers of organic products may also benefit from a more assured market than they may find for non-organic products. Non-certified organic agriculture offers no particular price or marketing advantages (unless non-certified organic produc-

---

[1] The IFOAM standards stipulate that standards-setting organizations may grant exceptions to the requirement that mineral fertilizers not be chemically treated, but this does not apply to fertilizers containing nitrogen.

tion is pursued as a prelude to certification). Therefore, it seems unlikely to be widely pursued in SSA except as a form of LEISA because of the advantages of LEISA approaches that have already been discussed in the previous section. Thus, we do not discuss non-certified organic agriculture further in this section, but focus on the potential impacts of certified organic agriculture in SSA.

Given the price premium available for organic products in developed country markets, increasing organic production is an attractive option for some producers of export commodities in SSA. Studies of organic farms in Europe and the USA have found that organic farms are able to earn comparable profits to those of conventional farms, despite somewhat lower yields in organic farming, as a result of the price premiums that organic farmers receive (Offermann and Nieberg, 2000; Scialabba and Hattam, 2002). Less evidence is available on the profitability of organic farming in developing countries, though a recent evaluation of organic agriculture projects in Latin America supported by the International Fund for Agricultural Development (IFAD) found that organic producers were able to earn substantially higher net returns than conventional producers in several cases, despite often higher production and certification costs (Damiani, 2002). Comparable evidence from SSA is not available, but these results suggest that profitable organic production is possible in SSA. Given that most farmers in SSA use little or no chemical inputs, the near-term yield disadvantage of organic production compared to conventional production seen in Europe and the USA may be less of a barrier to organic production in SSA. However, producers of some export crops in SSA do use significant quantities of chemical inputs and thus could face yield disadvantages of organic farming compared to their conventional practices.

Certified organic agriculture is still quite limited in SSA. In 2001, there were fewer than 40,000 certified organic farms in SSA, producing on about 190,000 ha (Parrott and van Elzakker, 2003). Organic projects are reportedly achieving good success in SSA, though they are not without problems (ibid.). Farmer organizations have obtained certification to export to high value markets in Europe and the USA, and are obtaining substantial price premiums.

The main constraints to expanding certified organic production in SSA are the difficulties of identifying and securing access to markets and dealing with the procedures and costs of certification. Larger and more commercially oriented producers are likely to enjoy significant advantages over most smallholders in addressing these constraints, unless smallholders are organized into effective farmer organizations and/or are assisted by donor funded projects. In addition, the commodities that offer attractive opportunities for organic production are generally high value export commodities, for which specialized niche markets can be established. Thus, the ability to dramatically expand certified organic production in SSA is likely to be fairly limited for some time, confined to areas of higher potential and favourable market access where high value export commodities have comparative advantage, and to regions and countries where efforts by donors, governments and non-government organizations (NGOs) to promote development of effective farmer organizations and certification schemes have a major presence.

Because organic product markets are still relatively thin niche markets, despite their rapid growth, and because producers in many countries are becoming interested in entering these markets, there is significant risk that premiums will decline, perhaps rapidly, as new producers enter the market. Organic producers and processors also face risks that they will fail to find a market for their products, even if their production is certified. Prices paid by importers of organic products are often less flexible than those paid by buyers of conventional commodities, creating the risk of excess supplies that are not purchased. There is also a risk of damage to the reputation of organic products if any scandals involving disreputable certification practices or poor enforcement of standards were to emerge, which could reduce the premiums that consumers are willing to pay and the overall demand for organic products. Such risks are likely to increase as new certification bodies begin to be accredited in developing countries and new suppliers enter the market.

On the production side, there are serious concerns about the sustainability of organic production in SSA, where problems of phosphorus deficiency and phosphorus fixing soils are common (Buresh *et al.*, 1997). Use of phosphate rock is a promising option for some areas of SSA, particularly in areas close to highly reactive deposits that can be readily used as a soil amendment. Unfortunately, the supply of rock phosphate deposits in SSA is spotty, and these deposits vary greatly in their reactivity. Outside of West Africa, there are few sedimentary deposits of rock phosphate of sufficient reactivity to be usable without processing (ibid.). Most deposits in eastern Africa (where the problem of phosphorus fixing soils, and hence phosphorus requirements for crop production, is greater) are of igneous origin, are costly to mine and require at least partial acidulation to be usable. Coupling these problems with the high cost of transporting phosphate rock relative to its value implies that unprocessed phosphate rock is unlikely to be an economically viable option for phosphorus replenishment for most farmers in SSA.

Depletion of other soil nutrients can also be a constraint to organic agriculture. For example, Deugd (2001) found that depletion of potassium in organic banana production systems of Costa Rica was a serious problem. A similar problem has been observed in studies estimating soil nutrient balances in Uganda (Wortmann and Kaizzi, 1998), where production commonly uses LEISA approaches, though is not certified organic.

Integrated soil fertility management

Integrated soil fertility management (ISFM) is an approach that seeks to combine the use of organic and inorganic approaches to soil fertility management, exploiting complementarities between these different approaches.[2] By doing so, it may be possible to obtain higher yields and returns than by using either inorganic fertilizer or LEISA technologies separately. Numerous studies have demonstrated the potential advantages of this approach in SSA (Bationo *et al.*, 1996; Palm *et al.*, 1997; Sanchez *et al.*, 1997; Vanlauwe *et al.*, 2001; IFDC undated; Breman and Debrah, 2003). LEISA technologies can improve fertilizer use efficiency by increasing the capacity of the soil to store and provide water and nutrients to crops, helping to control weeds and pests, and reducing nutrient losses through erosion and leaching. Inorganic fertilizers can increase the effectiveness of biological nitrogen fixation or other LEISA technologies by providing limiting nutrients (especially phosphorus and potassium because of low concentrations in organic material (Buresh *et al.*, 1997)). Even when the impacts of inorganic and organic approaches are only additive, these approaches may still be complementary in an economic sense because of limited ability to provide particular nutrients from any one source (Place *et al.*, 2003).

Despite growing evidence of the potential of the ISFM approach, this approach is not always successful. Application of organic materials can immobilize nitrogen and reduce yields, especially if the organic materials are of low quality (e.g. high in carbon relative to nitrogen content) (Palm *et al.*, 1997). The quality of organic materials available may vary greatly as a result of variations in soil fertility, livestock feeding practices (in the case of manure) and other factors. Much still needs to be learned about how and when different fertility management approaches can be profitably combined, especially on non-cereal crops (Place *et al.*, 2003). Beyond this, scaling out successes of ISFM will require addressing the constraints affecting both HEIA approaches (such as credit constraints and market development) and LEISA approaches (such as labour, land, biomass and knowledge constraints).

Selecting the right approach

Addressing the problem of soil fertility depletion in SSA requires a pragmatic approach, focusing on what is feasible and profitable for small farmers in diverse settings. Excessive promotion of any particular approach is likely to fail unless the approach is adapted to local circumstances and strengthens the capacity of smallholders to innovate and adapt. It is thus crucial that agricultural research

---

[2] LEISA approaches do not necessarily exclude use of inorganic fertilizers (e.g. Reijntjes *et al.*, 1992). The difference between ISFM and LEISA approaches is therefore more a matter of emphasis and philosophy about the desirability of using such external inputs.

and technical assistance programmes related to soil fertility be farmer-centred and demand driven, focused on identifying and teaching principles of sustainable land management and innovation, rather than a specific set of technologies. These programmes must be able to learn from and build on farmers' own experience in managing soil fertility, since farmers already have substantial knowledge in this area.

Although the approach should draw from farmers' experiences and concerns, basic and adaptive research are also needed to clarify the principles of profitable and sustainable land management, and identify recommendation domains for available technologies. Research and technical assistance programmes cannot be conducted everywhere, so information is needed to help guide the allocation of resources towards where they are likely to have the most payoff in increasing agricultural production and reducing poverty. For this, information is needed on the likely range of profitability of different soil fertility management technologies in major biophysical and socioeconomic domains in SSA, taking into account differences in agricultural potential, access to markets and roads, population density, and other factors.

In high potential areas with good market access, both HEIA and certified organic agriculture may have high potential. The potential for certified organic agriculture is greatest for high value export commodities. For such commodities, organic agriculture offers farmers potential for substantial improvement in incomes, as well as better land management. Realizing this potential, especially for small farmers, will depend critically upon development of effective farmer organizations and marketing institutions, including accredited organic certification programmes in SSA. A good start in this direction has already been made in some countries, but much more investment in such organizations and institutions is needed if organic agriculture is to achieve a marked impact in SSA.

Even when the potential of certified organic agriculture is fully realized, however, it is likely to still reach only a small proportion of the farmers in SSA. Most farmers in SSA will continue for some time to produce low value subsistence-oriented crops and livestock products in areas where agro-ecological conditions and/or market access are not suited to production of high value export commodities.

In areas with lower potential or less access to markets and infrastructure, greater emphasis on LEISA or ISFM approaches will be needed. Recent research has shown substantial potential of such approaches to improve productivity and farmers' incomes in these types of environments. The challenge is to scale up and out these approaches to much broader domains and larger numbers of farmers.

Recommendation domains for technological approaches must also take into account the key constraints facing different types of farmers, and the livelihood strategies being pursued in given domains. Soil fertility management approaches are more likely to be adopted if they complement the farmers' livelihood strategy; e.g. promoting use of manure from intensive livestock operations in vegetable production in peri-urban areas, or tree planting activities in densely populated remote areas to increase income (e.g. timber and honey) and relieve biomass constraints that inhibit recycling of organic materials to the soil.

Where profitable and sustainable technologies are available that have good potential in the recommendation domains, efforts are needed to identify and address the key constraints inhibiting their adoption. Rather than trying to address a long list of constraints everywhere in SSA, the idea is to focus first on key constraints only where these are binding. For example, effective farmer organizations and certification systems are likely to be key limiting constraints to expansion of organic agriculture, where it has strong potential. Short-term credit and market constraints are likely to be limiting constraints in areas where HEIA approaches are profitable. By contrast, labour constraints and land tenure issues may be more important limitations on LEISA and ISFM approaches, and thus will need to be addressed as a component of the strategy for soil fertility replenishment in less-favoured areas.

Conclusions

To successfully address the problem of soil fertility depletion and low productivity in SSA, approaches need to be profitable to farmers in the near term, of acceptable risk, feasible given the constraints faced by African smallholder farmers, and economically, environmentally and socially sustainable. Considering these factors, no single approach to the problem is likely to be successful throughout the diverse circumstances of SSA. Less dogmatic debate and more pragmatic actions are needed.

Organic farming approaches can play a constructive role in many African contexts in helping to address soil fertility and productivity constraints. However, the scope for certified organic farming in SSA appears to be limited to high value products, which can be relatively easily marketed for overseas export. The local markets for organic products are simply too small to sustain commercial production based on price premiums. If the understanding of organic farming is extended to non-certified LEISA technologies without the use of pesticides and fertilizers, a larger part of SSA agriculture can be classified as organic. In order to ensure the sustainability of agriculture in SSA, organic farming and LEISA technologies should be presented to farmers as two options amongst others. The use of inorganic fertilizers and even limited amounts of pesticides may in some cases be required to ensure a stable or increasing production. More research is needed on what approaches are likely to be profitable and feasible under different conditions and considering the diversity of farmers and consumers in SSA. If a dogmatic approach is taken to promoting organic agriculture, it is likely to meet with limited success and could be counterproductive.

References

Abbas, J.D. (1997). Gender asymmetries in intrahousehold resource allocation in sub-Saharan Africa: some policy implications for land and labor productivity. In: Haddad, L., Hoddinott, J. and Alderman, H. (eds) Intrahousehold Resource Allocation in Developing Countries: Models, Methods, and Policy. Johns Hopkins University Press, Baltimore, MD.

Adams, W.M. and Mortimore, M. (1997). Agricultural intensification and flexibility in the Nigerian Sahel. The Geographical Journal 163: 150-160.

Alföldi, T., Fliessbach, T., Geier, U., Kilcher, L., Niggli, U., Pfiffner, L., Stolze, M. and Willer, H. (2002). Organic agriculture and the environment. In: Scialabba, N. and Hattam, C. (eds) Organic Agriculture, Environment and Food Security. Environment and Natural Resources Series No. 4, Food and Agriculture Organization of the United Nations, Rome.

Altieri, M.A. (1995). Agro-ecology: The Science of Sustainable Agriculture. Westview Press, Boulder, CO.

Asrat, K., Idris, K. and Semegn, M. (1996). The 'flexibility' of indigenous SWC techniques: a case study of the Harerge Highlands, Ethiopia. In: Reij, C., Scoones, I. and Toulmin, C. (eds) Sustaining the Soil: Indigenous Soil and Water Conservation in Africa. Earthscan, London.

Atwood, D.A. (1990). Land registration in Africa: The impact on agricultural production. World Development 18(5): 659-671.

Avery, D.T. (1995). Saving the Planet with Pesticides and Plastic: The Environmental Triumph of High Yield Farming. The Hudson Institute, Indianapolis.

Avery, D.T. (1998). The hidden dangers in organic food. The American Prospect, Fall, 19-22.

Barrett, C.B., Place, F., Aboud, A. and Brown, D.R. (2002). The challenge of stimulating adoption of improved natural resource management practices in African agriculture. In: Barrett, C., Place, F. and Aboud, A.A. (eds) Natural Resources Management in African Agriculture. CABI International, Wallingford, UK.

Bationo, A., Rhodes, E., Smaling, E.M.A. and Visker, C. (1996). Technologies for restoring soil fertility. In: Mokwunye, A.U., de Jager, A. and Smaling, E.M.A. (eds) Restoring and Maintaining the Productivity of West African Soils: Key to Sustainable Development. Miscellaneous Fertilizer Studies No. 14, International Fertilizer Development Center, Muscle Shoals, Alabama.

Benin, S. (2003). Increasing land productivity in high versus low agricultural potential areas: the case of the Ethiopian highlands. Paper under review for Food Policy. IFPRI. Mimeo, Washington, DC.

Besley, T. (1995). Property rights and investment incentives: Theory and evidence from Ghana. Journal of Political Economy 103(5): 903-937.

Binswanger, H.P. (1980). Attitudes towards risk: experimental measurement in rural India. American Journal of Agricultural Economics 62(3): 395-407.

Binswanger, H.P. and Townsend, R.F. (2000). The growth performance of agriculture in sub-Saharan Africa. American Journal of Agricultural Economics 82(5): 1075-1086.

Borlaug, N. (1992). Small-scale agriculture in Africa: the myths and realities. Feeding the Future (Newsletter of the Sasakawa Africa Association).

Borlaug, N. (1994). Chemical fertilizer 'essential'. Letter to International Agricultural Development (Nov.–Dec.), p. 23.

Breman, H. and Debrah, S.K. (2003). Improving African food security. SAS Review 23(1): 153-170.

Bryceson, D. (2002). The Scramble in Africa: Reorienting Rural Livelihoods. World Development 30(5): 725-739.

Buresh, R.J., Smithson, P.C. and Hellums, D.T. (1997). Building soil phosphorus capital in Africa. In: Buresh, R.J., Sanchez, P.A. and Calhoun, F. (eds) Replenishing Soil Fertility in Africa. Soil Science Society of America, American Society of Agronomy, SSSA Special Publication Number 51, Madison, Wisconsin.

Clay, D.C., Kelly, V., Mpyisi, E. and Reardon, T. (2002). Input use and conservation investments among farm households in Rwanda: Patterns and determinants. In Barrett, C. Place, F. and Aboud, A.A. (eds) Natural Resources Management in African Agriculture. CABI, New York.

Cleaver, K.M. and Schreiber, G.A. (1994). Reversing the Spiral: The Population, Agriculture and Environment Nexus in Sub-Saharan Africa. The World Bank, Washington, DC.

Club du Sahel (1996). Etats des réflexions sur les transformations de l'agriculture dans le Sahel. Note de Synthèse. OECD, Paris. Available at http://isur.iep.free.fr/ docoursisur/PDM/data/origine/pdf/451.pdf

Damiani, O. (2002). Small farmers and organic agriculture: lessons learned from Latin America and the Caribbean. Office of Evaluation and Studies, International Fund for Agricultural Development, Rome.

Deininger K., Jin, S., Gebre Selassie, H.S., Adenew, B. and Nega, B. (2003). Tenure security and land-related investment: evidence form Ethiopia. World Bank Policy Research Working Paper 2991. The World Bank, Washington, DC.

Demeke, M. and Egziabher, T.G. (2003). Sustainable rural (regional) development in Ethiopia: the case of marginal woredas in Tigray region. Paper presented at the 7th Annual Conference of the Agricultural Economics Society of Ethiopia, 7–9 August, Addis Ababa. Mimeo, Addis Ababa University.

Deugd, M. (2001). Feasibility of production systems in Talamanca, Costa Rica. Centre of Rural Development Studies, Free University of Amsterdam, San José, Costa Rica.

Dorward, A., Kydd, J., Morrison, J. and Urey, I.. (2004). A policy agenda for pro-poor agricultural growth. World Development 32(1): 73-89.

Eicher, C.K. and Byerlee, D. (1997). Accelerating maize production: synthesis. In: Byerlee, D. and Eicher, C.K. (eds) Africa's Emerging Maize Revolution. Lynne Rienner Publishers, London.

FAOSTAT data (2004). Online at http://apps.fao.org/default.jsp (FAOSTAT is an online and multilingual database currently containing over 3 million time-series records covering international statistics in the agricultural area).

Gabre-Madhin, E. and Amha, W. (2003). Getting markets right in Ethiopia: Institutional changes to grain markets. International Food Policy Research Institute. Mimeo, Washington, DC.

Gavian, S. and Fafchamps, M. (1996). Land tenure and allocative efficiency in Niger. American Journal of Agricultural Economics 78(May): 460-471.

Gebremedhin, B. and Swinton, S.M. (2002). Sustainable management of private and communal lands in northern Ethiopia. In: Barrett, C.B., Place, F. and Aboud, A.A. (eds.) Natural Resources Management in African Agriculture: Understanding and Improving Current Practices. CAB International, Wallingford, UK.

Giller, K.E., Cadisch, G., Ehaliotis, C., Adams, E., Sakala, W.D. and Mafongoya, P.L. (1997). Building soil nitrogen capital in Africa. In: Buresh, R.J., Sanchez , P.A. and

Calhoun, F. (eds) Replenishing Soil Fertility in Africa. Soil Science Society of America, American Society of Agronomy, SSSA Special Publication Number 51, Madison, Wisconsin.

Gladwin, C.H., Buhr, K.L., Goldman, A., Hiebsch, C., Hildebrand, P.E., Kidder, G., Langham, M., Lee, D., Nkedi-Kizza, P., and Williams, D. (1997). Gender and soil fertility in Africa. In: Buresh, R.J., Sanchez, P.A. and Calhoun, F. (eds) Replenishing Soil Fertility in Africa. Soil Science Society of America, American Society of Agronomy, SSSA Special Publication Number 51, Madison, Wisconsin.

Grepperud, S. (1996). Population Pressure and Land Degradation: The Case of Ethiopia. J. Env. Econ. and Mgmt 30: 18-33.

Gruhn, P., Goletti, F., and Yudelman, M. (2000). Integrated Nutrient Management, Soil Fertility, and Sustainable Agriculture: Current Issues and Future Challenges. Food, Agriculture, and the Environment Discussion Paper 32, International Food Policy Research Institute, Washington, DC.

Haggblade, S. and Tembo, G. (2003). Conservation farming in Zambia. Environment and Production Technology Division Discussion Paper No. 108, International Food Policy Research Institute, Washington, DC.

Hagos, F. (2003). Poverty, institutions, peasant behavior and conservation investment in northern Ethiopia. Dissertation No. 2003:2, Department of Economics and Social Sciences, Agricultural University of Norway, Aas, Norway.

Hassan, A. (1996). Improved traditional planting pits in the Tahoua Department (Niger): an example of rapid adoption by farmers. In: Reij, C., Scoones, I. and Toulmin, C. (eds) Sustaining the Soil: Indigenous Soil and Water Conservation in Africa. Earthscan, London.

Herweg, K. (1993). Problems of acceptance and adaption of soil conservation in Ethiopia. Topics in Applied Resource Management 3: 391-411.

Høgh-Jensen, H. (personal communication). The Royal Veterinary and Agricultural University, Taastrup, Denmark.

Holden, S., Shiferaw, B. and Wik, M. (1998). Poverty, market imperfections and time preferences: of relevance for environmental policy? Environment and Development Economics 3: 105-130.

IFDC. Undated. International Fertilizer Development Center. The integrated soil fertility management project.

IFOAM (2002). International Federation of Organic Agriculture Movements. IFOAM Basic Standards for Organic Production and Processing. Available at http://www.ifoam.org

Joelibarison (2001). Summary economic analysis of SRI methods. Unpublished Cornell University memorandum, October. (Cited by Moser and Barrett (2002)).

Kaboré, D. and Reij, C. (2004). The emergence and spreading of an improved traditional soil and water conservation practice in Burkina Faso. Environment and Production Technology Division Discussion Paper No. 114, International Food Policy Research Institute, Washington, DC.

Kaizzi, C.K. (2002). The potential benefit of green manures and inorganic fertilizers in cereal production on contrasting soils in eastern Uganda. Ecology and Development Series No. 4, Center for Development Research (ZEF), University of Bonn, Germany.

Kirchmann, H. and Bergström, L. (2001). Do organic farming practices reduce nitrate leaching? Communications in Soil Science and Plant Analysis 32(7and8): 997-1028.

Leach, M. and Mearns, R. (1996). The Lie of the Land: Challenging Received Wisdom on the African Environment. International African Institute with James Currey, Oxford.

Lema, A.J. (1996). Cultivating the valleys: vinyungu farming in Tanzania. In: Reij, C., Scoones, I. and Toulmin, C. (eds) Sustaining the Soil: Indigenous Soil and Water Conservation in Africa. Earthscan, London.

Mäder, P., Fliessbach, A., Dubois, D., Gunst, L., Fried, P. and Niggli, U. (2002). Soil fertility and biodiversity in organic farming. Science 296: 1694-1697.

Mekuria, M. and Waddington, S.R. (2002). Initiatives to encourage farmer adoption of soil-fertility technologies for maize-based cropping systems in southern Africa. In: Barrett, C.B., Place, F. and Aboud, A.A. (eds) Natural Resources Management in African Agriculture: Understanding and Improving Current Practices. CAB International, Wallingford, UK.

Millenium Project Task Force on Hunger (2004). Halving hunger by 2015: a framework for action. Interim Report. Millenium Project, New York.

Moser, C.M. and Barrett, C.B. (2002). Labor, liquidity, learning, conformity and smallholder technology adoption: the case of SRI in Madagascar. Department of Applied Economics and Management, Cornell University. Mimeo, Ithaca, NY.

Nandwa, S.M. and Bekunda, M.A. (1998). Research on nutrient flows and balances in East and Southern Africa: state of the art. Agriculture, Ecosystems and Environment 71: 5-18.

Nkonya, P., Pender, J., Jagger, P., Sserunkuuma, D., Kaizzi, C.K. and Ssali, H. (2004). Strategies for sustainable land management and poverty reduction in Uganda. Research Report No. 133. International Food Policy Research Institute, Washington, DC.

Offermann, F. and Nieberg, H. (2000) Economic performance of organic farms in Europe. In: Organic Farming in Europe: Economics and Policy. Vol 5. University of Hohenheim, Stuttgart-Hohenheim.

Otsuka, K. and Place, F. (2001). Issues and theoretical framework. In: Otsuka, K. and Place, F. (eds) Land Tenure and Natural Resource Management: A Comparative Study of Agrarian Communities in Asia and Africa. Johns Hopkins, Baltimore.

Palm, C.A., Myers, R.J.K. and Nandwa, S.M. (1997). Combined use of organic and inorganic nutrient sources for soil fertility maintenance and replenishment. In: Buresh, R.J., Sanchez, P.A. and Calhoun, F. (eds) Replenishing Soil Fertility in Africa. Soil Science Society of America, American Society of Agronomy, SSSA Special Publication Number 51, Madison, Wisconsin.

Parrott, N. and van Elzakker, B. (2003). Organic and like-minded movements in Africa. International Federation of Organic Agriculture Movements, Germany. (Available at www.ifoam.org).

Pender, J. (1996). Discount rates and credit markets: Theory and evidence from rural India. Journal of Development Economics 50: 257-296.

Pender, J. and Gebremedhin, B. (2004). Impacts of policies and technologies in dryland agriculture: evidence from northern Ethiopia. In: Rao, S.C. (ed.) Challenges and Strategies for Dryland Agriculture. American Society of Agronomy and Crop Science Society of America, CSSA Special Publication 32, Madison, WI.

Pender, J. and Kerr, J. (1998). Determinants of farmers' indigenous soil and water conservation investments in India's semi-arid tropics. Agricultural Economics 19: 113-125.

Pender, J., Place, F. and Ehui, S. (1999). Strategies for sustainable agricultural development in the East African Highlands. Environment and Production Technology Division Discussion Paper No. 41, International Food Policy Research Institute, Washington, DC.

Pender, J., Gebremedhin, B., Benin, S. and Ehui, S. (2001). Strategies for sustainable development in the Ethiopian highlands. American Journal of Agricultural Economics 83(5): 1231-40.

Pender, J., Ssewanyana, S., Edward, K. and Nkonya, E. (2004). Linkages between poverty and land management in rural Uganda: evidence from the Uganda National Household Survey, 1999/2000. Environment and Production Technology Division Discussion Paper No. 122, International Food Policy Research Institute, Washington, DC.

Pinstrup-Andersen, P. and Pandya-Lorch, R. (1994). Alleviating Poverty, Intensifying Agriculture, and Effectively Managing Natural Resources. Food, Agriculture, and the Environment Discussion Paper 1. International Food Policy Research Institute, Washington, DC.

Place, F. and Hazell, P. (1993). Productivity effects of indigenous land tenure systems in sub-Saharan Africa. American Journal of Agricultural Economics 75(1): 10-19.

Place, F., Franzel, S., DeWolf, J., Rommelse, R., Kwesiga, F., Niang, A. and Jama, B. (2002a). Agroforestry for soil-fertility replenishment: evidence on adoption processes in Kenya and Zambia. In: Barrett, C.B., Place, F. and Aboud, A.A. (eds) Natural Resources Management in African Agriculture: Understanding and Improving Current Practices. CAB International, Wallingford, UK.

Place, F., Swallow, B.M., Wangila, J. and Barrett, C.B. (2002b). Lessons for natural resource management technology adoption and research. In: Barrett, C.B., Place, F. and Aboud, A.A. (eds) Natural Resources Management in African Agriculture: Understanding and Improving Current Practices. CAB International, Wallingford, UK.

Place, F., Barrett, C.B., Freeman, H.A., Ramisch, J.J. and Vanlauwe, B. (2003). Prospects for integrated soil fertility management using organic and inorganic inputs: evidence from smallholder African agricultural systems. Food Policy 28: 365-378.

Place, F., Franzel, S., Noordin, Q. and Jama, B. (2004). Improved fallows in Kenya: history, farmer practice, and impacts. Environment and Production Technology Division Discussion Paper No. 115, International Food Policy Research Institute, Washington, DC.

Platteau, J.P. (1996). The evolutionary theory of land rights as applied to sub-Saharan Africa: A critical assessment. Development and Change 27: 29-86.

Pretty, J. (1999). Can sustainable agriculture feed Africa? New evidence on progress, processes and impacts. Environment, Development and Sustainability 1: 253-274.

Quiñones, M.A., Borlaug, N.E. and Dowswell, C.R. (1997). A fertilizer-based Green Revolution for Africa. In: Buresh, R.J., Sanchez, P.A. and Calhoun, F. (eds) Replenishing Soil Fertility in Africa. Soil Science Society of America, American Society of Agronomy, SSSA Special Publication Number 51, Madison, Wisconsin.

Reardon, T. and Vosti, S. (1995). Links between rural poverty and the environment in developing countries: Asset categories and investment poverty. World Development 23(9): 1495-1506.

Reijntjes, C., Haverkort, B. and Waters-Bayer, A. (1992). Farming for the Future: An Introduction to Low-External Input and Sustainable Agriculture. MacMillan Press, London.

Rosegrant, M.W., Cline, S.A., Li, W., Sulser, T. and Valmonte-Santos, R.A. (2004). Looking ahead: long-term prospects for Africa's food and nutrition security. Plenary paper presented at the conference 'Assuring Food and Nutrition Security in Africa by 2020', April 1–3, Kampala. International Food Policy Research Institute, Washington, DC.

Sanchez, P.A. (2002). Soil fertility and hunger in Africa. Science 295: 2019-2020.

Sanchez, P.A., Shepherd, K.D., Soule, M.J., Place, F.M., Buresh, R.J., Izac, A.-M.N., Mokwunye, A.U., Kwesiga, F.R., Ndiritu, C.G. and Woomer, P.L. (1997). Soil fertility replenishment in Africa: an investment in natural resource capital. In: Buresh, R.J., Sanchez, P.A. and Calhoun, F. (eds) Replenishing Soil Fertility in Africa. Soil Science Society of America, American Society of Agronomy, SSSA Special Publication Number 51, Madison, Wisconsin.

Scherr, S.J. (1999). Soil Degradation: A Threat to Developing-Country Food Security by 2020? Food, Agriculture, and the Environment Discussion Paper 27, International Food Policy Research Institute, Washington, DC.

Scherr, S. (2000). A downward spiral? Research evidence on the relationship between poverty and natural resource degradation. Food Policy 25: 479-498.

Scialabba, N. and Hattam, C. (2002). General concepts and issues in organic agriculture. In: Scialabba, N. and Hattam, C. (eds) Organic Agriculture, Environment and Food Security. Environment and Natural Resources Series No. 4, Food and Agriculture Organization of the United Nations, Rome.

Scoones, I., Reij, C. and Toulmin, C. (1996). Sustaining the soil: indigenous soil and water conservation in Africa. In: Reij, C., Scoones, I. and Toulmin, C. (eds) Sustaining the Soil: Indigenous Soil and Water Conservation in Africa. Earthscan, London.

Smale, M. and Jayne, T. (2003). Maize in eastern and southern Africa: 'Seeds' of success in retrospect. Environment and Production Technology Division Discussion Paper No. 97, International Food Policy Research Institute, Washington, DC.

Tegene, A. and Wiebe, K. (2003). Resource quality and agricultural productivity in sub-Saharan Africa with implications for Ethiopia. Paper presented at the International Symposium on Contemporary Development Issues in Ethiopia, Ghion Hotel, Addis Ababa, Ethiopia, July 11–12.

Templeton, S. and Scherr, S. (1999). Effects of demographic and related microeconomic changes on land quality in hills and mountains of developing countries. World Development 27(6): 903-918.

Tiffen, M., Mortimore, M. and Gichuki, F. (1994). More People, Less Erosion: Environmental Recovery in Kenya. John Wiley and Sons, New York.

Toulmin, C. and Guèye, B. (2003). Transformations in West African agriculture and the role of family farms. Sahel and West Africa Club, OECD, Paris. http://www.sahel-club.org/doc/agri_doc_1.pdf

Uphoff, N. (2002). Opportunities for raising yields by changing management practices: the system of rice intensification in Madagascar. In: Uphoff, N. (ed.) Agroecological Innovations: Increasing Food Production with Participatory Development. Earthscan, London.

Vanlauwe, B., Diels, J., Sanginga, N. and Merckx, R. (2001). Integrated Plant Nutrient Management in Sub-Saharan Africa: From Concept to Practice. CABI, Wallingford, UK.

Viyas, V.S. (1983). Asian agriculture: achievements and challenges. Asian Development Review 1: 27-44.

Voortman, R.L., Sonneveld, B.G.J.S. and Keyzer, M.A. (2000). African land ecology: opportunities and constraints for agricultural development. CID Working Paper No. 37, Center for International Development, Harvard University, Cambridge, MA.

Woelcke, J., Berger, T. and Park, S. (2002). Land management and technology adoption in Uganda: An integrated bio-economic modeling approach, In: Nkonya, E., Sserunkuuma, D. and Pender, J. (eds) Policies for improved land management in Uganda: Second national workshop. EPTD workshop summary paper No. 12: 131-136.

Wood, S., Sebastian, K. and Scherr, S.J. (2000). Pilot Analysis of Global Ecosystems. International Food Policy Research Institute and World Resources Institute, Washington, DC.

Wortmann, C.S. and Kaizzi, C.K. (1998). Nutrient balances and expected effects of alternative practices in farming systems of Uganda. Agriculture, Ecosystems and Environment, 71: 115-129.

Yanggen, D., Kelly, V., Reardon, T. and Naseem, A. (1998). Incentives for fertilizer use in sub-Saharan Africa: a review of empirical evidence on fertilizer response and profitability. MSU International Development Working Paper No. 70, Department of Agricultural Economics, Michigan State University, East Lansing, MI.

Yesuf, M. (2004). Risk, time and land management under market imperfections: applications to Ethiopia. Economic Studies No. 139, Ph.D. Dissertation, Department of Economics, Göteborg University, Sweden.

9
Sustainable veterinary medical practices in organic farming: a global perspective

Mette Vaarst, Stephen Roderick, Denis K. Byarugaba, Sofie Kobayashi, Chris Rubaire-Akiiki and Hubert J. Karreman*

Introduction 243
The potential for organic livestock farming 244
Disease management in organic livestock production 247
The use and risks of antimicrobial drugs 248
  The use of antimicrobial drugs in veterinary treatment 248
  Antimicrobial residues 249
  Antimicrobial resistance patterns 249
  An organic approach to reducing antimicrobial drug usage 250
Vector-borne diseases and organic livestock farming 254
  Characteristics of vector-borne diseases 254
  Strategies for controlling tick-borne diseases 255
  Developing a holistic approach to the control of TBDs 256
  Context-based control strategies for TBDs 257
  Tick-resistant cattle breeds 258
  Approaches to the control of tsetse-borne trypanosomiasis 259
Disease control issues associated with land use and land tenure 261
  Pastoralism, nomads and transhumance 261
  Communal grazing and community participation 261
  Bio-security and closed herds 262
  Zero grazing and organic farming 263
Developing organic strategies to enhance animal health 264
  Breeding strategies and the role of indigenous breeds 264
  The use of local knowledge and traditional medicine 266
  Vaccination 268

---

* Corresponding author: Danish Institute of Agricultural Sciences, P.O. Box 50, Blichers Allé 20, DK-8830 Tjele, Denmark. E-mail: Mette.Vaarst@agrsci.dk

Moving from an 'organic approach' to 'organic animal production' 268
    'The hidden world of ecological farming' 268
    From non-certified to certified organic production...................................... 270
    Context-related plans for converting to organic livestock production 271
Conclusions and recommendation 272

Summary

Livestock production systems are the focus area of this chapter, where the prospects for an organic approach to veterinary treatment and disease control are discussed in particular. We have taken a case presentation approach to this topic by selecting some widely different farming systems to represent different challenges and opportunities for using and reducing veterinary medical products, as well as developing disease prevention and health-promoting strategies that meet the ideas of organic animal husbandry. The major challenge in organic livestock production systems is to 'think the organic principles' into a wide range of diverse systems under a wide range of circumstances and conditions, including systems which are not certified as 'organic' at the moment. We recommend that developing organic animal husbandry at all times requires a thorough analysis of the problems, opportunities and existing knowledge. All organic systems should allow animals to perform their natural behaviour as far as possible, and naturalness is an important principle also of organic livestock farming. We consider various organic approaches to breeding for disease resistance (use of indigenous breeds), the role of vaccination, traditional medicine and alternatives to biomedical treatments and other approaches to disease management. We have given particular emphasis on the need for a reduction in the use of antimicrobial veterinary drugs, as we can see some potential for a reduction of dependency on veterinary medicine, and – when successful – the associated problems of drug residues and resistance. The potential for the control of vector-borne diseases in the development of organic systems in tropical areas is also included in the discussion. In North Europe and large areas of the north-western world, production diseases related to high yield and performance dominate, whilst in the tropical regions the risk of infectious and epidemic diseases is a greater concern. The development of organic farming must always be careful not to threaten local and regional disease control programmes, particularly where the diseases are zoonotic in nature, are highly infectious or are of widespread economic importance, e.g Rinderpest and foot and mouth disease. According to experiences from the USA and Europe, effectiveness of organic approaches to health management are not always immediate. Whole communities can benefit from implementing and organizing an organic approach to disease prevention, e.g. in the case of communal grazing systems.

Introduction

Livestock production systems form a major part of farming and food production worldwide. A quarter of the world's land is used for grazing, and currently accounts for some 40% of the gross value of agricultural production (Steinfeld, 2004), one-third of this being livestock production in the so-called developing countries. Changes in consumption patterns will continue to result in an average net increase in the demand for animal food products (Steinfeld, 2004). Animals also play an important role within organic farming, and it is appropriate that any discussion of the global prospects for organic farming should incorporate issues associated with the role of animals. This discussion could extend to their role as forage consumers, as producers of milk, meat and fibre, their market potential, their role in recycling nutrients, the impact on the environment or their relative social and economic relevance. However, in this chapter we have restricted our focus to the prospects for an organic approach to veterinary treatment and disease control.

We have taken, in part, a case presentation approach to this discussion by selecting some widely different farming systems that represent different challenges and opportunities for using and reducing veterinary medical products, as well as developing disease prevention and health-promoting strategies that meet the ideals of organic animal husbandry. Presenting the discussion in this manner also helps us to illustrate the diversity and complexity confronting the development of organic livestock production. The major challenge here is to 'think the organic principles', in this case focusing on disease management, into the wide range of diverse systems that have developed under a wide range of circumstances and conditions. Not all of the examples and case studies are certified organic, but they are systems where there is either a potential or a need to adopt an organic approach.

Apart from excluding systems that continuously confine animals, we have not used animal welfare as selection criteria for our cases. It is assumed that the natural elements of the systems allow animals to exhibit, at least in part, their natural behaviour, and animal welfare is to a large extent defined on this basis. The preference for cattle within the cases and examples reflects the experiences of the authors. However, the general problems and conclusions that can be drawn from these examples can be also applied to other species and systems of production.

Throughout the discussion we consider various organic approaches to disease management with reference to such practices as breeding for disease resistance, the role of vaccination, traditional medicine and alternatives to biomedical treatments. We have given particular emphasis to the need for a reduction in the use of antimicrobial veterinary drugs. The potential for the control of vector-borne diseases is also discussed in some detail. Referring to specific systems, we include examination of the importance of land use, seasonal extremes, traditional knowledge and community participation.

The potential for organic livestock farming

Organic animal production methods have developed mainly in Western Europe and the USA, where they have primarily served as producers of niche products for consumers who give priority to environmental and animal welfare concerns. In some countries, organic livestock production offers a possibility to establish a niche production with a higher price, but in other situations this opportunity may not exist, or may not be appropriate. In such cases, the potential of organic farming may be viewed more in terms of the development of a method of farming, where organic principles are adopted into existing systems and can contribute to sustainability, environmentally friendly production, food security and food quality.

There is, of course, a huge diversity of farming systems across the globe, ranging from the purely commercial to those that barely achieve subsistence. In order to place the diversity of production and economic conditions into perspective, Table 9.1 describes key demographic and background information from countries from which the case studies have been drawn.

Table 9.1. Background information from the five countries chosen as case studies in this chapter.

Background information	Bhutan	Denmark	Kenya	Uganda	USA
Area (km^2)	47,000	42,430	569,140	197,100	9,156,960
Population	734,340^1	5,343,000	31,904,000	24,780,000	288,530,000
Population density (persons/(km^2)	46.8	125.9	56.1	125.7	31.5
Urban population,%	7.1	88.1	33.1	14.2	77.2
Children per woman, no.	5.3	1.7	4.4	7.1	2
Infant mortality (per 1000)	77	4	77	81	7
Life expectancy at birth (years)	63	77	49	46	78
Access to safe water,% of population	62	100	57	52	100
Land use					
- arable land	7.7^1	61.6^2	6.9	25^3	18.9
- forest and woodland	72.5^1	12.6^2	28.9	24^3	29.2
- other	19.8^1	25.8^2	64.2	18^3	51.9
Oil, kg used per person per year	NA	3,924.5	504.6	NA	7,936.9
Fertilizer use (kg/ha)	0	1,704	383	9	1,123

[1] Statistical yearbook of Bhutan 2003.
[2] Gyldenkærne et al., 2005.
[3] Statistical yearbook of Uganda. Other information from 'The World Guide' 2003/2004, Instituto del Tercer Mundo, New Internationalist Publications Ltd, UK.

Two of our case studies are in sub-Saharan Africa where about 70% of livestock is in the hands of rural farmers, many within mixed crop–livestock situations. De Leeuw *et al.* (1995) and others broadly classify African livestock systems into pastoralist (30% of cattle on the continent), agro-pastoralist (20%), crop-livestock farmers (40%) and crop farmers with livestock (10%). Resource-poor farmers may account for half of the rural population, yet they keep a very small share of the livestock wealth. While they contribute little to the aggregate output from livestock production, the few sheep, goats and poultry that they do own are generally very important to their welfare (de Leeuw *et al.*, 1995). The economic opportunity for formal conversion to organic production will vary significantly within and between these systems, as will the technical feasibility of adopting organic production methods.

Parrott *et al.* (Chapter 6) suggest that the certified organic farming sector is highly heterogeneous and that perceptions about organic farming vary considerably. In this chapter, non-certified organic farming is considered loosely as farming systems that pertain, either consciously or otherwise, to the basic agro-ecological principles associated with a low reliance on external resources. The driving forces for adopting an organic approach may be commercial, environmental, cultural, or legislative and might include other related or independent reasons, but may not necessarily justify or require certification as 'organic'. In this discussion we give equal credence to those systems that have no commercial reason for formal conversion and organic certification, but which offer possibilities for adopting an organic approach.

Boxes 9.1 and 9.2 present profiles of Uganda and Bhutan as examples of countries with a potential for developing organic livestock production. In Bhutan organic farming is not well developed, yet there is a significant emphasis on the use of local breeds of livestock – an important element of the organic approach to sustainability, disease prevention and health promotion. Uganda provides an example where traditional, low input livestock production proliferates within mixed farming systems. Here, a relatively strong organic sector has developed, but has focused on organic crop production systems. Examples from these, and other countries, are described later in the text to demonstrate some specific issues associated with animal health, welfare and disease control.

Box 9.1. The agricultural background of Uganda, including the development of the organic sector.

Livestock production in Uganda accounts for 7.5% of the total gross domestic product (GDP) and 17% of agricultural GDP. The national cattle herd ranges between 5 million and 6 million. Ninety per cent of this exists within mixed-farming smallholder and pastoral systems. Indigenous zebu cattle account for 70% of the cattle population, followed by indigenous Sanga (Ankole) (15%). Indigenous breeds also dominate the sheep, goat and poultry sectors.

Conventional agricultural practices have had limited penetration and adoption in sub-Saharan Africa, and Uganda in particular, with limited use of agrochemicals and the predominance of traditional production practices. This 'organic-by-default' production stems from a lack of choice, given poor access to alternatives. There is an underlying assumption that farming practices in many places are organic and that the agriculture systems are very suitable for conversion to organic farming. Therefore, part of the market-oriented promotion of organic agriculture has focused on the sale of smallholder farmers' produce on the international organic market to provide a premium farm gate price, thereby increasing farmers' incomes (sometimes with prices up to 50% higher and with no significant increase in production cost).

Recognition and promotion of organic agriculture in Uganda started in the early 1990s when non-governmental organizations (NGOs), adopted and initiated programmes under the label of 'sustainable agriculture', with the intention of arresting the degradation of smallholder farm lands and rejuvenating production to improve food security. Recently a network, the National Organic Agricultural Movement of Uganda (NOGAMU), has been formed by various involved parties to promote organic agriculture in Uganda and also to address policy issues. Beside the market-oriented driving force, the promotion of organic agriculture has also been based on natural resource conservation and improvements in the livelihoods of resource-poor farmers. Donor support has played a crucial role in assisting the sub-sector to gain a foothold in Uganda. Most organic agriculture promotion initiatives in Uganda have been in the crop sector, with few initiatives in the livestock sector. Uganda's natural environment provides good grazing conditions for cattle, sheep and goats. The natural freshwater lakes provide a favourable environment for freshwater fish (natural 'organic' fish from natural lakes with no inputs is an important export earner, accounting for US$ 87.4 m in 2002/3). The natural tropical forests provide a good environment for bee keeping. The major constraints to livestock production in Uganda, as with other tropical areas, are vectors and vector-borne diseases, such as ticks and tick-borne diseases, tsetse flies and trypanosomiasis. Control of these diseases has depended on heavy use of acaricides and trypanocides. Vaccines or more environmentally friendly methods of control are not very well developed.

Box 9.2. The agricultural background of Bhutan.

Bhutan is a landlocked mountainous country in the Eastern Himalaya, similar in size to Switzerland. Elevations range from 150 m to 8000 m above sea level, and climatic conditions range from hot and humid subtropical in the south (bordering India) to alpine climate in the high altitudes (bordering the Tibetan Plateau). The population of Bhutan is estimated at approximately three quarters of a million people, around 90% of whom are dependent on subsistence agriculture and livestock production. The traditional, self-sustaining farming systems integrate livestock production with crop production and forest products. Very little is produced for the market, though some produce is bartered between high and low altitude settlements. In addition to livestock products such as milk, meat, eggs and hair/wool, livestock contribute manure and draught power to the farming systems, and recycle nutrients between the forests and cropland. Due to the Buddhist culture and religion, the Bhutanese are extremely uncomfortable with the idea of animal slaughter on a commercial scale (Mack and Steane, 1996). Farming systems vary widely, and agro-climatic conditions, the ethnic background of the rural population, and access to market are important factors influencing the farming systems in practice (Roder *et al.*, 1992).

Bhutan has not developed an organic agricultural sector, but the Royal Government of Bhutan (RGOB) is committed to sustainable development (economic, ecological and social/cultural) (RGOB, undated). Livestock, especially ruminants, are an integral part of all farming systems. Three distinct types of large ruminant production systems exist in Bhutan: a transhumant Yak system, a migratory cattle system in the temperate/subtropical areas, and sedentary livestock rearing in rural settlement areas. Farmers who keep more cattle than needed to maintain the farming system traditionally let their cattle migrate between higher altitude areas, used in summer, and lower altitude areas, used in winter. With increasing numbers of cattle, extra pressure is put on the forests, threatening animal health as diseases are transmitted between regions.

Disease management in organic livestock production

Organic farming generally does not accept the use of synthetic chemical inputs. As a consequence, the role of veterinary medicine often dominates discussions of the relative merits or otherwise of organic livestock production. High standards of animal welfare are a basic premise of organic agriculture and, thus, animal suffering should be avoided. If necessary, this may involve medical treatment. This element of organic farming is firmly established in organic farming legislation. This places livestock in a unique position compared to other organic produce where there are more stringent restrictions on chemical and artificial inputs.

Most veterinary treatments in modern animal production systems are bio-medical, and this is the most widely accepted treatment approach in Western European and North American societies. Schillhorn van Veen (1997) views this approach, based on an understanding of pathophysiology and immunology, as being linked to 'a rational western society and its beliefs'. Yet with a few exceptions (such as mass vaccinations, insecticides, and some antibiotics and anthelmintics), he argues that this approach has not really reached the non-Western world. He points to the fact that, even today, there is much greater reliance on traditional methods of disease control amongst livestock keepers in developing countries. In this context, it is worth considering what the approaches to treatment of livestock are in other parts of the world, and how they are related to the disease patterns in these countries (see sub-section 'The use of local knowledge and traditional medicine').

A major goal of organic animal husbandry is improving animal health, preventing disease and ensuring animal welfare (Box 9.3). The organic understanding of animal welfare addresses aspects of naturalness and adopts the principle of precaution (Alrøe *et al.*, 2001; Vaarst *et al.*, 2003, 2004; Verhoog *et al.*, 2004), and these serve as guidelines for the organic approach to disease management. We consider it a very important strength and potential of organic animal farming that there is an explicit and strong focus on the health of the whole animal production system, the animals and their interactions with humans, other animals and the wider farm system. This is viewed as the primary way of reducing the risk of disease outbreaks and medicine use.

Box 9.3. A short description of the organic approach to animal health and disease, modified from Padel *et al.*, 2004.

An organic approach to disease management: keep the animals healthy!

Organic principles of disease management focus on prevention rather than treatment, and routine use of veterinary medicines is prohibited. Effective management should be employed so that animals can be kept in good health using two main approaches:
- Improving immunity and resistance in the animal population and in the individual animal through good nutrition, selection of robust breeds and resistant strains and reducing stress by allowing animals to exhibit natural behaviour; and
- Reducing the disease challenge through the operation of closed herds, low stocking rates, good hygiene and good grazing management.

The use and risks of antimicrobial drugs

The use of antimicrobial drugs in veterinary treatment

The appearance of antibiotic residues in food products and the development of antimicrobial resistance is an increasing problem in global agriculture. About half of all the antimicrobials produced today in the world are used in animal production, presenting real possibilities for drug residues in animal food products (WHO, 2001). In 1997, approximately 10,493 t of active antimicrobial substance were used in the EU, of which 3494 t (33%) were used therapeutically in veterinary medicine, and 1599 t (15%) in animal feed (Seveno *et al.*, 2002). Organic farming systems may substantially reduce this figure as they emphasise disease prevention and maintain explicit standards designed to reduce the use of medicine through breeding, feeding, housing, appropriate flock and herd sizes and active health care. In this section we use examples from Uganda to illustrate the problem of antibiotic residuals and resistance in Africa, and case studies from Denmark and the USA to demonstrate the possibilities for reduced veterinary inputs in Northern countries.

Antimicrobial residues

Contamination of the food supply by pharmaceutical and veterinary chemicals is an increasing problem for the livestock industry, both in terms of cost and public image. Even though the incidence of drug residues in food animals remains low, and the human health risks associated with these residues are small compared to other food-related hazards (Sundlof et al., 1991), public awareness is crucial to stem serious consequences.

Table 9.2 provides an example of problems of antimicrobial residues in the African context. It shows the prevalence of penicillin G residues in milk and tissues (meat, kidney and liver) from two districts of Mbarara and Masaka, where there is predominance of pastoral livestock production. These levels are extremely high and should be of great concern for consumers. In Uganda, antibiotics are widely available from many sources, and farmers are generally not educated in the use of veterinary medicine or the handling of treated animal food products for human consumption.

Table 9.2. Prevalence of penicillin G residues in milk and tissues (meat, kidney and liver) from two districts of Mbarara and Masaka in Uganda based on 837 samples.

District	Prevalence (%) of Penicillin G residues above the stated minimum residue limits (MRL)	
	Milk >4µg/L	Tissues >50µg/kg
Mbarara	10.6	23.1
Masaka	15	15

Source: Sasanya et al., 2004

Antimicrobial resistance patterns

Resistance to antimicrobials is a natural consequence of bacterial cell adaptation following exposure to antibiotics. Multiple use and misuse of antimicrobial agents in agriculture has increased the selective pressure for resistance in a wide range of bacterial groups (Seveno et al., 2002). Use of veterinary medicines is now accepted as the single most important factor responsible for increased antimicrobial resistance, which now poses a potential public health hazard.

In the UK, the presence of tetracycline-resistant *S. typhimurium* isolates from calves fell from 60% in 1970 to 8% in 1977, following the ban of tetracycline as a feed additive (Seveno et al., 2002). Indeed, the commensal gastrointestinal flora of healthy animals are a reservoir of resistance genes that can colonize the flora of humans. If the underlying resistant genes are horizontally transferred into human pathogenic bacteria, this can result in failure when treating a bacteriological infection as a consequence of antimicrobial resistance. Seveno et al. (2002) emphasize that antimicrobial resistance is now so widespread in the environment that it should be recognized as a phenomenon of global genetic ecology. Because livestock animals are carriers of food-borne pathogens such as *Salmonella*, *Yersinia* spp. and *Campylobacter* spp., these bacteria also undergo similar selection pressure due to the use of antimicrobial drugs (Aarestrup and Engberg, 2001).

Table 9.3 shows resistance patterns of clinical isolates from bovine mastitis in Uganda, as found in smallholder cattle systems. In a study of mastitis prevalence and management (Byarugaba, 2004), many farmers were found to be under-dosing when treating mastitis, e.g. using one intramammary tube to all four glands. Such practices may partly explain the very high levels of antimicrobial resistance. This clearly illustrates that a high level of antimicrobial resistance can develop on widely different backgrounds, here rather mis-use and under-use than over-use.

Table 9.3. Resistance patterns of bacteria isolates from bovine clinical mastitis in Uganda.

Antibiotic	Percentage resistance to important isolates					
	Staphylococcus aureus	Streptococcus spp.	E. coli	Klebsiella spp.	Pseudomonas auruginosa	Corynebacterium pyogenes
Gentamycin	26	0	9	21	30	0
Kanamycin	19	0	22	22	70	0
Chloramphenicol	23	21	35	32	80	20
Erythromycin	47	41	62	8	92	60
Neomycin	43	51	42	3	NA	NA
Streptomycin	39	40	63	66	90	100
Tetracycline	100	100	100	67	70	80
Cloxacillin	96	90	100	100	100	100
Ampicillin	100	76	100	100	97	60
Penicillin	95	90	NA	NA	NA	NA

Source: Byarugaba *et al.*, 2001

An organic approach to reducing antimicrobial drug usage

Given the problems and risks associated with the use of antimicrobial drugs, a much more thorough analysis of their necessity and use is needed. An organic approach to the use of medicine needs to continuously and critically analyse the necessity and relevance of using antimicrobials. The emphasis on reduced veterinary medicine in organic farming and the experiences of organic producers are therefore of relevance. There have been a number of studies from European organic farming that show the comparative success of disease control using minimal levels of antibiotics. Some of these are discussed below.

The high levels of antimicrobial resistance that may be found in some countries (e.g. USA and The Netherlands) can be explained by ready market access to such medicines, and subsequent mismanagement and inappropriate use. In the USA with an unregulated market for veterinary medicines, organic farming systems have completely abandoned the use of antibiotics and other chemical medicines, in order to create a clear distance and demarcation between organic and conventional farming systems. Health problems and associated suffering then have to be alleviated through other means (Karreman, 2004), which may include alternative treatment methods, good planning, and *in extremis* cullings (Karreman, 2004).

The background to the case study presented in Box 9.4 is the complete prohibition of antimicrobials from organic farming in the USA. There are certain parallels between this situation and situations where antimicrobials are not available, where farmers use alternative treatments (e.g. homeopathy) or, in some situations, rely upon traditional treatments, as discussed later in this chapter.

The case study in Box 9.5 takes this issue beyond mere treatment replacement. This case study, from Denmark in the early 1990s, briefly describes the combination of a non-medical approach with a high level of care taking. It shows how, when preventive measures inevitably break down and disease occurs, a mere replacement of a medical approach with a non-medical approach is insufficient and that 'care' is also an important element.

In contrast to these results more recent studies from Denmark showed the health situation in organic herds to be generally no better than in conventional herds (see Vaarst and Bennedsgaard, 2001; Bennedsgaard *et al.*, 2003). The results of the more recent studies might be explained by the higher proportion of newly converted organic farms within this study. These farms paid less detailed attention to the health situation of the herd, compared to the more well-established herds. In part, this was associated with the absence of 'mental conversion' following technical conversion to organic production. The 'mental conversion' experienced by the more well-established organic farmers motivated many to increase emphasis on care-taking routines, health promotion and disease prevention. These are concepts firmly rooted in the principles of disease management in organic farming. The inclusion of this case study illustrates the evolution of the ideas of health promotion.

In a second Danish project, 'Phasing out antibiotics from organic herds' (Vaarst *et al.*, 2005), 23 dairy farmers from one small dairy company set themselves the goal of completely eliminating the use of antimicrobial treatments in their herds. The preliminary results show a reduction of antimicrobial treatments to approximately half of the pre-project level, with no negative impact on the disease patterns in the herds, or on milk somatic cell counts. The only changes implemented by the farmers have been general improvements in housing systems and management practices. One of the important factors in this general improvement has been the use of a 'Farmer Field School' approach, where farmers support each other in group sessions and through common data analysis and discussions regarding the goals for each herd.

Box 9.4. US organic farming presented as a case of prohibited use of antimicrobials and chemical substances.

In organic farming in the USA, the use of synthetic medicine for treatment of livestock is prohibited. Most medicines commonly used by veterinarians and farmers in conventional livestock agriculture are synthetic and therefore prohibited, e.g. all antibiotics (penicillin, ampicillin, tetracyclines, sulfa drugs, ceftiofur, aminoglycosides, fluoroquinolones, etc.) and synthetic hormones (dexamethasone, isoflupredone, prostaglandins, gonadorelin and recombinant bovine somatotropin). Yet at the same time organic regulations clearly state that: 'The producer of an organic livestock operation must not withhold medical treatment from a sick animal in an effort to preserve its organic status. All appropriate medications must be used to restore an animal to health when methods acceptable to organic production fail. Livestock treated with a prohibited substance must be clearly identified and shall not be sold, labeled, or represented as organically produced'.
- 7CFR §205.238 (c) (7) USDA National Organic Program, Livestock health care practice standard

It is not clear what is considered 'appropriate medication', but in practice farmers try everything in their power to not use prohibited substances in order to keep the animals in the herd.

The USDA National Organic Program allows the use of all natural substances. Vitamins, minerals and other nutritional products used for the treatment of disease can be from any source (natural or synthetic) if already approved by the Food and Drug Administration (FDA) or American Association of Feed Control Officials (AAFCO). Calcium gluconate, electrolytes, dextrose, hypertonic saline and other metabolic stabilizers are allowed, but the list of synthetic substances allowed for use on organic livestock farms is very small in comparison to the vast amount of synthetic substances available to conventional farmers. The most commonly used and available complementary and alternative veterinary medicines (CAVM) include biologically derived medicines (serums, colostrum-whey derivatives), botanical medicines, and homeopathy. These are widely used on many organic farms in the USA depending on the farmer, veterinarian knowledge and acceptance of CAVM, as well as commercial availability. As an example, boric acid, used as a liquid solution, is employed as an antiseptic infused into udder quarters. Homeopathic remedies are quite widely used. On some organic farms, antimicrobials will be used, since knowledge, and availability, of CAVM may be minimal, with the consequence that the animal cannot stay in the herd. Some farmers operate 'split farms', having both an organic and a conventional herd at the same location. In such a situation a farmer may be more inclined to use prohibited material, such as antibiotics, since the requirement to remove a treated organic animal will not have as adverse an effect on overall production as it can be placed in the conventional herd.

The health consequences of the non-antibiotic policy in the US organic sector are minimal, at least in smaller herds. The incidence of pneumonia may be of concern, although this can, to a very large extent, be prevented by ensuring appropriate climatic conditions (low humidity, fresh air, space etc.) and good hygiene measures. With regard to mastitis, the prevalence of three-teated cows is not known. In general, on small organic farms, it seems that the milk somatic cell count easily stays within normal US legal limits (750,000 cells/ml); however, organic milk processors often demand somatic cell counts no higher than 400,000 cells/ml. For some farmers this can be a source of continual hardship and almost all organic farmers experience some difficulties to keeping counts below 400,000 cells/ml at certain times of the year (hot, humid summer months). One milk processor in California, selling only liquid milk, demands that organic milk be kept below a somatic cell count of 200,000 cells/ml.

Box 9.5. An example from organic dairy herds in Denmark in the early 1990s, here good udder health was explained by good management and consequent care-taking routines.

Danish research, conducted from 1991 to 1994, compared udder health between organic and conventional herds and found that the pattern of causative organisms of mastitis was similar. Generally, udder health was better in organic herds (see table below).

Udder health parameters	Organic herds		Conventional herds	
	Median	10-90% percentiles	Median	10-90% percentiles
Antibiotic mastitis treatments (% of lactations)	5	0–14	31	7–52
Milk somatic cell counts (% cows >500,000/ml)	14	3–26	19	12–32
Annual subclinical mastitis incidence (% cows)	28	11–44	43	20–65
Mean individual milk somatic cell count ($\times$ 1000)	240	148–452	347	213–613
Mean bulk milk somatic cell counts ($\times$ 1000)	210	90–350	315	200–550

> Over 3 years disease prevention routines were well described. No 'particular and distinct organic characteristics' could be identified. There was wide variation within both groups (conventional and organic). The only factor that characterized the majority of organic farms was the relatively high degree of 'care-taking routines', such as extra milking-out by hand between machine milking in a very early stage of mastitis cases, provision of extra bedding material in critical situations (e.g. after calving), and careful inspection of milk and udders. This higher degree of labour-intensive intervention was characteristic of many of the organic herds. The difference was not explained by 'being organic' – but rather by 'being good dairy herd managers', and having time enough to take care of the cows and undertake preventive (and health promoting) efforts as a part of the daily routines. It may be that basic organic farming ideas stimulate this effort in combination with the restrictive antibiotic policy. In interviews, many of these farmers described a 'conversion in the herd', which made them feel responsible for their own animals in a different way than before conversion, when they considered a veterinary treatment to be sufficient in order to solve a mastitis case.

Vector-borne diseases and organic livestock farming

Characteristics of vector-borne diseases

Many of the diseases encountered in commercial livestock production systems in Europe (such as mastitis and lameness in dairy cattle) are largely a consequence of management and the environment of the production system. In this section we consider the problem of vector-borne diseases in Africa, and evaluate the potential of an organic approach to these diseases. In some respects, the conventional control of these diseases has evolved along lines that conform more closely with an organic approach. This is partly a consequence of the nature of these diseases and their epidemiology, and partly a result of the social and economic environment in which they occur.

In many areas of the world vector-borne diseases, such as the tick-borne diseases (TBDs) (e.g East Coast Fever (ECF), babesiosis and anaplasmosis) and the tsetse-borne disease trypanosomiasis, proliferate. These diseases frequently have a wide-reaching economic impact. They are often a result of complex interactions involving the causative agents, the vector, the vertebrate host and the environment in which they exist. The interactions are influenced by a wide variety of factors ranging from climate, soil, vegetation and human activity that contribute to different epidemiological states. For tick-borne diseases, these include situations where the host animal population can exist either in a situation of endemic stability (where the infection is present in the area at a low level, which stimulates the immune defence of the animals), or instability. These situations are, in part, influenced by the susceptibility of the animal to the disease, the level of infection in the vector and cattle host, the management systems and land use pattern (Perry, 1994). For both tsetse and tick-borne disease situations, a 'mixed scenario' often exists where wildlife and/or resistant cattle are in contact with susceptible cattle.

As with other diseases, addressing problems of vector-borne diseases requires a holistic approach, this implies:

1. A thorough knowledge of vector distribution and population dynamics;
2. A better understanding of the diagnostic methods of these diseases; together with their epidemiology and pathogenesis;
3. An appraisal of the production environment (customary/acquired husbandry practices and control regimes and criteria for choice of technology); and
4. Knowledge of the economic losses associated with the diseases.

Strategies for controlling tick-borne diseases

Over the past 30 years there have been increases in the incidence of TBDs and there have been many situations where ticks have developed resistance to acaricides, thus reducing their effectiveness in tick control. This has occurred at a time when there has been a steady increase in demand for animal products and a concomitant demand for high performing livestock, resulting in the importation of susceptible exotic breeds. Control strategies remain vulnerable to disruption by drought, livestock movement, escalating costs of inputs and socio-economic and political instability. Any disruption of intensive control regimes can have disastrous effects (e.g. the liberation war in Zimbabwe: see Norval *et al.*, 1991). Mismanagement of acaricides in Uganda has been described by Okello-Onen *et al.* (2004), who distinguished six different shortcomings in the application of acaricides: delivery practices, choice, dilution rate, methods of application, frequency of application, storage and disposal. This illustrates how lack of information and training regarding the use of medicine can lead to incorrect application on many different levels.

In many control situations, there has been little consideration of the differences in the epidemiology of diseases in different agro-ecological zones or production systems. There is a lack of appropriate technology for tick control and TBDs in terms of meeting the needs, resource endowments and desires of farmers within specific production environments. There are usually conflicts between the recommended technology and the limitations within which farmers operate. Farmers' decisions on control measures are frequently based on subjective considerations; and unfortunately in Africa, such decisions are often speculative, with little consideration given to economic and environmental sustainability.

Developing a holistic approach to the control of TBDs: understanding the disease

Adequate scientific knowledge already exists to support changes in the philosophy and strategy underlying tick control. The general principles on which future strategies should be based have been summarized by Tatchell and Easton (1986).

- preserve enzootic stability, or re-establish, it through immunization against TBD;
- educate farmers and advisers to accept the benefits gained both from boosting immunity to TBD and achieving host resistance to ticks that would result from relaxed tick control regimes;
- assess the true benefits of using different cattle breeds, as some breeds are significantly more resistant to these diseases than others, and these should be an important focus, if the use of acaricides is to be reduced;
- modify animal husbandry practices, for example, by allowing the more susceptible cattle to be grazed and treated together;
- institute tick control regimes based on sound economic thresholds.

Whilst the basic approach of using routine acaracide treatments is unacceptable from an organic perspective, there are elements of the strategies described above that lend themselves to the philosophy of preventive disease management promoted in organic farming.

Control of TBDs requires a thorough understanding of their epidemiology: the presence of disease is itself not an *a priori* reason for introducing control measures. These measures should be selective and applied to the livestock sector most at risk. For example, in a highly susceptible population of animals the impact of disease can be devastating, whereas in a population constantly exposed to ticks and diseases, endemic stability can develop and the impact of the disease may be negligible.

Disease epidemiology is often complex, demanding a basic knowledge of tick–host relationships and environmental influences. This complexity is compounded, in East Africa at least, by the presence of a range of vertebrate hosts for both ticks and the disease agents, as well as diverse environments, climatic variability and extremes and a broad range of land use patterns ranging from small scale subsistence to large scale commercial farming.

A careful examination of the infections due to *Babesia* and *Theileria* parasites, the major tick-borne diseases in East and Central Africa, reveals major differences between their biology and the epidemiology of the two important diseases they cause. Such a comparison shows that:

1. There are dose-dependent host infection rates in *Theileria parva* infections but not in *Babesia* species infections;

2. There are high tick infection rates in *Theileria* but low rates in *Babesia*, especially *B. bovis*;
3. *Babesia* species' geographical distribution is wide, while that of *T. parva* is primarily localized within East and Central Africa;
4. *B. bovis* and *B. bigemina* can only affect bovines whereas *T. parva* has the ability to infect other ungulates;
5. There is transovarial transmission by one-host ticks of the *Babesia* species and trans-stadial transmission from 3-host ticks for *Theileria* species.

Large as these differences appear, they should not be allowed to mask the important similarities between the two parasites: (i) the ability of hosts to acquire immunity; (ii) the ability of the hosts to act as carriers; and (iii) the ability of their hosts to survive without disease in presence of pathogens in the vector ticks and in the blood of the hosts (endemic stability). These similarities make it possible to think of a common approach for disease control, or even possibly developing a generalized mathematical model of their epidemiology.

Rubaire-Akiiki *et al.* (2004 and 2005, submitted manuscripts) determined the epidemiological status of TBDs in the heavily populated highland area of Mbale and Sironko Districts in eastern Uganda and identified the cattle populations most at risk, according to spatial and temporal situations. Tick load was highest in the lowland agro-ecological zones (AEZ), where *R. appendiculatus* was the main species infesting the animals. However, the mean number of standard ticks on animals (ticks that had fed well and that would detach in the next 24 h) was highest in the upland AEZ.

Context-based control strategies for TBDs

Control strategies can be evaluated on their robustness and ability to perform over a wide range of biological and technical variables. However, the methods also need to be complementary, economically viable and sustainable. In this context, it is necessary to consider specific target groups or areas, depending on the disease epidemiology, production economics and whether resources are available in the farming system to support changes in livestock type and numbers. There is a need to take into account the farm circumstances and the variation in risk (e.g. for East Coast Fever (ECF)), both spatially and temporally, in order to design appropriate control strategies for any TBD. Control strategies for ECF will vary with the epidemiological situation as dictated by the production system.

Inappropriate control measures may be introduced if the impact of disease is wrongly assessed; not only will they be unnecessarily costly, but there is the risk that endemic stability, often a mainstay of disease control in developing countries, will be jeopardized or destroyed.

Blanket/all-purpose control measures that risk worsening a disease situation should be avoided. Integrated control is necessary as it is now clear that acaricides are not a feasible means of tick eradication in Africa (Young *et al.*, 1988). It has also become clear that the control of ticks and tick-borne diseases by intensive acaricide treatment of livestock is both expensive and epidemiologically unsound (Norval, 1983; Young *et al.*, 1988; Norval and Young 1990). Further, the Australian experience has shown that ticks develop resistance to acaricides, especially if there is over-reliance on acaricide use (Nolan, 1990). There is also evidence that production losses due to tick infestation per se are too small to justify intensive acaricide application on economic grounds in zebu and sanga cattle (Pregram *et al.*, 1989; Okello-Onen, 1994).

The existence of endemic stability in some areas implies that control can be selective, strategic and focused only on susceptible target cattle populations (Perry *et al.*, 1990). The strategy for control of tick-borne diseases may be to 'live with it' or to 'prevent it' depending on the individual farm or area, conditions of climate, type of livestock kept etc. 'Living with' TBD implies the creation of a balance of challenge and immunity within the management unit. Preventing TBD implies the prevention of any feeding by ticks on the animal. There are various factors that impinge on the dynamic system, which a manager attempts to create if she elects to 'live with' TBD. The use of a vaccine, for example, is an aid to 'living with' the disease.

Tick-resistant cattle breeds

Massive losses caused by ticks and TBD occur in susceptible breeds of cattle if unprotected. Local indigenous cattle kept completely tick-free become equally susceptible. Indigenous dual-purpose breeds are known to be highly resistant to ticks, but one should be cautious about overgeneralization or simplification. For example, the East Coast Fever situation is certainly not identical to that of babesiosis. Outside of Africa, tick-resistant cattle (cattle that are better able to limit the proportions of attaching ticks which survive to complete engorgement) can be used to control major tick species such as *Boophilus microplus* without causing outbreaks of TBDs. *Boophilus microplus*, being sessile and having one major host species (bovine) is among the simplest of pests to control; this has been done through taking advantage of the bovine tick resistance phenomenon. In addition the pathogens transmitted by this tick have characteristics that allow the population of disease organisms to reach a stable situation (endemic stability) without normally having clinical effects upon their hosts. In such situations it is possible to have both ticks and tick-borne haemoparasites in a cattle population without any measurable economic loss or disease.

In Africa, several other parasites (*T. parva*, *Cowdria ruminantum*) are transmitted by a variety of ticks. In many cases there is, as yet, no detailed understanding of their epidemiology. There is however increasing evidence that, in some breeds of indigenous cattle, endemic stability to theileriosis is possible.

The phenomena of host resistance to ticks and enzootic stability to TBD are well documented (Perry *et al.*, 1985; Tatchell, 1988; Latif and Pegram, 1992). For example, in Africa east of the Great Rift Valley, the tick species *R. pulchellus* feeds heavily on large wild mammals while overall population densities are usually limited by host resistance. In cattle, resistance levels may vary greatly between breeds and between animals.

Approaches to the control of tsetse-borne trypanosomiasis

Trypanosomiasis remains one of the most devastating diseases of animals and humans in sub-Saharan Africa, despite the development of numerous control methods. This disease is vector borne, and transmitted by the tsetse fly (*Glossina* spp.). Control efforts in the first half of the last century emphasized environmentally and socially unacceptable forms of pesticide control. Latterly, there has been greater emphasis by the research and development community on controlling the vector through non-chemical vector control, such as the use of odour-baited traps and targets, pour-on insecticides, trypano-tolerant breeds and sterile insect techniques (Holmes, 1997). There has also been increasing emphasis on rural development and community participation in particular.

Enormous amounts of drugs are used across the continent in the battle against trypanosomiasis. In some years as many as 30 million doses of the three effective drugs against tsetse-transmitted trypanosomiasis have been recorded as being used (discussed in Holmes, 1997). Increasing resistance to these drugs has been observed. Holmes (1997) associates this with the demise of governmentally funded veterinary services across Africa, followed by under-dosage and mismanagement of medical products. In some cases multi-resistance to all three previously efficient drugs has been proven (Peregrine, 1994). In addition to this threat of resistance, reliance on trypanocidal drugs is unsustainable because they are often only sporadically available.

Apart from using insecticides, vector control can be achieved by using traps, targets and bait technology. All of these control methods have disadvantages and none has proved to be sustainable. There is growing interest in integrated control, which can be at three levels: integration with rural development, integration with other disease control measures and integration of various tsetse and trypanosomiasis control measures. An integrated control approach may improve the effectiveness and sustainability of the individual control measures. This offers the possibility to introduce control approaches that are in accordance to organic livestock farming approaches. Holmes (1997) points to the necessity of aiming at effective suppression rather than eradication, but such an approach requires a more detailed evaluation of the relationship between tsetse population densities and the occurrence of disease than is presently available.

Box 9.6 describes a Kenyan pastoral livestock system, where the seasonal movements of animals has been used as a traditional means of control, in which tsetse trapping has been attempted, but chemical means remain the key element (Roderick *et al.*, 1999, 2000).

Box 9.6. An example of the dilemma facing pastoralists in tsetse-infested areas.

African pastoralism and trypanosomiasis: an example from Kenya

The Maasai of Olkiramatian Group Ranch in south-west Kenya keep livestock under a constant threat of tsetse-borne trypanosomiasis. The traditional approach to control is based upon a system of transhumance. This involved only allowing animals into tsetse-infested areas during the dry season when tsetse populations are at their lowest, to graze 'standing hay'. During the wet season, when these grazing areas become heavily infested with migrating fly populations, livestock were moved to tsetse-free areas. Critical periods occurred during prolonged drought, when animals were forced to graze in the dense vegetative habitats favoured by tsetse. Susceptibility to disease is increased by the fact these critical periods occur when the animals are under-nourished.

Land pressure and land tenure changes over recent decades have put pressure on the sustainability and effectiveness of this approach, and there has been an increasing reliance on trypanocidal drugs. There have also been attempts by outside agencies, using community participation approaches, to focus on controlling the vector by means of tsetse trapping. Although Dransfield *et al.* (1990) report remarkable success in controlling tsetse populations, there has been a general failure to sustain this type of control programme. This failure may in part, be due to the organisational complexity involved in operating a tsetse-trapping programme, compared to administering individual veterinary treatments. The latter requires the owner of individual animals to take responsibility for his/her own animals whereas control based on tsetse trapping requires a community effort, exemplifying the conflict between public and private goods (see Hardin, 1968). As a consequence, individually administered veterinary drugs remain the preferred option for livestock producers. Similar experiences of the response to community based tsetse control (i.e. a preference for veterinary drugs) have been recorded by Catley *et al.* (2002).

This case study illustrates an interesting dilemma for the development of organic approaches to disease management in communally operated farming systems. These issues are discussed in more detail in the next section.

Disease control issues associated with land use and land tenure

Pastoralism, nomads and transhumance

Livestock farming is based on transhumance in large parts of the world, including many African, Asian and South American countries. In Greenland, Norway and Switzerland it is estimated that approximately 41% of the sheep and 21% of the cattle are produced in transhumant systems (Eckert and Hertzberg, 1994). Globally, approximately 50–100 million people and 120 million cattle are estimated to live a nomadic life. There are particular characteristics of such a lifestyle that are worthy of specific consideration when evaluating the potential for organic approaches to animal health management.

Macpherson (1995) identifies some consequences of this lifestyle for livestock diseases: many transhumant people migrate through areas where there are no veterinary or medical services, no education, no abattoirs and no safe water supply. Yet, at the same time migration also substantially reduces the risk of the build up of many faecally transmitted protozoal and parasitic diseases. Research in the 1990s changed the rather stereotypic perceptions of pastoral communities as being 'irrational' to that of being 'rational entrepreneurs facing the challenge of non-equilibrium environments' (de Leeuw *et al.*, 1995). However, the same lifestyle can increase the risk of contact with geographically limited, or seasonally related diseases, such as trypanosomiasis (see previous section), may increase contact between domestic and wild animals, and increases the risk of certain virus diseases, including Rinderpest, malignant catarrhal fever, foot and mouth disease, African horse sickness and rabies. Bourn and Blench (1999) include a discussion of these issues in their extensive review of the potential for co-existence of livestock and wildlife in Africa.

Communal grazing and community participation

Apart from the mobile production systems discussed above there are also significant areas of communal grazing associated with sedentary systems. These are not restricted to grazing ruminants. Some animals, such as poultry, and in some cases pigs (traditionally 'scavenger animals') use common grazing or scavenging areas. In these situations there are disease risks associated with contact between neighbouring herds and flocks, and preventive disease approaches may therefore require community participation and collaboration. Further, formal conversion to organic production requires animals to be fed diets that are largely organically produced. Again, in this respect, conversion at the communal level is required if these systems wish to undergo formal conversion to organic production. This is perhaps less of an issue in the case of pastoral communities, where the pasture is the main form of feed, and that pasture is natural, and does not receive chemical inputs.

Low stocking rates (livestock units per unit area) are one of the mainstays of organic approaches to disease prevention, as this enables a reduction in disease pressure and transmission and the potential for breaking disease cycles. However, in communally managed systems there is a potential for conflict between this and the production goals of individual farmers or pastoralists. Again, this is a scenario of conflict between public and private goods, as described earlier for community-based tsetse trapping. Added to this potential conflict is the problem of a decline in the area available for communal grazing. Steinfeld (2004) identifies a number of pressures on such systems, as the availability of range land decreases, due to arable land encroachment, land degradation, conflict, etc. These pressures reduce the scope for further increases in production from increasing herd numbers and can lead to pressures to reduce livestock numbers, which in turn can give rise to conflicts.

Livestock projects that encompass, interactive, community participation and aim at self-mobilization are most likely to result in sustained benefits for livestock keepers (Catley and Leyland, 2001). The same authors reach this conclusion also by comparing negative experiences with community-based tsetse fly control (discussed in Case Study 6) with the more positive experiences with Rinderpest vaccination using community animal health workers in the Horn of Africa.

Bio-security and closed herds

The level of disease risk and need for protection or acceptance of disease exposure is directly linked with the specific epidemiological situation in a country or a region. Geographic differences, statutory approaches to livestock disease and even cultural practices regarding bio-security and disease control vary significantly between countries. In Europe, the UK and Finland could be taken as two extreme examples: in the UK, the sheep and cattle industry struggle with over 25 different endemic diseases that the farmer has to either live with or protect animals against, without any statutory or communal control programmes. By contrast, in Finland, the majority of these diseases have been eradicated or are carefully controlled by communal or statutory efforts, so that the risk of an individual farm introducing one of these diseases is minimal. In these two widely different situations, the 'organic' approach to disease control and health promotion is bound to differ significantly as well.

The principle of closed herds is an important component of achieving manageable disease control as a basis for a preventive approach in organic farming. Obviously the potential for adoption of this varies considerably between systems and regions. The most commercial farming systems are perhaps best suited to limiting movement of animals between herds. Traditional farming systems are more likely to involve, or even rely on, movement between herds and flocks, e.g. the sharing and loaning of animals between families. Additionally, marketing tends to be less formal and less controlled.

The informal trade of live animals between regions, countries and continents can be significant. King and Mukasa-Mugerwa (2002) discuss how Dinka cattle from Sudan are marketed in northern Uganda, where demand far outweighs supply. Despite a shortfall of supply in southern Uganda, there is concern that movement of animals further south may spread disease, particularly Rinderpest. Interestingly, at the time as this report was published livestock populations in southern Sudan had increased as a consequence of improved Rinderpest control and community participation in disease control. The war in southern Sudan and poor infrastructure meant that the marketing structure had not improved.

Trading may also involve returning unsold animals to an otherwise closed herd, thus introducing a biosecurity risk. Informal trading is not unique to the Southern hemisphere. The informal trading of livestock in the UK, through cattle auctions, contributed greatly to the spread of foot and mouth disease in 2001.

Zero grazing and organic farming

The multiple enterprises that frequently characterize smallholder systems throughout the world provide real opportunities for the development of integrated and sustainable production, particularly with regard to nutrient recycling. However, the opportunities for an organic approach to disease management are more contentious than those presented by livestock in pastoral situations. Zero-grazed systems in many parts of the world frequently have small herds or flocks of animals, kept partly in confinement. In Uganda, for example, these systems rely on so-called exotic breeds (normally Holstein-Friesian type), either pure or in cross-breeds. These breeds are highly susceptible to the prevalent diseases and therefore depend on veterinary inputs. Further, they rely on feed beyond that available from natural pasture. The lack of access to exercise, natural pasture, and in many cases an absence of social contact with other cattle mean that these farming systems do not conform well with the organic approach. Whilst the systems themselves are not well-suited to organic livestock production, there is still a need to reduce the use of antimicrobial medicines and to place more emphasis on disease prevention, particularly for production diseases such as mastitis in dairy cattle.

Developing organic strategies to enhance animal health and livestock production

Breeding strategies and the role of indigenous breeds

The use of environmentally adapted breeds is an important feature of organic production, although the extent to which this occurs tends to depend on the type of enterprise: beef and sheep systems tend to use indigenous breeds more than dairy and egg production. There is, of course, a tremendous genetic pool of indigenous breeds available, which are widely used, and this has positive ramifications for the development of organic systems at a global level.

Indigenous breeds tend to be more associated with subsistence and resource-poor communities in climatically disadvantaged areas, where adapted breeds are the only or best sustainable option. Many indigenous breeds have the further strength of having disease tolerant characteristics, such as the trypanotolerant N'dama cattle of West Africa (de Leeuw *et al.*, 1995; Holmes, 1997) and the Red Maasai helminth-tolerant sheep breed of East Africa. However, such species may only be partially disease tolerant, and in areas of heavy tsetse challenges, even trypanotolerant breeds can succumb to infection. Increasing recognition is being paid to the role that disease-tolerant breeds can play in integrated control strategies (see the section on vector-borne disease). While some breeds may show high levels of resistance to disease, this may be at the expense of productivity.

Bishop and Woolliams (2004) provide evidence of the benefits of selection for resistance to disease as a means of reducing dependence on chemical inputs. They cite research showing that, after 6 years of continuous selection for tick resistance, a herd can be created which effectively requires no intervention with acaricides, and state that:

> An important sociological aspect of the sustainability of a livestock production system is the extent to which it depends on external inputs, e.g. chemicals, feedstuffs, breeding material, etc. Another sociological aspect is the recognition and enhancement of the value or cultural identity of the indigenous animal genetic resources, especially when they contribute to disease management or reduce risks. Such recognition further empowers local communities. Therefore, when considering genetic solutions to sustainable livestock production, genetic management strategies that utilise indigenous animal genetic resources and reduce the reliance on external chemical inputs are those most likely to be successful and sustainable.

They illustrate this point by comparing disease-resistant animals with more productive ones (comparing the Red Maasai sheep breed with the Dorper breed). Even if the former have a comparatively low production potential, they perform better when judged according to their contribution to the efficiency of the whole system.

New breeding technologies can rapidly increase the rate and success of breeding programmes. However, from the perspective of an organic approach to health management, these developments need to be balanced with ethical considerations. Cristofori *et al.* (2005) describe how artificial insemination using local cattle breeds in Niger has allowed a more 'modern' breed selection programme. Crossbreeding programmes using artificial insemination with exotic breeds are widely used in developing countries, with the result that local breeds may be diluted and eventually become extinct. Exotic breeds (often Holstein-Friesian types, Ayrshire or Jersey) frequently lack resistance to local diseases and climatic conditions and, in the absence of high-quality feed and management, produce poorly and lack strength.

FAO has formulated a Global Strategy for the Management of Farm Animal Genetic Resources (FAO, Rome, 1999) on the basis of a pilot project on 'Conservation and Use of Animal Genetic resources in Asia'. One of the 11 countries participating in the pilot project was Bhutan (see Box 9.2), where animal disease control is given a high priority. For example, migration and vaccination check posts are established on the main migratory passes between epidemic zones of the country. The largest share of the government's budget for veterinary medicines is spent on anthelmintics and FMD accounts for the greatest proportion of expenditure on vaccines (RGOB, undated).

Veterinary services are free of charge in Bhutan (as are health services and education) through veterinary hospitals and livestock extension centres. Regional veterinary laboratories are responsible for technical backstopping of field activities (RGOB, undated). However, the lack of trained personnel, the long distances and the inaccessibility of remote areas still constrain the effectiveness of these strategies (Gyamtsho, 1996). On average a livestock extension centre covers 585 households. Farmers mostly perceive diseases as a penance for bad deeds or as part of a natural cycle of bad luck (Thinlay *et al.*, 2000). Sustainable disease control solution in Bhutan however, is achieved not primarily through vaccines, veterinary medicine or a focus on disease control, but through forming an animal population, which fits into the farming conditions in the country, with animals that can survive and prosper under the local circumstances. Explicit breeding strategies aiming at preserving the traditional breeds are viewed as a sustainable solution (see Box 9.7).

Box 9.7. Cattle breeding strategies in Bhutan supporting disease resistance and optimal adoption to living conditions.

Livestock, especially ruminants, are an integral part of farming systems in Bhutan (see Box 9.2). Farmers breed cattle for milk production and traction. The indigenous cattle breed is a *Bos indicus* named Siri or Nublang, and constitutes 70% of the cattle population. Within this breed no selection takes place except for survival, and the breed has a higher resistance to tick-borne diseases than exotic breeds. The Siri is crossed with Mithun (*Bos frontalis*) acquired from India, and the F1 (Jatsa/Jatsam) to produce highly valued combined draught and milk animals. Traditional breeding practice implies backcrossing to Siri for five generations, and the various crosses are all recognized until the F5 which is considered a pure Siri, and the cycle starts again. Bhutanese living in the transhumant yak herding system practise similar crossbreeding between yak and Siri, or between yak and the Tibetan *Bos taurus*, Goleng. These traditional breeding practices still prevail despite increasing possibilities of crossbreeding with Jersey and Brown Swiss for milk production. This is probably because the exotic breeds are not suited for draught power, are more susceptible to tick-borne diseases, and are not well adapted to grazing and browsing in the forests. Bhutanese farmers are generally considered cautious and reluctant to adopt new technologies (e.g. more than 90% rice is traditional, rather than high yielding varieties (Thinlay *et al.*, 2000)).

An FAO project on Animal Genetic Resources has raised awareness on the importance of conserving the local breeds. The project identified and classified the local breeds and developed a strategy for their conservation and use. During the project it became evident that Bhutanese farmers have a good knowledge of breeding and are well aware of the importance of conserving the local breed, which they need for crossbreeding with yak and Mithun. The Royal Government now recognizes farmers' preference for Mithun crosses, and has started producing semen from Mithun at their breeding farm, and will be able to offer Mithun semen at the AI centres.

The use of local knowledge and traditional medicine

The absence of biomedical treatments for livestock disease in many parts of the world, was highlighted earlier in the chapter. In this section we consider the role of other approaches, based on local knowledge and tradition.

Local knowledge of animals, diseases, treatment and control is widespread throughout the world, and particularly so in areas where traditional methods of production are still common, e.g in pastoral livestock areas. Acknowledgement of the value of this knowledge and active support and encouragement from the institutional environment can empower local farmers to develop and use traditional knowledge to solve disease problems in their own herds in a cost effective and sustainable way. However, this local knowledge has often become lost as a result of the widespread use of veterinary drugs. Reducing the emphasis from chemical approaches to health management allows greater opportunity to explore and rediscover this knowledge. In this section we examine how this may complement the adoption of organic methods.

In Uganda, the recent 'Livestock System Research Programme' built up knowledge from the existing farming systems, to which relevant solutions could be found. One of the approaches developed within the framework of this project has been the 'Farmer Field Training Groups', which closely resemble the

'Farmer Field Schools' that have been widely used throughout the world and are based on farmers learning about farming in groups.

In the LSRP project, one of the emphases was on ticks and tick-borne diseases. Farmers expressed that they had 'become proud of what they previously did' in using non-medical methods of tick control, e.g. hand-picking of ticks. These methods worked well and supported the endemic stability in the area, but the farmers had wanted medicine because it was a more 'modern way' of farming. This project also emphasized the use of local feed sources, e.g. from the household, and the condition of the animals significantly improved during the period of these training and learning groups as result of implementing these practices. The important starting point in this project was the focus on the local situation, the farm and community conditions, opportunities and goals of the farmers.

Adolph *et al.* (1996) emphasize the importance of ethno-veterinary knowledge in Sudan, how this knowledge should be central to livestock programmes and how other inputs should be seen as complementary to this knowledge. They also highlight the importance of gender, and the need to involve women in these programmes, as they frequently have specialist practical knowledge and experience that is not always recognized.

When knowledge exists, it has the potential of becoming an important element in the process of developing organic livestock farming practices. Traditional treatment approaches have been categorized as 'folk medicine' as defined by Kleinmann (1980), where traditional knowledge is passed from one person to the next. In contrast to 'professional medical schools' (e.g. so-called 'alternative' treatment methods having a theory and practice like classical homoeopathy and acupuncture), these systems of knowledge are not supported by specific, coherent theories of health and disease. Although some practices are primarily 'spiritual' or 'belief-oriented' in character, others are more based on proven experience and knowledge about the effects of certain substances and plants.

Perceptions of disease and what causes, defines and cures diseases vary widely between cultures and these differences need to be articulated and discussed by the partners involved in solving a certain disease problem or implementing an organic approach to diseases. Johnsen (1997) describes the perception and treatment of disease among the Maasai of East Africa who, for example, determine the seriousness of diseases according to whether they are 'hot' or 'cold'. The linguistic merging between the categories 'trees' and 'medicine' is so total that it is very difficult to explain to outsiders. This reflects the Maasai's perception that all trees are potential medicines: even small herbs are considered to be 'small trees'. In other cases, there are clear connections between the perceptions of local cattle keepers and veterinarians about diseases. Catley *et al.* (2002) describe how Boran cattle keepers in Kenya had developed local characterization of two diseases known as *gandi* and *buku* which was similar to veterinary knowledge of chronic trypanosomiasis and haemorrhagic trypanosomiasis respectively. Alawa *et al.* (2002) describe ethnoveterinary medical practices for ruminants in the sub-humid zone of northern Nigeria. Approximately 75 substances were used for 20 different disease complexes and herdsmen readily identified signs of diseases. There are many other recorded examples, including those used in north Europe and North America (Vaarst 1995; Karreman, 2004).

Vaccination

Vaccines may be considered 'artificial', 'synthetic' or 'chemical' ways of handling disease, and the risks and opportunities of developing organic animal husbandry practices without them must be discussed. Decisions regarding the use of vaccination need to involve risk and impact assessment. For some diseases, such as the clostridial diseases affecting sheep, the incidence is inconsistent and does not follow a predictable pattern, and decisions may be most influenced by the relative risk averseness of the farmer. For contagious disease, control decisions have to take into account the potential local, regional and international impact, in terms of the risk to other animals and to trade. It is imperative that the promotion and adoption of an organic approach does not jeopardize regional, national and international vaccination programmes aiming to eradicate a given disease problem.

Moving from an 'organic approach' to 'organic animal production'

'The hidden world of ecological farming'

In Chapter 6 of this book, Parrott *et al.* discuss the issue of non-certified organic farming and present certified organic farming as 'the tip of the iceberg'. They identify four main areas of organic farming, three of which constitute a 'hidden world of ecological farming':

1. 'Explicit organic approaches', where IFOAM membership etc. shows an explicit interest in organic approaches to farming.
2. 'Like-minded approaches', which comprise approaches very similar to those of organic farming. They mention bio-dynamic movements in some countries and the permaculture movement in Zimbabwe.
3. 'Low external input sustainable agriculture', understood as participatory approaches to farming system development promoting e.g. optimal use of local resources and processes. In the examples given, much focus is on soil fertility and organic or green manure.

4. 'Traditional farming', often understood as 'agriculture without chemicals', also referred to as 'organic by default'.

When applying any of these definitions to systems that have livestock, it is necessary to incorporate livestock-specific organic aims and principles. Perhaps the most important of these should relate to the health and welfare status of the animals within these 'ecological farming' scenarios. It is proposed that these be defined by assessment of two criteria:

- the risks and severity of the prevalent diseases and the actual and potential opportunities to reduce the impact of these diseases using organic principles of disease management; and
- the actual and potential opportunity for animals to fulfil their basic behavioural needs.

The aim in this chapter has been to consider the strengths, weaknesses, opportunities and threats of adopting organic principles in order to reduce dependence on veterinary medicines within a broad range of systems and for a number of animal health scenario. In this context we have identified both existing organic approaches and the potential for the adoption of organic health management within a number of very different situations.

In the case of traditional pastoral systems in Africa the management system itself has been identified as having an 'organic approach', in that it is a relatively natural environment allowing animals access to natural feeds and freedom of movement, etc. Other elements also exist that are likely to aid the preventive approach to animal health, such as traditional knowledge and well-adapted breeds. Parrott *et al.* (Chapter 6) suggest that 'organic farming on extensive grasslands may not require very many changes in management practice' (related to conversion from 'traditional' to 'organic farming'). This may be true in some areas but, as discussed above, it may be the case that animals grow up under very 'natural conditions' on large, extensive grassland areas but, at the same time, rely on a high input of veterinary medicine. Another system, the zero-grazed method of animal management, is often represented as operating within categories 1–3 of the ecological farming described above. Yet, these systems generally are highly dependent on medicinal inputs and imported feeds, little thought has been given to their potential for adopting preventive animal health practices and to providing a natural environment for the animals. Our selection of the Bhutanese livestock system as an example, is due to its strong emphasis on one specific element associated with an organic approach, the use of adapted breeds as part of a more sustainable health programme.

From non-certified to certified organic production

The development from 'an organic approach' to a more formal 'organic agricultural system' will be very much driven by particular circumstances. It will depend on what the driving forces are for change, the characteristics of the production system, etc. However, given the diversity of such systems globally, a framework is required to help evaluate the potential for conversion and help implement conversion where appropriate. This framework is now well defined within the IFOAM principles. For animal-based systems, these should include:

1. The whole animal production system should be based on non-medical health promotion and disease prevention initiatives with bio-medical treatments being used only to avoid suffering.
2. Closed flocks and herds and minimal transport should form the basis of activities to reduce the spread of diseases.
3. There should be attempts to ensure harmony between animal species, herd/flock size, the production level of the animals and the land area. Likewise, the natural harmony of the individual animals should be emphasized through feeding to achieve physiological balance using natural animal feeds and not forcing growth or yield.
4. The breeds for organic production should be suited to the prevailing climatic conditions, living in a stable herd/flock and be tolerant of, or resistant to, common diseases.
5. Animal welfare should be promoted by giving animals access to conditions that allow them to express their natural behaviour. This should be combined with a management regime that is not too intrusive but which responds rapidly when intervention is required.
6. Chemical/synthetic medicines should only be used after a thorough analysis of the local situation and in combination with health-promoting initiatives.

It is accepted that in many situations achieving all of these aims will be either undesirable or unachievable for a range of practical and economic reasons. Even, in such situations, it remains possible to move towards a general organic approach to issues such as reducing the reliance on veterinary drugs. The traditional approaches to disease management described by Schillhorn van Veen (1997), based on ecological management, breeding for resistance, development of immunity and the use of plant and animal extracts as therapy, can be clearly recognized as an organic approach. However, the sustainability and success of such strategies is likely to be less effective if they are adopted individually and in isolation.

Context-related plans for converting to organic livestock production

As emphasized throughout this discussion, organic animal production (whether certified or not) should be based on basic organic principles, which have been thought through, and analysed, in the context of local conditions, opportunities, threats and the existing animal production systems. The EU-funded network 'Sustaining Animal Health and Food Safety in Organic Farming' gave the following recommendations to the principles developed by IFOAM for the organic livestock sector (Padel and Alrøe, 2005). They are proposed as guidelines for assessing the impact of developing organic livestock production:

1. Define 'preventive use of drugs' as 'drugs used without a thorough analysis of the conditions, the risk and prognosis of disease in the area'. This will prohibit blind routine use of medicine and blanket therapy, but still allow partial application of the approach to infectious and environmentally related diseases.
2. Analyse each region in relation to the possibilities for implementing organic livestock farming, including:
   a. The provision of veterinary services and government extension services;
   b. Farming culture and types of traditional production;
   c. Identification of the major disease risks and planning at farm, and if relevant community level, in order to minimize the use of medical inputs;
   d. Market opportunities and cost-benefit;
   e. Societal impact;
   f. Environmental impact;
   g. Food quality and safety.
3. Develop animal health plans, including farm-level plans and communal activities (including resource sharing, biosecurity and, if possible, plans for handling disease using non-medical disease treatment methods).
   a. Focus on animal welfare, and the possible positive and negative consequences of implementing an organic approach to the farming system, health promotion, and disease management.
   b. A plan for development of breeds suited to organic farming systems, under specific local conditions should be developed, including explicit breeding goals and concrete suggestions to fulfil these goals.
   c. Depending on the local possibilities for advisory services etc., a plan for practical support for the conversion process throughout the entire transition period should be developed. This should prevent individual farmers from finding themselves in the situation where they lack the necessary knowledge and experience. Farmer groups, for example a Farmer Field School approach focusing on conversion, could be a relevant part of this.
   d. Education, training and improving understanding of organic principles must be an integrated part of the conversion process, and farmers as well as advisors and extension officers must be included in this training.

e. Explore the possibilities for traditional or non-medical disease treatment methods in the area.

Conclusions and recommendations

- The starting point and basis for developing organic animal husbandry in any country with little experience in organic animal production must be a thorough analysis of the problems, opportunities and existing knowledge.
- Conversion to organic animal husbandry, with a focus on disease prevention strategies, can reduce dependency on veterinary medicines. When successful, the associated problems of drug residues and resistance will also be reduced. Although organic approaches to animal health can be effective in this respect, they require a multi-disciplinary and holistic approach, preferably extending to the whole farm system.
- An important part of the organic animal production system is the principle of 'naturalness'. Many livestock systems, e.g. pastoral and other systems in tropical countries, fulfil this criteria. At the same time, they present real challenges in terms of disease control, because of endemic diseases, large flocks/herds and how they use the land areas under the given conditions. Other systems in tropical countries, such as smallholder crop–livestock systems have potential for conversion to organic production from a perspective of efficient resource utilization. However, the livestock elements of these systems are not readily converted, as they do not fulfil naturalness criteria, often use breeds which are not resistant to diseases nor well suited for outdoor life, and which are frequently heavily dependent on veterinary medicine, e.g. zero-grazed systems.
- In north Europe and large areas of the industrialized world, production diseases related to high yield and performance dominate, whilst in the tropical regions the risk of infectious and epidemic diseases are a greater concern. Disease patterns are also influenced by factors such as the economic and political environment, culture, history, climate and vegetation. Therefore, the experiences and technologies from one region cannot be directly transferred to other regions of the world. Some experiences and approaches can, however, be relevant sources of inspiration to other situations.
- Experiences from the USA and Europe show that the effectiveness of organic approaches to health management is not always immediate, and that the human element associated with animal care is an evolving process.
- The use of indigenous breeds that have adapted to the prevailing environmental and disease conditions should provide the cornerstone of an organic approach to animal health.
- The involvement of whole communities in the implementation and organization of an organic approach to disease prevention becomes more

important in situations where there is significant societal involvement in the farming systems, e.g. communal grazing systems. These situations are more prevalent in the areas where subsistence farming proliferates.
- The development of organic farming in any particular region must be mindful of, and should not threaten, local and regional disease control programmes, particularly where the diseases are zoonotic in nature, are highly infectious or are of widespread economic importance e.g. Rinderpest and foot and mouth disease.

References

Aarestrup, F.M. and Engberg, J. (2001). Antimicrobial resistance of thermophilic *Campylo-bacter*. Vet. Res. 32: 311-321.
Adolph, A., Blakeway, S., Linquist, B.J. (1996). Ethno-veterinary knowledge of the Dinka and Nuer in Southern Sudan. Report to Community-based Animal Health and Participatory Epidemiology Unit (CAPE), OAU.
Alawa, J.P., Jokthan, G.E. and Akut, K. (2002). Ethnoveterinary medical practice for ruminants in the subhumid zone of northern Nigeria. Preventive Veterinary Medicine 54: 79-90.
Alrøe, H.F., Vaarst, M. and Kristensen, E.S. (2001). Does organic farming face distinctive livestock welfare issues? Conceptual analysis. Journal of Agricultural and Environmental Ethics 14: 275-279.
Bennedsgaard, T.W., Thamsborg, S.M., Vaarst, M. and Enevoldsen, C. (2003). Eleven years with organic dairy production in Denmark – herd health and production in relation to time of conversion and with comparison to conventional production. Livestock Prod. Sci., 80: 121-131.
Bishop, S.C. and Woolliams, J.A. (2004). Genetic approaches and technologies for improving the sustainability of livestock production. J. Sci. Food Agric. 84: 911–919.
Bourn, D. and Blench, R. (eds) (1999). Can Livestock and Wildlife Co-exist? An Interdisciplinary Approach. Published by ODI (Overseas Development Institute) and ERGO (The Environmental Research Group, Oxford).
Byarugaba, D.K. (2004). A view on antimicrobial resistance in developing countries and responsible risk factors. International Journal of Antimicrobial Agents 24(2): 105-110.
Byarugaba, D.K., Olilia, D., Azuba, R.M., Kaddu-Mulindwa, D.H., Tumwikirize, W., Ezati, E., Khalid, T. and Mpairwe, Y. (2001). Development of Sustainable Strategies for the Management of Antimicrobial Resistance in Man and Animals at District and National Level in Uganda. Feasibility Report, May 2001, Makerere University.
Catley, A,T. and Leyland, T. (2001). Community participation and the delivery of veterinary services in Africa. Preventive Veterinary Medicine 49: 95-113.
Catley, A., Irungu, P., Simiyu, K., Dadye, J., Mwakio, W., Kiragu, J. and Nyamwaro, S.O. (2002). Participatory investigations of bovine trypanosomiasis in Tana River District, Kenya. Medical and Veterinary Entomology 16(1): 55-66.
Cristofori, F., Issa, M., Yenikoye, A., Trucchi, G., Uuaranta, G., Chanono, M., Semita, C., Marichatou, H. and Mattoni, M. (2005). Artificial Insemination Using Local Cattle Breeds in Niger. Tropical Animal Health and Production 37(2): 167-172.

De Leeuw, P.N., McDermott, J.J. and Lebbie, S.H.B. (1995). Monitoring of livestock health and production in sub-Saharan Africa. Preventive Veterinary Medicine 25, 1995, 195-212.

Dransfield, R.D., Brightwell, R., Kyorku, C. and Williams, B. (1990). Control of tsetse fly (Diptera: Glossinidae) populations using traps at Nguruman, south-west Kenya. Bulletin of Entomological Research 80: 265-276.

Eckert, J. and Hertzberg, H. (1994). Parasite control in transhumant situations. Vet. Parasit. 54: 103-125.

Gyamtsho, P. (1996). Assessment of the condition and potential for improvement of high altitude rangelands of Bhutan. PhD thesis, Swiss Federal Institute of Technology, Zürich. 249 pp.

Gyldenkærne, S., Münier, B.M., Olsen, J.E., Olesen, S.E., Petersen, B.M. and Christensen, B.T. (2005). Opgørelse af CO_2-emissioner fra arealanvendelse og ændringer i arealanvendelse LULUCF. Metodebeskrivelse samt opgørelse for 1990–2003 [Evaluations of CO_2-emissions from the use of areas and the changes. Land Use, Land Use Change and Forestry. Methods and evaluations 1990-2003]. Work report from National Environmental Research Institute No. 213, in press.

Hardin, G. (1968). The tragedy of the commons. Science 162: 1243-1248.

Holmes, P.H. (1997). New approaches to the integrated control of trypanosomosis. Veterinary Parasitology 71, 1997, 121-135.

Johnsen, N. (1997). Masaai Medicine. Practising health and therapy in Ngorongoro Conservation Area, Tanzania. PhD Dissertation 1997. Institute of Anthropology, University of Copenhagen.

Karreman, H. (2004). Treating dairy cows naturally: thoughts and strategies. Paradise Publications, USA, 268 pp.

King, A. and Mukasa Mugerwa, E. (2002). Livestock marketing in Southern Sudan, with particular reference to the cattle trade between southern Sudan and Uganda. Report to Community-based Animal Health and Participatory Epidemiology Unit (CAPE), OAU.

Kleinmann, A. (1980). Patients and Healers in the Context of Culture. An Exploration of the Borderland between Anthropology, Medicine, and Psychiatry. University of California Press, USA.

Latif, A.A. and Pegram, R.G. (1992). Naturally acquired host-resistance in tick control in Africa. Insect Sci. and Appl., Vol. 13.

Mack, S. and Steane, D. (1996). Livestock breeding and research. Issues and Options paper. Animal Production and Health Division, FAO, Rome.

Macpherson, C.N.L. (1995). The effect of transhumance on the epidemiology of animal diseases. Preventive Veterinary Medicine 25: 213-224.

Nolan, J. (1990). Acaricide resistance in single and multi-host ticks and strategies for its control. Parasitologia 32: 145-153.

Norval, R.A.I. (1983). Arguments against intensive dipping. Zimbabwe Veterinary Journal, 14: 19-25.

Norval, R.A.I and Young, A.S. (1990). Problems in tick control and its modification after immunization. In: Young, A.S, Mutugi, J.J. and Maritim, A.C. (eds) Progress towards Control of East Coast Fever (Theileriosis) in Kenya. Proceedings of an East Coast fever planning workshop held at the National Veterinary Research Centre, Muguga, 17-18 July 1989, KARI, pp. 88-94.

Norval, R.A.I., Lawrence, J.A., Young, A.S., Perry, B.D., Dolan, T.T. and Scott, J. (1991). *Theileria parva*: influence of vector, parasite and host relationships on the nature and distribution of theileriosis in southern Africa. Parasitology 102: 247-356.

Okello-Onen, J. (1994). Cattle tick population dynamics and the impact of tick control on the productivity of indigenous cattle under ranch conditions in Uganda. PhD Thesis Makerere University, Kampala. 263 pp.

Okello-Onen, J., Ssekitto, C.M.B. and Mway, W. (2004). Factors affecting the sustainability of tick and tick-borne disease control in Uganda and malpractice associated with acaricide use. Uganda Journal of Agricultural Sciences, 9, no. 1, 663-666.

Padel S. and Alrøe, H.F. (2005). Report of the working group discussion of the draft IFOAM principle in relation to animals at the 4th SAFO workshop. In: Proceedings, 4th SAFO workshop, 33-35.

Padel, S., Schmidt, O. and Lund, V. (2004). Organic Livestock Standards. In: Vaarst, M., Roderick, S., Lund, V. and Lockeretz, W. (eds) Animal Health and Welfare in Organic Agriculture. CABI Publishing, UK, pp. 57-72.

Pregram, R.G., Lemche, J., Chizyuka, H.G.B., Sutherst, R.W., Floyd, R.B., Kerr, J.D. and McCosker, P.J. (1989). Effects of tick control on liveweight gain of cattle in central Zambia. Veterinary and Medical Entomology 3: 313-320.

Peregrine, A.S. (1994). Chemotherapy and delivery systems: haemoparasites. Vet. Parasit., 54: 223-248.

Perry, B.D. (1994). The role of epidemiology and economics in the control of tick-borne diseases of livestock in Africa. The Kenya Veterinarian Special Issue 18(2): 7-10.

Perry, B.D., Musisi, F.L., Pegram, R.G. and Schels, H.F. (1985) Assessment of enzootic stability to tick-borne diseases in Zambia. World Anim. Rev., 56: 24-32.

Perry, B.D., Grandin, B.E., Mukhebi, A.W., Young, A.S., Chabari, F., Maloo, S., Thorpe, W., Delehanty, J., Mutungi, J.J. and Murethi, J. (1990). Techniques to identify target populations for immunization and to assess the impact of controlling East Coast Fever in Kenya. Editors A.S. Young, J.J. Mutugi and A.C. Maritin, KARI.

RGOB (n.d.): National Policy Document and Work Plan on use and maintenance of domestic animal diversity. Thimphu, Bhutan [mid 1990s] 19 pp.

Roderick, S., Stevenson, P., Ndung'u, J. (1999). Factors influencing the production of milk from pastoral cattle herds in Kenya. Animal Science 68: 201-209.

Roder, W., Calvert, O. and Dorji, Y. (1992). Shifting cultivation systems practiced in Bhutan. Agroforestry Systems 19: 149-158.

Roderick, S., Stevenson, P., Mwendia, C., Okech, G. (2000). The use of trypanocides and antibiotics by Maasai pastoralists. Tropical Animal Health and Production 32(6): 361-374.

Rubaire-Akiiki, C., Okello-Onen, J., Nasinyama, G.W., Vaarst, M., Kabagambe, E.K., Mwayi, W., Musunga, D. and Wandukwa, W. (2004). The effect of agro-ecological zone and grazing system on the incidence of East Coast Fever in calves in the smallholder dairy farms in Mbale and Sironko Districts, Eastern Uganda (submitted manuscript).

Rubaire-Akiiki, C., Okello-Onen, J., Nasinyama, G.W., Vaarst, M., Kabagambe, E.K., Mwayi, W., Musunga, D. and Wandukwa, W. (2005). The prevalence of serum antibodies to tick-borne infections in Mbale District, Uganda: The effect of agro-ecological zone, grazing management and age of cattle. Journal of Insect Science 4: 8. Also online: insectscience.org/4.8

Sasanya, J.J., Enyaru, J., Ejobi, F. and Olila, D. (2004). Penicillin G residues in Bovine Milk and Tissues from Mbarara and Masaka Districts. In: Proceedings of the Livestock Systems Research Programme annual Scientific Workshop, 1–3 June 2004, Kampala.

Schillhorn van Veen, T.W (1997). Sense or nonsense? Traditional methods of animal parasitic disease control. Veterinary Parasitology 71: 177-194.

Seveno, N.A., Kallifadas, D., Smalla, K., Dirk van Elsas, J., Collard, J.M., Karagouni, A.D. and Wellington, E.M.H. (2002). Occurrence and reservoirs of antibiotic resistance genes in the environment. Reviews in Medical Micobiology 13(1): 15-27.

Steinfeld, H. (2004). The livestock revolution – a global veterinary mission. Veterinary Parasitology 125, 2004, 19-41.

Sundlof, S.F., Riveire, J.E. and Craigmill, A.L. (1991). Food Animal Residue Avoidance Databank Trade Name File: Food Animal Drugs, 7th edition. Institute of Food and Agricultural Sciences, Gainesville, FL.

Tatchell, R.J. (1988). Astudy of the effect of tick infestation on liveweight gain of cattle in the Sudan. Tropical Pest Management 34: 165-167.

Tatchell, R.J. and Easton, E. (1986). Tcik (Acari: Ixodidae) ecological studies in Tanzania. Bulletin of Entomological Research 76: 229-246.

Thinlay, X., Finck, M.R., Bordeos, A.C. and Zeigler, R.S. (2000). Effects and possible causes of an unprecedented rice blast epidemic on the traditional farming system in Bhutan. Agriculture Ecosystems and Environment 78: 237-248.

Vaarst, M. (1995). Sundhed og sygdomshåndtering i danske økologiske malkekvægbesætninger. PhD thesis, Clinical Institute, Royal Veterinary and Agricultural University, Copenhagen.

Vaarst, M. and Bennedsgaard, T.W. (2001). Reduced medication in organic farming with emphasis on organic dairy production. Acta Vet. Suppl., 95: 51-57.

Vaarst, M., Thamsborg, S.M., Bennedsgaard, T.W., Houe, H., Enevoldsen, C., Aarestrup, F. and De Snoo, A. (2003). Organic dairy farmers decision making in the first two years after conversion in relation to mastitis. Livestock Prod. Sci., 80: 109-120.

Vaarst, M., Martini, A., Bennedsgaard, T.W. and Hektoen, L. (2004). Approaches to the treatment of diseased animals. In: Vaarst, M., Roderick, S., Lund, V. and Lockeretz, W. (eds) Animal Health and Welfare in Organic Agriculture. CABI Publishing, UK, 279-303.

Vaarst, M., Byarugaba, D.K., Nissen, T.B., Klaas, I., Bennedsgaard, T.W. and Østergaard, S. (2005). Farmer Field School approach for learning and exchanging knowledge on mastitis and health promotion. Action research projects in Uganda and Denmark. Proceedings, IDF Conference, Maastricht, The Netherlands, 12–15 June 2005.

Van den Bogard, A. (2001). Antimicrobial resistance, pre-harvest food safety and veterinary responsibility. Veterinary Quarterly 23: 99.

Verhoog, H., Alrøe, H.F. and Lund, V. (2004). Animal Welfare, Ethics and Organic Farming. In: Vaarst, M., Roderick, S., Lund, V. and Lockeretz, W. (eds) Animal Health and Welfare in Organic Agriculture. CABI Publishing, UK, pp. 73-94.

WHO (2001). The use of antimicrobials outside human medicine. http: www.who.int/emc/diseases/zoo/antimicrobial.htm

Young, A.S., Groocock, C.M. and Kariuki, D.P. (1988). Integrated control of ticks and tick-borne diseases of cattle in Africa. Parasitology 96: 403-432.

10
The impact of organic farming on food security in a regional and global perspective

Niels Halberg,* Timothy B. Sulser, Henning Høgh-Jensen, Mark W. Rosegrant and Marie Trydeman Knudsen

Introduction 278
Overview of existing food supply and security projections 281
    Global food production and food security today 281
    Projections for global food production and demand 282
    Important factors impacting on scenarios 285
Significant factors determining the effect of OF 287
    The characteristics of existing agricultural systems 287
    Significance of agro-ecological conditions 292
    The influence of socio-economic conditions 296
    Different types of organic farming systems', impact on food security 298
Modelling consequences of large scale conversion to OF for food security 301
    The IMPACT model and methodology for predicting food security 302
    Definition of the scenarios for large scale conversion to organic farming ... 303
    Results of the scenario modelling 307
Discussion 313
Conclusions 316

Summary

The spread of organic and agro-ecological farming (OF) methods in developing countries has raised a debate whether a large scale adoption of OF will increase or decrease global food security. This will however depend on a number of socio-economic factors together with the relative yield levels of OF versus conventional farming systems. Relative yields again depend on a number of

---

* Corresponding author: Danish Institute of Agricultural Sciences, P.O. Box 50, DK-8830 Tjele, Denmark. E-mail: Niels.Halberg@agrsci.dk

agro-ecological factors and the characteristics of farming systems before conversion.

In areas with intensive high-input agriculture, conversion to OF will most often lead to a reduction in crop yields per ha by 20–45% in crop rotations integrated with leguminous forage crops. In many areas with low input agricultural systems farmers have little incentive or access to use chemical fertilizer and pesticides, and yields may increase when agro-ecological principles are introduced.

While present food production in theory is sufficient to cover the energy and protein needs of the global population there are still more than 740 million food insecure people, the majority of whom live in South Asia and Africa South of Sahara (SSA). This number will only decrease over the next 20 years if the present policies are changed. The food policy model IMPACT was used to project possible effects on food security of a large scale conversion to OF in Europe/North America (E/NA) and SSA. Results indicate that a conversion of approximately 50% of E/NA agricultural area will have a 6–10% impact on world prices on (non-meat) agricultural commodities under the assumptions of 35% lower OF yields after conversion and 50% higher yield growth rates compared with conventional crop yields. The indirect effect on food security in SSA would be small as the up scaling experiences from case studies into scenarios for conversion of 50% of agricultural area in SSA results in increased self-sufficiency and decreased net food import to the region. Given the assumption of higher relative yields in most organic crops compared with existing low input agriculture, there is potential for improving local food security in SSA if non-certified OF is supported by capacity building and research. More knowledge is needed, however, to confirm that these optimistic results of non-certified OF apply to large areas in SSA and other regions with low input agriculture.

Introduction

The practices of organic farming (OF) are spreading in developing countries both in the form of certified, export-oriented production as well as through farming systems building on agro-ecological principles (Altieri, 1995) and locally available resources as an explicit strategy, as reviewed by Parrot *et al.* (Chapter 6). While still being a niche production in most countries the rapid increase of (certified) OF has raised questions regarding the productivity and the longer-term consequences of not supplying mineral fertilizers or using improved varieties if they have been genetically modified (GMOs). Proponents of the productivity-focused understanding of sustainability (food sufficiency school) claim that limitations on the use of fertilizer and pesticides are problematic because it will reduce the total food production. On the other hand, a number of case studies have shown that farmers have indeed improved their food security after the introduction of OF methods (Scialabba and Hattam, 2002; Pretty *et al.*, 2003). However, it is not always clear whether this improvement was an effect of using OF or agro-ecological methods *per se* or simply an effect of enhanced extension on basic traditional farming systems using state-of-the-art participatory extension methods. In some cases the increased food security is an effect of higher income and asset building due to price premiums on certified organic products. Whereas local evidence is mixed, the impacts on food supply and security at the regional and global levels are unknown, raising the question if organic farming would feed or starve the world if it became more widespread (see also Vasilikiotis, 2000; Pretty and Hine, 2001; Stockdale *et al.*, 2001; Tiffen and Bunch, 2002; Surridge, 2004).

The issue builds on the general assumption that the yields in OF will continue to be low because of limited inputs, which could be problematic from the point of view of global food production. This is not surprising because OF has developed in Europe and North America as an explicit criticism of the development of conventional farming relying on speicialization and high external inputs. Many farmers in some developing countries have benefited immensely from the Green Revolution, which relies on high input intensity. Therefore, as seen from this perspective, the idea of a deliberate cut off from the productivity-enhancing, high-input intensive production systems seems destructive for the development of a rational agricultural system and thus for food security (Borlaug, 1994; Reason Foundation, 2000).

However, Green Revolution agriculture has shown mixed impact in resource poor environments in, for example, sub-Saharan Africa (SSA) and has bypassed a significant fraction of the poorest farmers. There have also been negative ecological, social and economic side effects of high external input agriculture. In some parts of Asia Green Revolution agriculture, combined with inappropriate policies such as water and fertilizer policies and trade protection for rice, has resulted in falling ground water tables, reduced agro-biodiversity and the degradation of natural resources (Pingali and Rosegrant, 2000). Farmers in numerous developing countries report human health hazards through pesticide poisoning (Pretty, 1995). Others have become indebted as they are not able to pay back loans, which they required for purchasing seeds, pesticides and fertilizers.

As discussed by Hauser (Box 3.4) and Parrot *et al.* (Chapter 6) organic farming may improve the asset building and food security in smallholder families in a variety of ways, such as, higher prices on cash crops, reduced risks through farming system diversification, and occasional potential for higher yields in organic grown crops among others. But there may be a limit as to how much yields can improve in OF or low external input sustainable agriculture (LEISA, http://www.leisa.info/) systems in the long run compared to the potential increases in conventional systems given the same access to extension. It is also not clear how representative the positive case stories of OF are for the majority of the resource poor farmers in, for example, sub-Saharan Africa, given the low organic matter content, low pH, and nutrient status – especially lack of phosphorus – in many tropical soils (see Chapter 8). This aspect in combination with the projections for population growth, and the resulting global pressure on land use, water, biodiversity and fish resources leads some critics to conclude that promoting OF is a luxurious idea of the privileged rich consumers in developed countries. Only few attempts exist so far to give aggregated evaluations of increasing uptake of OF on national or regional food supply and food-security in developing countries. OF in the tropics is multifaceted and comprises many farming systems in different agro-ecological settings. The effects of increased uptake of organic farming on food supply and food security will depend on geographical and socio-economic conditions and on the scale considered.

Moreover, recent food crises in some parts of India and Argentina, for example, show that food insecurity is not only a result of low agricultural productivity, but have a distinct socio-economic and political dimension. Often people experience food shortages because they are economically not in the position to maintain access to adequate amounts of high quality food. Food insecurity and poverty are interlinked, which is an important consideration in the discussion about the food security impacts of organic agriculture that are too often production and productivity biased. The working group preparing for the UN conference Earth Summit 2002 defined food security as:

> the peoples right to define their own policies and strategies for the sustainable production, distribution and consumption of food that guarantees the right to food for the entire population, on the basis of small and medium sized production, respecting their own cultures and the diversity of peasant, fishing and indigenous forms of agricultural production, marketing and management of rural areas, in which women play a fundamental role. (McHarry *et al.*, 2002)

Thus, the overall question of whether organic farming can feed the world is too simplified to allow for a thorough inquiry and to be policy relevant. On the other hand, it may be of relevance to policy makers and development bodies to consider the regional or global relations between large scale promotion of OF and LEISA systems and their effects on the food supply and malnutrition.

The objective of this chapter is to analyse and discuss the following topics:

- Under what circumstances and to what extent is OF a beneficial solution to food insecurity and low agricultural productivity?
- What could be the consequences of a high percentage OF in developed and developing countries for the medium and long-term food supply seen in relation to global developments in food consumption and demography?

Overview of existing food supply and security projections

Global food production and food security today

Global food production increased during the last decade and 'global food production at present would be sufficient to provide everyone with his or her minimum calorie needs if available food was distributed according to need' (von Braun *et al.*, 2003). However, recent inventories of the world food situation show that there are still 742 million people without enough to eat and that this number has increased during the 1990s except in China where food security on average has improved (FAO, 2003). Moreover, new methods for measuring food insecurity based on household surveys indicate that more families in sub-Saharan Africa are food insecure than previously estimated using national statistics. Food-security is not simply a question of global food production since the access to sufficient food may be compromised by a number of factors including private socio-economic conditions such as resource endowments, income, and access to land and knowledge. In many cases, food security depends more on socio-economic conditions than on agro-climatic ones, and on access to food rather than the production or physical availability of food (Smith *et al.*, 2000; FAO, 2003). FAO (2003) states that producers have satisfied effective market demands in the past, and it is likely that they will continue to do so. But effective markets do not represent the total need for food and other agricultural products, because hundreds of millions of people lack the money to buy what they need. FAO (2003) predicts that even by 2030 hundreds of millions of poor people will remain undernourished unless local food production is given higher priority and inequality of access to food is reduced. Furthermore, three-quarters of undernourished people are the destitute population of poor agricultural regions, indicating that when attempting to increase food security, focus should be on local food production in poor agricultural regions, where financial resources for inputs are low. An evaluation of the consequences of increased hectarage with OF for food-security must address the different factors contributing to food security for people in different situations such as urban poor, landless rural poor, and small-scale farmers in different agro-ecological settings among others.

In many developing countries different trends seem to co-exist. At the same time parts of the agricultural sector are become increasingly involved in the globalization process through either growing cash crops for exports or intensive export-oriented livestock production, while other parts become or stay isolated from the markets, such as the resource poor in areas with poor soils and infrastructure. The fact that some countries like India have food exports while still having food shortages adds to this picture.

Projections for global food production and demand

The International Food Policy Research Institute (IFPRI) makes projections of global food security to the year 2020 and beyond based on the International Model for Policy Analysis of Agricultural Commodities and Trade (IMPACT) (Rosegrant et al., 2001; Runge et al., 2003). Through the years the IMPACT model has been developed and extended to incorporate an expanded set of commodities and critical production details such as water usage in what is called the IMPACT-WATER model (Rosegrant et al., 2005). The most current baseline set of projections from IMPACT-WATER will be used here as a baseline to inform comparisons (Rosegrant et al., 2002).

While developing countries will continue to drive increases in the global demand for cereals, growth in their demand for cereals has begun to slow down. With slowing population growth rates and increasing diversification of diets away from cereals due to rising prosperity and changing dietary preferences, annual growth in cereal demand in the developing world is projected to decline to 1.9% between 1995 and 2020 from 3.8% in 1967–82. Nevertheless, the absolute increase in the demand for cereals during 1995–2020 is expected to be as large as the increase in demand during the preceding 20 years. Developing countries in Asia, because of their larger and more urbanized populations and rapid economic growth, will account for half of the increase in global demand for cereals, with China alone accounting for one-quarter.

Global demand for meat will grow much faster than that for cereals. Worldwide, demand for meat is projected to increase by more than 58% between 1995 and 2020, with most of the increase occurring in developing countries. China alone will account for more than 40% of this increase, compared to India's 4%. Although demand for meat will double in South Asia, Southeast Asia and sub-Saharan Africa, per capita consumption of meat will remain far below levels in the developed world indicating considerable potential for even more consumption. Poultry will account for 40% of the global increase in demand for meat, followed by pork's 32% and beef's 23%. Cereal crops will increasingly be grown for animal feed to fuel the explosive rise in demand for meat rather than for direct human consumption. As a result, maize will rise in importance, at the expense of wheat and rice. By 2020, maize will become the world's leading cereal, accounting for 40% of global cereal production compared to 25 and 20% for wheat and rice respectively.

Growth in demands for other staple food commodities will also be strong in developing countries. In many parts of sub-Saharan Africa, roots and tubers, especially cassava, sweet potatoes and yams, are a major source of sustenance. In the late 1990s, they accounted for 20% of calories consumed in the region, with an even higher concentration in the diets of the poor. In much of Asia and Latin America, roots and tubers provide an important, supplemental source of carbohydrates, vitamins and amino acids in food systems that are dominated by other commodities. These patterns will continue, with total roots and tubers demand in the developing world increasing by 49% (212 million t) between 1995 and 2020. Sub-Saharan Africa is projected to account for 42% of this increase, indicating that roots and tubers will continue to be an important part of the diet in that region. Asia will also account for a significant amount of the total increase, with both East and South Asia accounting for 16% each. Improved yields will be necessary to drive roots and tubers production increases throughout the developing world, and the area planted to roots and tubers will actually shrink significantly in the developed world. Area expansion will remain important in sub-Saharan Africa increasing by a projected 60% between 1995 and 2020.

Cereal production in some developing countries will not keep pace with increases in demand. In parts of Asia, almost all the available land is already under cultivation, urban land conversions are encroaching on prime agricultural land, and land degradation is becoming an increasingly serious problem (FAO, 2003). Sub-Saharan Africa and Latin America have more potential for area expansion, with the area under cereal production projected to expand by over 20 million ha in sub-Saharan Africa and by 18 million ha in Latin America during 1995–2020.

Increases in cereal production will be highly dependent on increases in productivity. But increases in crop yields are slowing across all cereals and all regions, with the notable exception of sub-Saharan Africa, where yields are projected to recover from past stagnation. Yield growth rates in most of the world have been slowing since the early 1980s. In the developed world, the slowdown was primarily policy-induced, as North American and European governments drew down cereal stocks and scaled back farm-price support programmes in favour of direct payments to farmers, while in Eastern Europe and the former Soviet Union economic collapse and subsequent economic reforms further depressed productivity. Factors contributing to the slowdown in cereal productivity growth in developing countries, particularly in Asia, include high levels of input use, which means that it takes increasing input requirements to sustain yield gains, slowing public investment in crop research and irrigation infrastructure, and growing water shortages as irrigation development slows and non-agricultural water demands divert water from agriculture. These forces are expected to further slow annual cereal yield growth rates from 1.6% between 1982 and 1995 to 1.2% between 1995 and 2020 (see Figure 10.1).

By 2020, with developing countries unable to meet fully their cereal demands from their own production, international trade will become even more vital for providing food to many regions of the globe. Fortunately, cereal producers in the Americas and in Europe appear ready and able to meet this demand. The United States will become an even more dominant force in agricultural markets and Europe will continue to be a major agricultural exporter. Net cereal imports by developing countries will more than double to 2020 with Asian nations, particularly China, boosting their imports enormously. However, countries that falter economically, leaving them unable to muster enough foreign exchange to pay for adequate food imports, will become increasingly vulnerable.

Sharp decreases in food prices over the last two decades were a great benefit to the poor, who spend a large share of income on food. But international cereal prices are projected to decline only slightly during the next two decades, a significant break from past trends. This tighter predicted future price scenario indicates that additional shocks to the agricultural sector, particularly shortfalls in meeting agricultural water and other input demands, could put serious upward pressure on food prices.

The prospects for reducing child malnutrition are mixed, and in some regions poor. Overall, the number of malnourished children is expected to continue its gradual decrease, from 162 million in 1995 to 127 million in 2020. The number of China's malnourished children will halve while India will experience slower improvement and will remain home to a third of all malnourished children in the developing world. Sub-Saharan Africa, with its combination of high population growth and lagging economic performance, will be caught in an increasingly perilous situation with the number of malnourished children forecast to increase by 6 million or 18% compared to 1997.

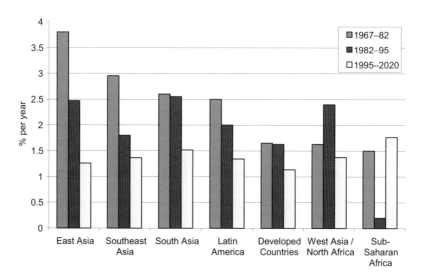

Figure 10.1. Yield growth rates by region for all cereals, 1967–82, 1982–95, and projected 1995–2020. (Source: FAOSTAT (FAO, 2004) and IMPACT-WATER model projections, June 2005.)

Important factors impacting on scenarios for food supply and agricultural development

Year-to-year variability in food production and prices often sends incorrect signals about the fundamental trends in food security. But it is inappropriate to make judgements about long-term food security based on short-term trends in global markets. Indeed, year-to-year variability in prices and production – and the influence that this variability has on the amount and type of attention devoted to the global food situation – may in fact contribute to long-term food problems by encouraging complacency during periods of strong harvests. In order to understand the future of food supply and demand and food security, it is essential instead to focus on long-term forces, such as income growth, population growth, rates of urbanization, and technological change in agriculture driven by investments in agricultural research, irrigation and roads. The natural resource base is a key supporting system for food production, and future trends and policies in soil and water quality and availability will play a major role in determining the sustainability of food security.

Future food security will be determined largely by the interplay of a number of factors such as political and socio-economic stability, technological progress, agricultural policies, growth per capita and national incomes, poverty reduction, women's education, drinking-water quality and increased climate variation (FAO, 2003). In relation to objectives of this chapter the interesting question is what role organic farming methods may play in different technological, political, economic and environmental change scenarios.

As described above, the past three decades have seen major shifts in human diets. Meat products have provided an increasing share of human diets, with poultry expanding fastest and pig production to a smaller extent. This development is primarily caused by higher incomes in parts of Asia and Latin America. There is a continued shift in livestock production methods away from more traditional mixed farming systems and grazing systems towards more intensive and industrial methods where the livestock require diets with a high concentration of energy and protein (FAO, 2003). Livestock are the world's largest user of agricultural land and about 20% of the world's arable area is used for the production of livestock feed, which is mostly concentrates (Hendy *et al.*, 1995). This development is already influencing and changing agriculture and it is one major reason for the increased trade in soybean and feed-grains. Since 1990 a strong growth in global trade in soybeans and soybean products has been seen, mainly caused by an increasing import of soybeans to China (USDA, 2004). The increased areas with soybeans in Argentina and Brazil, which are the major soybean exporters along with USA (Chapter 1; USDA, 2004), are possible due to a combination of replacing traditional farming systems in the Pampas (see Chapter 1) and cultivating previously undisturbed landscapes. While the conversion of natural lands has negative consequences for biodiversity, it is positive seen from a food sufficiency point of view and contributes to alleviate the pressure on other crops.

Will hogs and poultry compete with poor people for the daily cereal and protein meal in the next decade? Or will the increased livestock feed use be compensated for by increased land use and higher yields? This is important for the assessment of the consequences of a large scale conversion to OF because speicialized and industrialized meat production is not in accordance with OF.

In spite of the increased demand for cereals and meat over the last two to three decades relative prices on the major staple foods traded globally – such as maize and rice – have fallen (FAO, 2003). And according to FAO's and IFPRI's food and population projections the risk that food prices will rise significantly is not high.

The impact of this increased demand for livestock feeds on food prices and availability for rural and urban poor in developing countries is not straightforward because the prices for staples at local markets is not directly linked to the world market prices. Staple foods for many poor people are not world market commodities but are more regional, non-traded crops such as millet, cassava and bananas in Africa. The prices and availability of staple food at local rural markets in many developing countries are significantly influenced by poor marketing channels and lack of sufficient infrastructure. Thus, in some countries like Malawi, a regional food deficit has been observed even in years with relatively good maize yields in other regions of the country because the incentives and means to transport the excess harvest were not sufficient. On the other hand, the increasing urban population in many developing countries may be more directly influenced by the global demand for food commodities, and because they obviously depend more on purchased food than the rural poor, even small increases in prices for staples may have significant impact on their ability to satisfy family nutritional needs.

The changing global climate due to increased average temperatures will have implications for agriculture because of shifting rain patterns, increased evaporation in combination with the water scarcity arising from overuse of irrigation water and competition from other sectors (FAO, 2003).

The effect of climate changes on the global food production was modelled by Parry *et al.* (2004), who found that there is a risk that large populations in developing countries will become food-insecure because of reduced yields in dry and semiarid regions, where the majority of the population growth will also happen. Anecdotal evidence from East Africa suggests that weather insecurity increases especially in the low potential areas. In areas with bimodal rainfall patterns farmers report that the start of both short and long rainy seasons have been delayed in recent years. In eastern Uganda, for instance, increasing rainfall variability has disrupted farmers' traditional cropping calendars with serious implications for food availability at household level. According to Parry *et al.* (2004) the overall food production may still be sufficient on a global scale, but this is an effect of increasing yields in developed countries counteracting the decreasing yields in the South. And this may not give food security for poor populations in developing countries that will not have the purchasing power to access the surplus in the North. The study does not address the adaptive capacity of smallholder farmers and the possible counteracting effects of agro-ecological methods, which may ameliorate the negative consequences through increased soil fertility, water retention capacity and water harvesting techniques (Rodale Institute, 1999; IFAD, 2003; Pretty *et al.*, 2003).

Significant factors determining the effect of OF on food supply and food security

The question of whether a conversion to organic farming methods will improve or reduce food production and food security for the poor, both urban and rural, depends on a number of factors. Some factors relate to the agricultural system before conversion and some to the overall conditions in the food system in question, as illustrated in Figure 10.2.

The following section will discuss a number of these issues to the extent that information is available in the literature. Figure 10.2 is a conceptual description of the qualitative relations which will to a large extent determine the consequence of large scale conversion to OF under different conditions. Not all of these factors will be dealt with.

The characteristics of existing agricultural systems

It is a basic assumption that the effects of promoting OF will differ significantly between intensive farming systems, mostly in developed countries but also found in developing countries in the form of high-yielding, irrigated smallholder plots, large scale commercial farms, and plantations among other forms and smallholder, low-input, traditional farming systems.

The effect of OF methods in low input farming systems

In many low potential areas of sub-Saharan Africa the principles of Green Revolution agriculture have rarely been adopted by the smallholder farmers. This situation is not surprising as most Green Revolution technologies and practices were developed with the assumption that smallholder farmers physically and financially have access to mineral fertilizers, synthetic pesticides, improved crop varieties, and irrigation from ground and surface water sources or relatively favourable rainfed conditions. In reality, however, the majority of the poor in sub-Saharan Africa lack access to such inputs and infrastructure, which makes it difficult to adopt such technologies. Some farmers who had experimented with Green Revolution technologies in low potential areas increased the production risks due to many economic, social and political uncertainties. In Ethiopia and Mozambique attempts to increase yields through Green Revolution approaches have shown mixed impact (Jiggins *et al.*, 1996; Howard *et al.*, 2003). Also, in non-Green Revolution crops such as maize, which historically received particular attention from local governments, early successes in improving yields have since stagnated (Byerlee and Eicher, 1997). In the 1967–82 period, northern and eastern sub-Saharan Africa along with Nigeria saw remarkable growth in maize yields of around 3% annually (FAO 2004). The initial impressive growth was due to adoption of improved seed varieties, whereas the subsequent stagnation arose from the limited use of complementary soil fertility-maintaining inputs and the lack of institutional support for further advances (Byerlee and Eicher, 1997; Rosegrant *et al.*, 2001). Such experiences have promoted emerging low external input movements, especially in sub-Saharan Africa as an alternative to Green Revolution agriculture.

The relatively small group of smallholder farmers that will have the chance to benefit from price premiums in a certified production scheme will improve their overall situation by having higher income (Giovannucci, 2005), which they may invest in their farm or non-farm activities. These benefits will usually weigh higher than questions of whether the yields could have been higher if they used fertilizer and pesticides. However, this cannot benefit the majority of smallholders in developing countries. As reviewed in Chapters 5 and 8 there are a number of examples of increasing yields and productivity in traditional farming systems when introducing improved methods based on agro-ecological principles such as LEISA. Thus, the short term effects of introducing agro-ecological principles in many areas of especially Africa and parts of Asia can be an increase in yields and improvement of local food-security as reviewed by Pretty and Hine (2001). This is then compared with the lower yields of some farming systems using little external input. But in the long run the question is, how much is it possible to increase and sustain yields without fertilizer and pesticides? As discussed by Abalu and Hassan (1998), the much needed increase in agricultural productivity in southern Africa is unlikely to come from a green revolution comparable to the development in Asia because of lack of roads and other infrastructure and poor market accessibility. Therefore, the strategy for intensification should build primarily on:

> internal sources for soil fertility maintenance and enhancement, such as Nitrogen-fixing and mycorrhiza associations and crop residue use since low doses of chemical fertilisers will be much more profitable under such conditions.
>
> (Abalu and Hassan, 1998, p. 486)

Abalu and Hassam interestingly call this an 'agro-ecological specific' approach and calls for flexible solutions. Therefore, they also recommend increased use of fertilizers in order to secure the sustainability of the improved use of internal resources because locally produced organic matter is unlikely to replenish nutrients mined with crops.

In sum, the impact of introducing methods of organic and agro-ecological farming systems in areas with traditional farming can be higher and more staple yields, although this has yet to be shown on a large scale. To promote wide adoption, these methods should have an immediate beneficial effect, be affordable to the smallholder family and not compete too much for labour and capital. A wide adoption will only happen under a pro-poor or pro-smallholder policy.

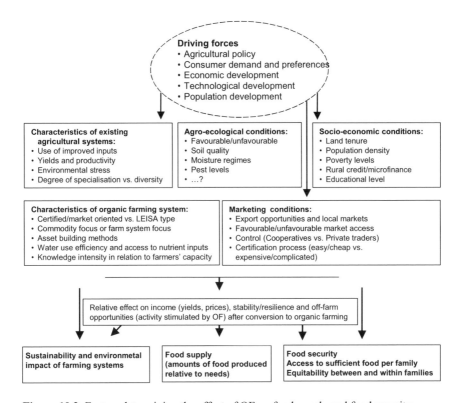

Figure 10.2. Factors determining the effect of OF on food supply and food security.

The effects of OF in high yielding farming systems

The yields in major cereals such as maize, wheat and rice have increased significantly over the last three to four decades in Europe, North America and parts of Latin America and Asia due to a combination of improved varieties and cultivation practices including mechanization and the use of chemical fertilizer and pesticides (see Chapter 1). In high potential areas in developing countries, Green Revolution agriculture has had great success and the doubling of cereal yields in developing countries has mainly been attributed to high yielding varieties. Good results with Green Revolution packages have been achieved especially in South and East Asia. However, projections of future yield increases following the same model indicate a low yield growth in the next decades (see above). Compared to high-yielding crops cultivated with the use of fertilizer and pesticides most organic crops yield less per hectare due to a combination of lower nutrient supply and yield reductions from weeds, fungi and insects. De Ponti and Pinstrup-Andersen (2005) collected 272 comparative data sets from 28 countries in Europe (135 entries), North America (106), Africa (11), Asia (9), Australia and New Zealand (8), and Latin America (3) and found that on average organic yields are 80% of those under conventional agriculture. In 60% of the cases organic yields are 75% or more of conventional yields. In 40% of the cases organic yields are 85% or more of conventional yields. The African results represent only one location in Zambia where organic yields in a number of legume crops were higher that conventional. Thus, the effect of large scale conversion to OF in high yielding environments in Europe and North America would be a reduction in especially cereal and feed crop production per hectare. Moreover, the feed conversion factor in organic meat production is less effective, which is why around 10% more cereals are needed per kg pig (Hermansen et al., 2004) produced and probably more for poultry. Organic milk production based on grass-clover meadows is however efficient (Kristensen and Mogensen, 1999; Zollitsch et al., 2003).

The result of large scale conversion to OF is not simply a scaling up of relative yields in individual crops because the choice of crop rotations and land use is different in organic farming. An increased area with legumes (pulses and fodder legumes) will partly compensate for the reduced nitrogen supply and yields in cereals. In a simulation of a 100% conversion to organic farming in Denmark, the combined effect of approximately 20% reduced cereal yields per hectare – assuming 15% improved yields compared to the recorded organic yields – and the need for more grass-clover in the crop rotation resulted in a reduction in national cereal yield of more than 50% (Bichel Committee, 1999). In this 'Bichel'-study, Danish milk production was assumed to be constant in the organic scenario, but the organic pig production would be only 45% of Danish conventional pig production (1996-level) if no feed import was allowed. If 25% of the feed needed for organic pig production could be imported, the production level could be 90% of conventional. In all scenarios the domestic cereal harvest would be used domestically for bread and feed leaving no surplus for export compared with an export of 2 million t of cereal in 1996. However, it should be noted that this cereal export was lower than the feed import to Denmark in 1996. The import of protein feed in 1996 was also higher than the projected import in the mentioned organic scenario (Bichel Committee, 1999, Table 5.3).

Penning de Vries *et al.* (1995) modelled the consequences for global food supply of different scenarios for crop yields including an organic scenario with low yields (33% of conventional yields). The results were aggregated at regional levels and showed a dramatic negative impact on food security in most of Africa and Asia if an affluent diet high in meat was consumed. The study did not consider socio-economic conditions or feed-back mechanisms at the local scale such as increased food-security and reduced risks for smallholders through the employment of improved agri-ecological methods. The reason for the low relative organic yields per hectare used by Penning de Vries *et al.* (1995) is not clear but does not seem realistic compared with the empirical data mentioned above.

The simple conclusion would be that the relatively rich consumers in the North and parts of Asia would increase their demand for imported feeds and have less food surplus for export after a large scale conversion to OF with negative impact on the food availability of others with less purchasing power. But this inference is only valid under an assumption of 'all-other-things-being-equal', which is not likely. Especially, the decoupling of European agricultural payment schemes may result in further abandoning of agricultural land with low productivity for conventional farming. Contrary to this development scenario of separation between intensively-used and marginalized land, organic farming may potentially support another scenario of less intensive land use on a larger area. This question revitalizes a 20-year-old discussion of the pros and cons of two different perspectives of agricultural development: separation and intensification vs. integration and extensification (Weinschenck, 1986; de Wit *et al.*, 1987; Halberg *et al.*, 1995).

This is reinforced by recent projections of future land use in European agriculture (Rounsevell *et al.*, 2005). The authors modelled the long term land use changes in Europe following different generic policy options such as a 'global economic and fossil fuel intensive world' (A1FI) and a global respectively 'regional environmental world' (B1, B2 scenarios). Their results indicate that over the next two to four decades large agricultural areas will be abandoned due to low increases in demand, technological developments (yield increases) in crops and a change in meat demand favouring pigs and poultry at the cost of (European) beef production. In the A1FI scenario the area used for crops and grassland in EU will be reduced by 15–20% compared with present land use. This will happen mostly in the so-called less favoured areas, that is, the areas less suited to high input agriculture. In the B scenarios this development is assumed to be partly offset by policy measures aimed at securing biodiversity in grasslands and (in B2 only) 'reducing crop productivity by encouraging extensification and organic production' (Rounsevell *et al.*, 2005). Thus, these results indicate that there would be abundant land in Europe from a food sufficiency point of view even if productivity is reduced for a while due to conversion to OF.

Moreover, reducing the area under 'set-aside schemes' in Europe and North America could increase the total grain production, but it is difficult to say how much without a detailed analysis of which areas have been taken out so far. In some areas the croplands with lower productivity were taken our first and these lands would not fully compensate for reduced yields in highly productive areas converted to OF. For these reasons total crop production in Europe under an organic scenario might not be as low as predicted from simple field level comparisons when comparing with the conventional scenario that may develop under the present agricultural policy. These complicated interactions point toward the need for a 'Bichel'-study on a European or American scale.

Significance of agro-ecological conditions

The degree to which organic farming methods can give high and sustainable yields will partly depend on the agro-ecological conditions such as soil quality (e.g. nutrients and moisture content), climate (e.g. rainfall patterns and growth seasons) and pest levels in a given area. In Europe the moderate levels of pests (insects and fungi) can often be controlled through crop rotation, resistant varieties and other mitigation efforts. But still, in southern Europe high value crops such as fruit and grapes are regularly treated with 'organic' fungicides such as copper sulphate. Under tropical conditions with high risks of pests and poor soils in large areas it may seem impossible to grow organic food without great losses in yields per hectare. Critics of the promotion of OF in tropical countries have claimed that the sudden and rapid outbreaks of crop diseases and pest infestations are too high a risk for growing crops without the possibility for using chemical control. According to evidence presented in Chapter 6 this seems not to be a significant problem, however, and even in cotton, where normally large quantities of pesticides are used, organic production has been practised, for example, in Uganda for more than 10 years. The recent IFAD study (Giovannucci, 2005) concludes the same from India and China. A number of methods have been developed with the aim to reduce the vulnerability of the smallholder farming systems to pests and climatic variation especially by increasing the number of crop varieties and using cash crops. These are described in detail elsewhere (Altieri, 1995; Scialabba and Hattam, 2002; Pretty *et al.*, 2003). A number of case studies have shown reduced levels of pests in organic crops (Altieri *et al.*, 1998). Likewise, agro-ecological methods have been developed with the aim of improving poor soils under a variety of organic farming systems (including LEISA), as discussed in Chapter 6. The irregularity of rainfall is an important factor for yields and food security in arid and semi-arid regions. A few case-studies suggest that increased SOM in combination with a high crop diversity in organic farms reduce the negative consequences of erratic rainfall (see Chapter 8).

Consequences of conversion to organic farming depend on the agro-ecological conditions, especially for staple food crops. The two key aspects for our purpose are: (i) the relative yield of organic vs. conventional with fertilizer use; and (ii) the yield growth over time. Under conditions of sufficient water, yields are to a large extent determined by the amount of plant available nutrients in the soil from shortly after germination and to when the reproductive phase starts. For many crops this is a fairly short period, which the farmer can influence in different ways via management.

A conversion to organic farming under *fertile* (good agro-ecological conditions and high fertilizer use) conditions will lead to a reduction in yield because the amounts of plant-available nutrients will mostly be lower in the short period of maximum plant needs. This is because nutrients in the organic system often are carbon-borne and thus must be released via mineralization. This agrees with situations in Denmark (Halberg and Kristensen, 1997), Indonesia (Martawijaya, 2004) and Bangladesh (Hossain *et al.*, 2002; Rasul and Thapa, 2004). Halberg and Kristensen (1997) found that *yield differences* between organic and conventional crops in Denmark were higher the better the agro-ecological conditions were. On sandy loam soils with good climatic conditions and on irrigated sandy soils, that is, the best potential crop yields in Denmark where nutrient supply is not limiting, the effect of lower nutrient supply and probably also the reduced control of pests and weeds in organic systems resulted in larger differences between organic and conventional crop yields than under less favourable growth conditions. The potential crop growth also explained interactions between growth season and yield differences (organic vs. conventional), which have also been found in other studies (Lockeretz *et al.*, 1981; Stanhill, 1990). The yield difference on the poorest Danish soils was relatively smaller. In a similar way, under relatively *infertile* conditions the conversion may lead to a gain in yield if the amounts of plant-available nutrients can be improved due to additions of organic materials which enhance the biological soil activity and thus the flux of nutrients that the crops potentially can tap. This agrees with situations in Western Africa (Pieri, 1992; Vanlauwe *et al.*, 2001; Yamoah *et al.*, 2002) where the use of crop residues or green manure often enhances the effects of fertilizers (Figure 10.3).

These synergies are often also observed after conversion to organic farming as farmers change their practices such as returning the residues to the soil and an increased use of green manures (Carperter, 2003). This illustrates that there may not always be one single cause–effect relation explaining the consequences on yield following conversion to organic farming. This is also well-known from Danish conditions as the use of cover crops for example has a surprisingly strong contributing effect on soil organic matter according to a simulation study (Foereid and Høgh-Jensen, 2004).

Regarding the variability of yields, a huge proportion of the production in the developing world takes place under marginal conditions, i.e. conditions with erratic and unreliable rainfall. Such conditions interfere with the classical dose–response relations and therefore the yields of organic farming cannot be deduced simply from fertilizer response curves. The variability in crop yields may restrain farmers from using fertilizers, and this can be assessed using the so-called 'adaptability analysis approach' (Hildebrand and Russell, 1996). An analysis of maize yield response to fertilizer in 20 farmers' fields in Malawi demonstrates that nitrogen is a limiting factor (Figure 10.4). But the yield response differed with some unspecific crop growth potential denoted 'Environmental index'. This index is calculated for each site from the mean yields over all treatments and is assumed to express the difference in overall growth conditions analogous to 'agro-ecological conditions'. Thus, Figure 10.4 illustrates that on some farmers' fields with low environmental index the yield increase from fertilizer application is only marginal (<0.5 t/ha) while in others it is large (situations with higher 'Environmental index').

Based on a quick view of Figure 10.4, it seems that most farmers would want to apply fertilizer under fertile conditions but that under low fertile conditions, the investment will not pay. A continuation of this example into a profitability analysis showed that the farmers not using fertilizers are making a profit across all environments whereas the use of fertilizers only pays off in the most fertile environments and only in the 20% most fertilised conditions gives a profit better that the unfertilized (data not shown).

Moreover, a risk assessment shows that the use of fertilizers increase the fluctuations in yield in absolute terms. If the farmer, for example, accepts a risk of 15% for a lower yield then this involves a decrease of 1 t/ha for the unfertilized compared to 2 t/ha for the fertilized crops (data not shown). Whether this is an acceptable risk depends on the socio-economic conditions especially the farmers' ability to invest cash (see later). The risk associated with investment in Natural Resource Management (NRM; e.g. improving nutrient recycling and use of agro-forestry) was recently confirmed in a much larger study by CIMMYT in Malawi, that concluded that only the most fertile soils would ensure a yield response that could pay for the investment of fertilizer (Benson, 1998).

It should be noted that the effect of agro-ecological conditions on the yield difference between OF and conventional practices is interacting with the characteristics of existing agricultural systems in the same areas. If farmers are not taking full profit from good rainfall or soils because of poor planting techniques and unclean seed and if fertilizer is not available at reasonable prices, the consequence of promoting organic farming methods in high potential areas may be even a yield increase.

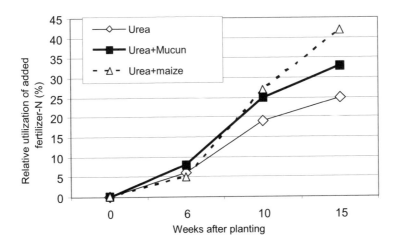

Figure 10.3. Relative utilization of fertilizer nitrogen by crops in combination with organic residues (adapted from Vanlauwe *et al.*, 2001). (Permission to reprint obtained from SSSA.)

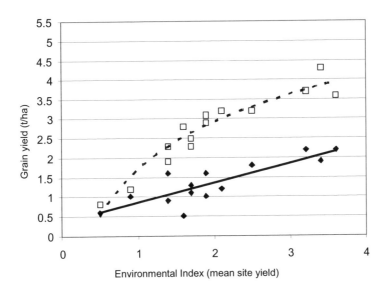

Figure 10.4. Maize grain yield of landrace varieties across 20 farmers' fields under unfertilized (full line) or fertilized (dotted line) conditions (adapted from Hildebrand and Russell, 1996).

Here the point is that the potential benefit and the practical and economic possibilities for implementing organic farming methods depend on a number of agro-ecological conditions such as soil types, soil fertility, climate (including rainfall patterns), and household specific constraints on factors of production such as labour and human capacity. The proper choice and implementation of organic farming methods depend on local circumstances and smallholder farmers should be more directly involved in selecting and adapting such methods (Uphoff, 2002). Therefore, assessments of large-scale uptake of non-certified organic farming based on LEISA and other agro-ecological methods cannot simply be based on a specific model for organic farming. Rather, ideally what is to be assessed is the scaling up of a development principle, of locally developed or adapted agro-ecological methods, which proves advantageous for smallholder farmers.

In conclusion, there is the potential that widespread adoption of agro-ecological farming methods in areas with low-potential agro-ecological conditions might actually increase both yields and the stability of yields.

The influence of socio-economic conditions

As shown in Figure 10.2 a number of socio-economic issues will impact the feasibility of introducing agro-ecological methods in any given community of smallholder farmers. These factors include land tenure, risks and credit, labour availability and human capacity in the form of education and skills. The willingness to invest in soil fertility management and agro-forestry techniques may depend on farmers' confidence in specific land tenure systems (Abalu and Hassan, 1998), which has been discussed in more detail in Chapter 8.

Besides the agro-ecological conditions mentioned above, changes in yields or occurring yield manifestations are often based in social or socio-economic conditions as Egelyng (2000) illustrates with several examples. Access to and use of irrigation water is one such social cause (Egelyng, 2000; Rockström and Falkenmark, 2000; Falkenmark, 2004). Access and use of commercial fertilizers is another. Subsidies of fertilizers and specific crops are further influencing the land use pattern and thus the possibility for incorporating leguminous crops into the system. The priorities in breeding investment can give further disadvantages to specific crops and to crop diversity. Thus, some of the principles on which organic farming is based, such as leguminous crops and crop diversity, can be disadvantaged in reality.

In the previous section it was demonstrated that under marginal conditions, evaluation of the consequences of conversion to organic farming must consider how organic practices influence the risk associated with particular investments. The principal investment will most often focus on labour. Labour is becoming a critical factor in some of the countries that are acutely suffering the HIV/AIDS epidemic. One of the worst hit regions is sub-Saharan Africa where an estimated

7.5% of the population is HIV-infected (www.unaids.org), many of whom are women. At the same time, there are an increasing number of female headed households. The burden on members of such households is high as they face serious trade-offs between labour investments into agriculture, education and off-farm employment. The use of some organic methods such as green manures is labour demanding to a degree, which may be an obstacle for adoption without development of labour-saving technologies. These aspects are currently investigated in a study carried out in Uganda (Bajunirwe *et al.*, 2004).

Gender is an important element to consider in relation to organic farming, not only when considering the labour inputs available in the household in relation to the technologies, but also in relation to knowledge. Extension systems often target males (Gilbert *et al.*, 2002), although many farmers are female and many households are female-headed. The main reason for women being less productive than men in relation to maize yields appears to be that women have less access to chemical inputs and technical know-how than men, and it is stressed that when women have equal access to such inputs their productivity matches that of men (Gilbert *et al.*, 2002; Gladwin, 2002). Gladwin (2002) therefore recommends that women's access to chemical inputs, among other assets, must be improved.

However, although this is true in a general sense, the use of purchased inputs may not always be a good livelihood strategy. Furthermore, it is important to emphasize results showing that women's productivity in relation to other crops such as pigeonpea in some cases can be even higher than that of men without access to special inputs (Odgaard *et al.*, 2003). In an agricultural setting where institutions support crop diversification, women farmers may not at all be disadvantaged.

Integrated approaches have been developed to a limited extent and appear to be well-received by farmers (Snapp and Silim, 2002; Snapp *et al.*, 2002; Odgaard *et al.*, 2003). Such approaches often focus on intercropping with complementary crops, whether in resource use or functionality (see Zethner *et al.*, 2003). In other cases, apparently productive technologies, like the Rice Intensification System, spread without a proper scientific understanding of the underlying mechanisms which caused disputes among agronomists (Surridge, 2004) regarding the complementarity between soil fertility and water management. Such cases underline that science does not yet have all the answers for dealing with complex biological systems.

Organic farming building on agro-ecological methods is often considered *knowledge intensive* as opposed to input intensive. While it is certainly a sympathetic idea to develop productive agricultural systems that build on improving the management of local resources, recycling of nutrients, and enhancing biological control through deliberate use of diversity in crop species, this ideal of maximizing the systems interactions may also be its own enemy. Only 5% of the projects reviewed by Pretty *et al.* (2003) have focused on the introduction of new elements into the farming system. The idea of integrated farming systems has been promoted in many developing countries over the last two decades but with quite different success in terms of uptake and persistence. An important factor for this may be the educational level and capacity of the involved farmers. Some ideas of farming systems are knowledge intensive in the sense that farmers need a variety of skills for successfully implementing and sustaining the integrated systems. Technical skills include both knowledge in when and how to plant crops and feed animals and observational skills linked to continuous decision making. New participatory learning methods linked to extension intend to improve on this aspect.

Many poor and traditional farmers have limited school education and limited experience with modern farming methods. Therefore, the idea of involving them in a process of developing locally adapted agro-ecological methods is at best a long and ambitious scheme that should build on patience and a longer time-frame than is available to many development projects. The risk is that the farmers will not be able to manage properly the diverse and knowledge-intensive farming systems and will give up when something is not going as predicted by the experts. Several years of promoting integrated farming systems among poor farmers in Cambodia through a number of NGOs seems to demonstrate the limitations of farmers' management skills if leaving them to fend for themselves too early (Nou, pers. comm.).

However, it is important not to assume *a priori* that new knowledge related to a certain technology cannot be taken up rapidly among farmers. The expansion of the quite knowledge-intensive coffee crop within two decades after the Second World War in East Africa is one example (Knight, 1974), as is the rapid adaptation of export-oriented production of fruits and vegetables among small-scale growers in highland central Guatemala (Hamilton and Fischer, 2003). Several papers in Uphoff (2002) also describe successful developments of diverse agro-ecological faming systems among smallholders. In general, a primary condition for success with introduction of agro-ecological methods is to involve farmers in the selection and adaptation of the relevant techniques (Onduru *et al.*, 2002; Uphoff, 2002; Halberg and Larsen, 2003).

Any investment in inputs, such as fertilizer, improved seeds, planting material, fish ponds, and the like that demand cash payments will compete strongly with the smallholder families' other cash needs for school fees and medical treatment among others. This is one of the rationales for introducing LEISA and other low input systems. But even non-certified organic systems introducing new crops or tools should consider wisely the need for well-designed rural credit systems (Zeller and Meyer, 2002).

Different types of organic farming systems' impact on food security

The different characteristics of OF and LEISA systems promoted in developing countries have different implications for the food produced and the way these systems support food security. As described in Chapter 6 some projects promoting certified OF focus on a single product or crop for the global market and do not address the whole farming system as such. Therefore, the participating farmers do not necessarily improve or convert the rest of their farms and the benefit in terms of food security is due to the higher external income from the certified crop. The implications of this for the distribution of assets and increased food security within the family may depend on the type of products sold and the degree to which the women in the households get part of the revenue.

For the purpose of discussing organic farming in relation to food security we assume that the distinction between certified and non-certified organic farming equals their market orientation, because the only reason for certification is to obtain a price premium. With this distinction it may be anticipated that the price premium for organic products and thus the motivation for certified production is mostly possible for farmers relatively close to good infrastructure that will allow export of the products. This follows empirical evidence from adoption patterns of other types of improved farming methods, such as milk-production based on concentrate use (Staal *et al.*, 2002) and high input crop production (see in Chapter 5). On the other hand, a number of examples of organic products exist where the involved farmers actually are situated quite a distance from key trading centres. For example, the large cotton project in Uganda described in Chapter 6 and the dried fruits project described in Chapter 1 are located in the provinces rather some distances from the air shipment facilities. This is possible when the products are not fresh but storable.

In other projects, such as the ones promoted under the broad terms 'Agro-ecology' (Altieri *et al.*, 1998), 'LEISA' or 'sustainable agriculture' (Pretty *et al.*, 2003), the focus is on diversification and interactions between a number of crop and livestock elements on smallholder farms. The basic hypothesis and aim of such projects is to increase overall productivity and resilience in the whole farming system while avoiding the economic risks of using purchased inputs.

Based on their review of around 200 projects promoting sustainable agriculture that include different agro-ecological methods, Pretty *et al.* (2003) distinguish a number of major types of interventions including intensification of a single component of the farm, introduction of new enterprises (crops, fish in rice fields, legumes as intercrops) or improved use of water and land to increase crop production. The majority of projects have focused on the yield improvement of staple cereals by introducing better seeds/varieties, integrated pest management methods (IPM) or using legumes. On average these interventions have doubled the yields per hectare for the involved farmers with the highest relative increases for the poorest farmers with the lowest yields.

Projects focusing on improved use of water and water harvesting (15% of projects reviewed covering 25% of farmers) were also successful in terms of increasing yields sustainable for poor farmers. While these types of non-certified organic farming may improve the food security in the particular farms, efficiently supporting a self-sufficient livelihood, these projects are often not market oriented. Pretty *et al.* (2003, p. 223) state:

> Although these initiatives are reporting significant increases in food production, some as yield improvements, and some as increases in cropping intensity or diversity of produce, few are reporting surpluses of food being sold to local markets.

It is thus an open question whether these types of farming systems' improvements can contribute significantly to large-scale food supply for rural non-farmers and urban populations.

As discussed above, certified organic farming usually improves the income of farmers and thus the asset building in terms of capital. Even though this may not increase food production on the farm itself, the food security of the family will improve due to the increased purchasing power (see also Chapter 6). This is not the case for non-certified organic systems, where increased food security should come from a higher and more resilient yield from the crops and only to a limited extent a higher output for sale.

The effects of introducing OF methods on food production, security, and access for the poor will depend on a complex interaction between agro-ecological and socio-economic conditions, the characteristics of existing farming systems, and the type of OF developed. As discussed above this means that large-scale conversion to OF in high yielding regions will reduce yields per hectare by roughly 20% on a crop level and possibly more on a crop rotation level. Some of this yield loss may be compensated for by counteracting the present process of externalization of agriculture in large parts of Europe due to economic conditions and the de-coupling of EU agricultural support. But there will most likely be an effect on world food prices if such a general yield decrease in the developed world were to happen.

Contrary to this, in large regions of the developing world where the majority of the food insecure lives, the introduction of agro-ecological methods as described above would be able to raise average yields and thus food-security in many regions (see Chapter 8 for a discussion of specific problems with P-deficient soils). For five-member households with around 1.5 ha, data show that it is possible to secure food for the year relying on non-certified organic farming but the yields of course depend on rainfall patterns and soil types. On most such subsistence farms little will be left for marketing. Therefore, it is still a question what the implications of this would be for the landless, or nearly so, and the rising populations of urban poor in developing countries. If the smallholder farmers are willing and able to hire more labour for such tasks as management of crops or green manure, this may benefit the poorest.

It should be noted that a systematic introduction of locally adapted agro-ecological methods would not be an obstacle for the later use of more input-intensive methods in the future if this proved to be advantageous. Finally, the most serious question related to large-scale conversion to organic farming is the effect on global food prices and the impact on food security.

Modelling consequences of large scale conversion to OF for food security

The discussion above on factors important for understanding the effect of OF on yields and food security show that there is no simple answer to the question of the consequences of promoting OF for food supply in a regional and global context. There are a lot of apparent success stories of the positive effects of OF and LEISA at a local scale and for the smallholder farmers involved in these studies. As described in Chapter 6, the positive effects can be on the income level due to certified export crops or on household food security due to improved yields primarily for self-sufficiency. Outcomes for national, regional and global food supply and, particularly, for food consumers, who increasingly reside in urban areas, are less certain. The question of the sustainability and socio-economic consequences of a large-scale promotion of LEISA and OF methods should be addressed based on policy simulations based on a global and regional food supply, demand and trade framework, such as IFPRI's IMPACT model.

Organic farming can provide a wide range of ecosystem services. They include enhanced organic matter content of the soil, reduced water pollution, and enhanced biodiversity. As discussed above, the actual relative yield level of organic farming varies between regions and farming systems: in high input farming systems of Europe and North America OF in general gives lower yields. Contrary to this, there is evidence of positive effect on yields when shifting low-input farming systems in developing countries to organic systems using agro-ecological methods such as enhanced crop diversity and use of organic fertilizer and green manures. In many developing countries, especially in Africa but also in parts of Latin America and Asia, OF is increasingly promoted in attempts to improve food security for resource-poor farmers who in practice cannot access fertilizer and pesticides. This should be taken into account when modelling the consequences of large-scale uptake of OF methods. The results of large scale OF in Asia would probably be a combination of yield reductions in high-yielding areas and smaller yield changes in semi-arid areas and on land where high dependence on chemical fertilizer cannot sustain yields in the long run. Given the significant links between food production and prices on the global market there is a risk that conversion to OF in areas with high yields will put pressure on the global food market and increase prices and availability of food for poor people in developing countries.

Our working hypotheses for this modelling exercise were therefore: (i) large scale conversion to organic farming in high input regions will have important negative impacts on food security in poor countries due to increasing global food prices; and (ii) large scale conversion in areas dominated by low input agriculture will increase local food security

The first set of scenarios will test the effect of conversion to OF in SSA and the sensitivity of different assumptions regarding yields and yield growth rates. Another set of scenarios will look at the effect of converting high intensity areas to OF on global food prices and subsequently the indirect effect on food security in resource poor areas as described in the following. A critical distinction to highlight between the two sets of scenarios is that the OF practices in SSA are not focused on certified organic production as in the high intensity areas of Europe and North America. The low-input scenarios include the potential to employ more general agro-ecological approaches that make limited and prudent use of input, which may not be allowed in certified OF.

The IMPACT model and methodology for predicting food security

IFPRI's IMPACT model offers a methodology for analysing baseline and alternative scenarios for global food demand, supply, trade, income and population. IMPACT covers 36 countries and regions and 33 commodity groups, including all cereals, soybeans, roots and tubers, meats, milk, eggs, oils, oilcakes, meals, vegetables, fruits, sugars and fish. IMPACT is a representation of a competitive world agricultural market for crops and livestock. In addition, this study employed an enhanced version of the basic model, IMPACT-WATER, which includes components that embed the core model in the context of water demand and availability. It is specified as a set of country or regional sub-models, within each of which supply, demand and prices for agricultural commodities are determined. The country and regional agricultural sub-models are linked through trade, a specification that highlights the inter-dependence of countries and commodities in the global agricultural markets. The model uses a system of supply and demand elasticities incorporated into a series of linear and nonlinear equations, to approximate the underlying production and demand functions. World agricultural commodity prices are determined annually at levels that clear international markets. Demand is a function of prices, income and population growth. Growth in crop production in each country is determined by crop prices and the rate of productivity growth. Future productivity growth is estimated by its component sources, including crop management research, conventional plant breeding, wide-crossing and hybridization breeding, and advances in biotechnology. Other sources of growth considered include private sector agricultural research and development, agricultural extension and education, markets, infrastructure and irrigation.

A wide range of factors with potentially significant impacts on future developments in the world food situation can be modelled based on IMPACT. They include: population and income growth, the rate of growth in crop and livestock yield and production, feed ratios for livestock, agricultural research, irrigation, and other investments, price policies for commodities, and elasticities of supply and demand. For any specification of these underlying factors, IMPACT generates projections for crop area and livestock numbers, yield, production, demand for food, feed and other uses, prices and trade. A base year of 1995 (a 3-year average of 1994–97) is used and projections are made to the year 2025. A presentation of the so-called baseline scenario, which is the projected consequences for world food production and food security if present policies continue, is present above. In the following section the alternative scenarios are compared with the baseline

Definition of the scenarios for large scale conversion to organic farming

In order to test the potential effect of OF on food security, two main scenarios were established that modelled large-scale conversion in selected regions dominated by high input and low input agricultural systems, respectively.

Conversion of high input regions

For the high input conversion scenario four of the IMPACT regions were selected to cover most of Europe and North America. As discussed, the relative yields of organic cereal crops in high input areas will depend on the supply of nutrients from either green manure or animal manure. Therefore, the most efficient organic crop production is integrated with ruminant production, which may use grass–clover pastures and other types of green manure. The scenario for high input regions was therefore defined as the agricultural area that may be assumed integrated with dairy and other ruminant production. Andersen *et al.* (2005) established a typology of agricultural sectors for all the farms in the EU and defined:

> grazing livestock farms as farms with more than 50% of the value of production from cattle, sheep or goat products. Together these farms account for 27% of all farms in EU-15, they hold 44% of the Utilised Agricultural Area (UAA) and 84% of the livestock units (LU) of grazing livestock.

From this it cannot be assumed that the grazing livestock farms also hold 44% of the cereal land; probably less, because they have more grassland in their rotations and some sub-types have in fact almost only grasslands. However, a scenario for large scale conversion of ruminant systems would imply that dairy and 50% of beef production over time would be better integrated in cereal and other cash crop rotations in order to better utilize the soil fertility building effect of grasslands in crop rotation. This was a key assumption in the Bichel scenarios

(Bichel Committee, 1999) and is also tested in organic demonstration farms in different European countries.

Table 10.1 shows a summary of the assumptions concerning percentage of area converted, and the proportional yields in the first years after full conversion. The yields were estimated as a percentage of the yields used in the IMPACT baseline scenario based on a review of organic and conventional yield comparisons and the authors' own assessment based on the evidence presented above. Thus, while De Ponti and Pinstrup-Andersen (2005) found that organic crops yield on average 80% of conventional, we find that this is most realistic in controlled experiments, which may to some extent favour the organic crops and possibly leave out total crop failures due to excessive pests or weeds. In order to get a more conservative estimate we used the lower relative yield (65%) based on statistical information from Danish organic farms and previous data from longitudinal studies on 30 farms (Halberg and Kristensen, 1997).

Table 10.1. Definition of scenarios for large scale conversion to OF in high input regions[1].

Yield growth rates	Conservative	Optimistic
% crop area converted[2]	40–60	40–60
OF Yields, percent of conventional[3]	60–100	60–100
Relative yield growth rate, OF vs. Baseline,%[4]	100	125–200
% OF Livestock, ruminants		
Dairy cattle	50–100	50–100
Beef cattle	50–100	50–100
Sheep and goats	50–100	50–100
% OF Livestock, non-ruminants	0	0

Conventional = Intensive farming

1) Countries: Europe, the USA, Canada were chosen as examples. Not intended to cover all high input regions.
2) Depending on crop types and regions, with 40% maize and cereal area in the USA and Western Europe.
3) Depending on crop types and regions e.g. cereals, maize 65% except 100% in Eastern Europe, potatoes 60%, milk yield 80–100%.
4) Values 125, 150, 200 tested in different scenarios. Baseline corresponds to the default parameterization of IMPACT, see text for explanation.

The relative yield is a crucial assumption with great importance for the results. In order to test the effect of relative crop yields (organic vs. conventional), the assumed relative growth rates in organic crop yields compared with the yield growth rates in the IMPACT baseline scenario varied between a factor of one and two. For example, in the 125% scenario, it was assumed that organic yields per ha improve 25% faster than conventional yields over the first 15 years due to intensified research, specific crop breeding for organic varieties and improved farm management including fertility-building crop rotations. The objective was to assess which level of relative growth rate was necessary to offset any negative impact on world food prices of large-scale conversion; thus any evaluation of the likeliness of these growth rates is left to others.

We focused on the results in terms of world prices and as indicators of how the conversion impacts food security, we used the parameters demand, and availability of selected food commodities and childhood malnutrition in SSA. This region has the larger part of food-insecure people and is the region most difficult to improve food production per capita (Runge *et al.*, 2003). SSA is used as a critical case to test the effect of changing world prices of key food commodities on the regional food security.

Conversion of Low Input areas

For the low input scenario the sub-Saharan African region (SSA) was selected (comprising all African countries in eastern, south-eastern, southern, south-western and West Africa south of Sahara) because the majority of food insecure will be localized there according to the IMPACT baseline scenario and because this region has experienced low growth in yields and use of fertilizer over the last four decades (Chapter 1). Table 10.2 shows the assumptions behind the scenarios concerning relative yields and yield growth rates, which were based on the available literature as reviewed in the section above. It was assumed that 50% of the areas in SSA would be suitable for conversion into organic farming systems. Since valid comparative data on yields in organic vs. low input conventional systems in Africa are scarce, the study aimed at testing the sensitivity of different assumptions concerning yields and growth in yields. In a 'pessimistic' scenario the organic yields are anticipated to be following the same yield increase as in the baseline scenario from a starting point with only moderately higher yields in OF. In the more optimistic scenarios, it was assumed that research into agro-ecological farming methods and suitable plant breeding could increase yields at a higher rate than in the baseline scenario in the first 15 years and levelling off thereafter.

In both sets of scenarios it was assumed that price elasticities were identical for conventional and organic commodities, i.e. no price premiums were paid for the organic commodities. Thus the question of whether to assume conversion to certified or non-certified organic production is not important for these scenarios.

Table 10.2. Definition of scenarios for large scale conversion to OF in low input regions.[1]

Yield growth rates	Conservative	Optimistic
% crop area converted	50	50
OF Yields, percent of conventional[2]	80–120	80–120
Relative yield growth rate, OF vs. Baseline[3],%	100	125
% OF Livestock, ruminants		
Dairy cattle	50	50
Beef cattle	50	50
Sheep and goats	50	50
% OF Livestock, non-ruminants		
Pigs	0	0
Poultry only	50	50

Conventional = low input/traditional farming

1) Countries: sub-Saharan Africa, including Nigeria.
2) Depending on crop types: wheat, potatoes 80%; maize, other coarse grains, roots and tubers 120%.
3) Baseline corresponds to the default parameterization of IMPACT, see text for explanation.

Many examples of successfully improving yields in local farming systems after conversion to organic practices build on diversification and use of intercropping as explained previously. However, the IMPACT model is developed in order to simulate food production, trade and security on regional and global level and only operates with major enterprises/commodities on a country or regional level. The model is, therefore, not suitable for investigating differences in farming systems at farm or local level and the introduction of increased diversity and mixed crops on farming systems level have been simulated only roughly by assuming that these effects translate into the relative yield levels.

Combined effect of the conversion in high and low input areas together

The two scenario types were then combined in order to test if the negative impact from yield reductions due to conversion in intensive areas could be counterbalanced by the positive effects of OF in low-input areas – and to what degree given the assumptions of conversion rates and yield developments. Rather than postulating a certain relative yield and yield growth rates, it was attempted to find the OF yield growth rates, which would reduce any negative global impact on food security to near zero. These figures could then feed into a discussion of the probability of attaining such growth rates in OF.

Results of the scenario modelling

Conversion of high input regions

The total production in the high intensity areas is reduced by around 10% for most crops if the yield growth rate is assumed to be equal in the organic scenario and the baseline scenario (Table 10.3). If the yields of organic crops grow 50% higher than the conventional the difference in overall production in 2020 between the organic and the baseline scenario is less than 10%. This reduction is a combination of an almost identical level of yield decrease in North America and EU, while the production is expected to increase in former Eastern Europe compared with the baseline scenario.

Table 10.3. Relative production under large-scale conversion to organic farming in Europe and North America and resulting world prices, and food demand in sub-Saharan Africa modelled for the year 2020. Results presented as percentage of projected results of IFPRI's baseline scenario.

Scenario	Intensive	Intensive	Intensive
Relative Yield Growth Rate (% of baseline)	100	150	200
Production in Europe and North America			
Wheat	92	95	97
Maize	90	92	94
Other coarse grain	92	95	97
Soybean	87	89	92
World prices			
Wheat	111	107	103
Maize	112	109	106
Other coarse grain	113	109	105
Sweet potato and yam	114	110	106
Cassava	109	106	103
Soybean	108	106	104
Food demand in SSA			
Wheat	94	96	98
Maize	97	97	98
Other coarse grain	96	97	98
Sweet potato and yam	100	100	100
Cassava	101	101	100
Soybean	95	95	96
Food security in SSA			
Food availability (Kcal/capita)	98	99	99
Total malnourished children	101	101	101

The lower production translates into higher world prices in the major crop commodities with the highest relative prices when yield growth rates are assumed identical in organic and conventional crops. In the scenario with 50% higher yield growth rate in organic crops, the increases in world prices are projected to be between 6 and 10% for major commodities. The relative higher world prices reduce the demand for maize and other cereals in SSA by 3–6% and soybean by 5% while other commodities are not influenced. Production in SSA of these commodities increases slightly, by 1 to 2%, in response to these changes. The overall food security situation in terms of food availability and number of malnourished children in SSA worsens only slightly with large scale conversion to OF in Europe and North America.

In Figure 10.5, the impact of the various scenarios for large-scale conversion to OF in Europe and North America on the development trends in net trade of major commodities in SSA is presented. The negative values mean that there will be a net import to the region of these crops. The baseline scenario projects a 13% increase in the amount of soybeans imported to SSA towards the year 2020 and 17% for 'other coarse grain'. Compared with the baseline scenario, the projected increases in import patterns were essentially dampened across the board for all commodities in the OF scenarios with the largest effects in the 100% scenario where yield differences were largest. This is a result of a combination of decreased production in high input areas (Europe and North America) and increases in world prices, which reduces the import demand in SSA. Also, though the decrease in food availability on the world market actually stimulates a slight increase in food production in SSA of 1 to 3% in the major commodities and roots and tubers, the concurrent shift in food security (Table 10.3) gives an indication that the global food production system would likely be experiencing a fair amount of dynamic shifts on a global scale. Though the scenarios modelled here result in a reduction in 2020 imports when compared to the baseline projections, the scenario import levels would be twice to three times 1995 levels for maize and wheat and around six to eight times 1995 levels for soybeans. Net trade in 'other coarse grain' is the most impacted by the changes in OF scenarios. The absolute magnitudes of both 'other coarse grain' and soybean, however, are actually quite small compared with the other major grains. Also, net trade is the result of the large market forces of supply and demand, so relatively small shifts in either can show up as large shifts in net trade.

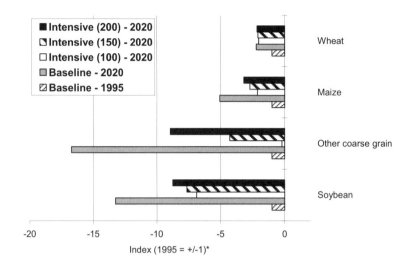

Figure 10.5. Changes in net trade for sub-Saharan Africa under three scenarios of large-scale conversion to organic farming in Europe and North America for the year 2020 compared with IFPRI's baseline projections (see text for scenario explanations). Results indexed to 1995 in IFPRI's baseline scenario (1995 = +/-1; *less than zero means a net import.

Conversion of low input regions

The scenarios focusing on conversion of half of SSA to OF resulted in a rather different picture for SSA in 2020 (see Table 10.4). Production in SSA declines for soybean and wheat by about 5 and 10%, respectively, while maize, other coarse grains, and roots and tubers see an increase somewhere between 4 and 8% above the baseline projections. The positive yield development had little effect on world prices – except on roots and tubers – due to the relatively small size of the total grain production in SSA compared with world markets. Food consumption (demand) in 2020 remains essentially equal to that of the baseline projections in SSA, which lead to a similar result in terms of food security in the region.

As shown in Figure 10.6, the large-scale conversion of SSA to OF resulted in a lower projected import compared with the baseline, which projects a fivefold increase in the maize imports into SSA from 1995 to 2020. In the organic scenarios only a two- to threefold increase is projected in maize import. This is due to the combination of assumptions of increased production along with increased or equal yield growth rates over that time period. Opposed to this, imports of soybean in 2020 dramatically increase in these scenarios compared with the baseline projections while wheat imports essentially match those in the baseline projections. Other coarse grains are again dramatically affected by the changes in modelled scenarios, with SSA actually becoming a net exporter in the most optimistic scenario. As mentioned above in the scenarios of OF conversion in Europe and North America, the absolute magnitudes of soybeans and other coarse grains are relatively small compared to the major grains, and net trade numbers belie the insignificance of some of the these shifts.

Table 10.4. Relative production under large-scale conversion to organic farming in sub-Saharan Africa (SSA) and resulting world prices, and food demand in SSA modelled for the year 2020. Results presented as percentage of projected results of IFPRI's baseline scenario, average over 3 years.

Scenario	Low Input	Low Input
Relative yield growth rate (% of baseline)	100	125
Production in SSA		
Wheat	89	92
Maize	105	108
Other coarse grain	106	109
Sweet potato and yam	104	107
Cassava	104	105
Soybean	95	98
World prices		
Wheat	100	100
Maize	99	98
Other coarse grain	98	96
Sweet potato and yam	96	94
Cassava	92	89
Soybean	100	100
Food demand in SSA		
Wheat	100	100
Maize	100	100
Other coarse grain	101	101
Sweet potato and yam	100	100
Cassava	100	100
Soybean	100	100
Food security in SSA		
Food availability (Kcal/capita)	100	100
Total malnourished children	100	100

Combined effect of conversion in high and low input regions

The two branches of scenarios – OF conversion in high input areas of Europe and North America and low input areas of SSA – were combined into a joint modelling of large-scale conversion in both regions. As shown in Figure 10.7, the combined scenario with baseline or 150% yield growth rates increased world prices on major cereals and soybean between 4 and 10% in 2020. The combination of the two scenarios had impacts on food security that were dampened slightly compared to the Europe and North America only scenario. The effect on number of malnourished children in SSA was a slight increase between 0.50 and 1.25% by 2020 (Figure 10.8), while the developing world as a whole was essentially the same on balance.

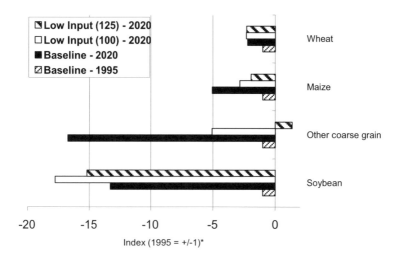

Figure 10.6. Changes in net trade for sub-Saharan Africa under large-scale conversion to organic farming in sub-Saharan Africa for the year 2020 compared with IFPRI's baseline projections (see text for scenario explanations). Results indexed to 1995 in IFPRI's baseline scenario (1995 = +/-1; *less than zero means a net importer while greater than zero means net exporter).

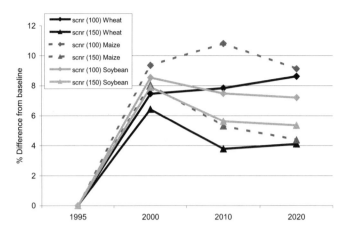

Figure 10.7. Projected changes in world prices on major cereals and soybeans from IFPRI baseline scenario over 25 years after large-scale conversion to organic farming in Europe, North America and sub-Saharan Africa. Relative crop yield growth identical to baseline (=100) or 50% higher for initial period (=150).

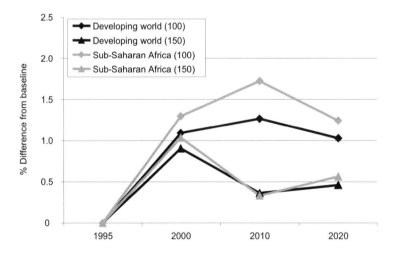

Figure 10.8. Projected changes in number of malnourished children in sub-Saharan Africa and the developing world as a whole relative to the IFPRI baseline scenario after the combined large-scale conversion to organic farming in Europe, North America, and sub-Saharan Africa. Relative crop yield growth identical to baseline (=100) or 50% higher for initial period (=150).

Discussion

The modelling shows that large scale conversion to OF is possible without severe negative effects on global food availability compared with the baseline projection of food production over the next 2–3 decades. These negative impacts could also potentially be mitigated through a series of focused policy initiatives. There may even be a positive effect on food security from OF in SSA but the model could not detect this as other than increased possibility for exports or reduced imports to the SSA region. Thus, in order to improve the overall production and demand for food and the food security in SSA compared with the baseline projection, the partial introduction of modern organic farming in SSA would have to give yearly yield increases of at least 50% higher than what is expected in the baseline scenario.

However, as stated by Rosegrant *et al.* (2001), the global food production is not a sufficient condition alone for eliminating food insecurity, which is why increased local production in areas with food shortage can be a better solution than increasing food production for export in high input regions. According to the model, though, the most likely outcome of increased regional production of maize and other food commodities in the low input conversion scenarios is reduced import, or even increased export, while the amount consumed in the SSA regions increases only little. This result probably builds on the model's assumption that all staple food produced is subject to potential long-distance trade, which means that increased production in areas with food-insecure, resource-poor people may not automatically benefit the disadvantaged. Thus, if the purchasing power is higher in other regions with unmet demands the increased production will be directed there and may not benefit the resource poor. However, this result may be too simple if the increased production is actually happening in smallholder farms in areas with poor infrastructure and marketing channels. This may be considered the case in the low input scenarios, where the idea is to improve smallholder farms by using local resources and OF. In this case, the improved yield may in fact benefit the rural resource poor and increase food security. Since the model does not distinguish between different food markets in the poor regions, it is not able to predict these differences. This will be improved in future versions of the IMPACT model.

The effect of reducing the subsidized export of food surplus from Europe and North America to developing countries may have a beneficial effect on domestic market-oriented agricultural production in Africa, but this was not included in the model.

IMPACT cannot directly model actual crop rotations or benefits of increased crop–livestock interactions, such as manure use, or the increased soil fertility from legumes or grass–clover. These important attributes of OF have, however, been included in the modelling through the proportions of OF in the scenarios and the assumptions regarding relative yield and yield growth.

In conclusion, it does not appear large-scale conversion to OF would lead to severe negative impact on global food supply and food security in a developing region such as SSA. Focus should then be on the potential combination of environmental benefits of conversion in high input areas with the global economic benefits on reducing food surplus in the North and the potential increase in local food availability and food security by improved organic farming in the South.

The consequences for food production and security of large-scale introduction of organic farming differ between regions and between different income groups within regions. They are the result of complex relations between ecological, socio-cultural and agronomic factors and the choice of parameters to express these relations will influence the results of modelling exercises. The studies of links between food security and OF differ in the way they focus on different aspects of the issue and the assumptions in much the same way as the different interpretations of the concept of sustainability, as discussed by Douglass (1984) 20 years ago. Thompson (1996) distinguished between two main schools of sustainability in agriculture (based on Douglass, 1984), the food efficiency school and the functional integrity school. As examples of the first approach to sustainability, Penning de Vries et al. (1995) projected standard estimates of (low) organic yields on a regional and global scale and estimated food security by dividing the estimated total amount of food produced (in grain equivalents) with the food need (number of people times a standard food ration). No consideration was given to socio-economic conditions or feed-back mechanisms at the local scale such as increased food-security in traditional smallholder systems when introducing modern agro-ecological methods. The reduced risks and possibilities for improved resilience of farming systems through LEISA for smallholders were not considered. Other authors also strongly argue against organic farming because of their belief that this farming system cannot feed the world's population (Borlaug, 1994; Reason Foundation, 2000; Centre for Global Food Issues, 2002). The scenarios modelled here with IMPACT are, by and large, within the food sufficiency tradition simply because the model does not account for either differences in farming systems and their environmental impacts or for different impacts on various social groups within a region. Future model development will attempt to distinguish food security effects by poverty levels and between rural and urban populations.

Contrary to this is the functional integrity school, which focuses on the benefits at local scale of increased reliance on recycling of organic matter and nutrients, diverse crop rotations, crop mixtures, agro-forestry and use of biodiversity and biological pest-control methods to increase and sustain yields as illustrated by 'agro-ecology' (Altieri, 1995) and LEISA. This school of agricultural development most often is linked with ideas of soil fertility management that focus on the use of local resources to improve soil OM, nutrient status, water holding capacity, and other critical soil quality dimensions, thus improving the farming system's reproductive capacity and self regulatory mechanisms. This may be understood as improving the functional integrity of the farming system and thus its sustainability. Proponents of this approach to sustainability such as Pretty *et al.* (2003) focus on the examples of increased yields in organic systems and thus see no risk for food security from opposing chemical fertilizers and pesticides. The documentation for this is discussed above. However, regardless of how sustainable the LEISA approach is for subsistence farmers it is seldom described how this type of farming system is capable of producing a surplus of food to supply the increasing number of urban poor in developing countries. While it may be argued that an important part of the urban population may supplement their income and food supply from urban or peri-urban farming, a significant surplus of food needs to come from rural areas in a situation with 40–50% of developing countries populations living in cities. And it still has not been demonstrated how the majority of such large urban populations will be able to access enough food in a situation with mostly OF or LEISA type farming systems. This question should be in focus in the development of a model for detailed food security projections related to OF.

However, there is an urgent need to leave such antipoles and build a more pragmatic approach to the use of agro-ecological technologies for the improvement of poor farmers' yields and livelihoods. Moreover, our findings suggest that increased conversion to extensive agricultural production such as OF in Europe and North America could be supported with little negative impact on global food security, especially if combined with food security focused policy initiatives. Whether this is a desirable development is outside the scope of this paper. But, as recommended by Dabbert *et al.* (2004), there may be complementary advantages of supporting further development of OF in Europe such as decreasing overproduction, reduction of environmental impacts, and development of multifunctional agriculture including maintenance of semi-natural grasslands and biodiversity. In order to identify the trade-offs between costs and benefits of such a development it is recommendable to perform a regional conversion scenario analysis, much in the style of the Bichel Committee (1999).

It is important to stress that in this modelling exercise we have not attempted to make direct claims as to the exact impacts of OF on yields in any given production system. Instead, scenarios were developed to conservatively represent a range of potential outcomes presented in existing literature, thus using different assumptions for relative yield growth rates. From there, the results generated by the IMPACT model represent a fairly optimistic up-scaling of the potential of OF in these different regions. The hope was to present an outer bound of the possibilities available through the innovations in an agro-ecological approach to improving farming systems. It is left to others to judge what yield levels and yield growth rates will be realistic after large scale conversion to OF in high and low input areas.

Conclusions

1. In high yielding regions with near to economic optimal inputs of fertilizer and pesticides the yields of organic farming are between 15–35% lower than present yields when comparing single crops and possibly at the low end (35%) when including crop failures and the needs for green manure in crop rotations. The rationale for introducing certified organic farming in Europe and North America is based on demand for environmental improvements and perceptions of increased health and food security combined with higher animal welfare in OF among groups of consumers. A large scale conversion to OF in these parts of the world would reduce the total food production but it is possible that the difference will be lower than assumed from a direct comparison of crop to crop yield differences because of counteracting changes in future land use. The impact on global food prices and the food security of poor people in developing countries would be limited due to counteracting effects (among others a shift towards other staple foods and improved local food production).

2. Modern non-certified organic farming is a potentially sustainable approach to agricultural development in areas with low yields due to poor access to inputs or low yield potential because it involves lower economic risk than comparative interventions based on purchased inputs and may increase farm-level resilience against climatic fluctuations. But concepts of modern OF should be adjusted to local agro-ecological conditions, for example allowing for the use of P fertilizer where this is unavoidable. Moreover, there is a need to consider closely the potential advantages of improved varieties even when these have been bred using selected biotechnological methods.

3. There is abundant evidence that locally adapted agro-ecological methods can improve the yields and sustainability of smallholder farms and increase local food security. Organic farming, LEISA, sustainable agriculture and other farms are different flavours of the same basic idea of building on locally available resources. This includes improving soil fertility through increased use of green manures etc., and enhanced diversity in the choice of crops both specially (crop mixtures etc.) and in time (multipurpose agro-forestry etc.). Significant questions remain as to the degree to which these technologies can achieve similar success when upscaled.

4. Whether coming from the productivity and growth school that promotes agricultural intensification through increased use of fertilizers, pesticides and biotechnology or from the so-called true organic movement arguing against any use of artificial input, there is a need to critically assess the suitability of such preconceived ideas in any given context. In any case, the successful introduction and up-scaling of OF necessitates increased research and extension based on local farming systems. Ideas for research in such organic agriculture are presented in the following chapter. The results presented here provide a pioneering attempt to assess organic farming impacts within a broader global food supply and demand context. Additional research of particular relevance to the analysis in this chapter is to further assess the underlying assumptions to be used in organic farming scenarios, including additional review of evidence on organic farming productivity impacts and the use of crop models to further assess the performance in a range of environments and agro-ecological zones.

References

Abalu, G. and Hassan, R. (1998). Agricultural productivity and natural resource use in southern Africa. Food Policy 23: 477-490.

Altieri, M. (1995). Agro-ecology. The scientific basis of agriculture. Intermediate Technology Publications, London.

Altieri, M.A., Rosset, P. and Thrupp, L.A. (1998). The potential of agro-ecology to combat hunger in the developing world. 2020 Brief 55, IFPRI.

Andersen, E., Elbersen, B. and Godeschalk, F. (2005). Assessing multifunctionality of European livestock systems. In: Floor Brouwer (ed.) Sustaining agriculture and the rural economy: Governance, policy and multifunctionality. Edward Elgar, New York.

Bajunirwe, F., Massaquoi, I., Asiimwe, S., Kamya, M.R., Arts, E.J. and Whalen, C.C. (2004). Effectiveness of nevirapine and zidovudine in a pilot program for the prevention of mother-to-child transmission of HIV-1 in Uganda. African Health Sciences 4(3): 146-154.

Benson, T.D. (1998). Developing fliexible fertilizer recommendations for smallholder maize production in Malawi. In: Waddington, S.R., Murwira, H.K., Kumwenda, J.D.T., Hikwa, D. and Tagwira, F. (eds) Soil Fertility Research for Maize-based Farming Systems in Malawi and Zimbabwe. Proceedings of the Soil Fert Net Results and Planning Workshop held 7-11 July 1997 at Africa University, Mutara, Zimbabwe, 275-285. Soil Fert Net and CIMMIT-Zimbabwe.

Bichel Committee (1999). Report from the Bichel Committee - Organic Scenarios for Denmark The Bichel Committee. Report from the main committee. Danish Environmental Protection Agency, Ministry of Food, Agriculture and Fisheries, Copenhagen. 112 pp. Available online (November 2004) at http://www.mst.dk/homepage/ (look for publications database and seek 'bichel').

Borlaug, N.E. (1994). Chemical fertilizer 'essential'. International Agricultural Development (Reading, Berkshire, UK) 14(6): 23.

Byerlee, D. and Eicher, C.K. (eds) (1997). Africa's emerging maize revolution. Lynne Reiner, Boulder, Colorado.

Carperter, D. (2003). An investigation into the transition from technological to ecological rice farming among resource poor farmers from the Philippine island of Bohol. Agriculture and Human Values 20: 165-176.

Centre for Global Food Issues (2002). British Crop Group Speaker Claims Organic Farming on a Global Basis Would be an Environmental Catastrophe. Available on-line (November 2004): http://www.cgfi.org/materials/articles/2002/nov_15pr_02.htm

Dabbert, S., Häring, A.M. and Zanoli, R. (2004). Organic Farming: Policies and Prospects. Zed Books, London and New York.

de Wit, C.T., Huisman, H. and Rabbinge, R. (1987). Agriculture and its environment: Are there other ways? Agricultural Systems 23: 211-236.

Douglass, Gordon K. (1984). The meanings of agricultural sustainability. In: Agricultural Sustainability in a Changing World Order. Westview Press, Boulder, Colorado, pp. 3-29.

Egelyng, H. (2000). Managing agricultural biotechnology for sustainable development: the case of semi-arid India. International Journal of Environmental Studies 2: 342-354.

Falkenmark, M. (2004). No freshwater security without major shift in thinking. Ten-year message from the Stockholm water symposia. Stockholm Water Symposium, 32.

FAO (2003). World Agriculture: towards 2015/2030 – an FAO perspective. Earthscan, Food and Agriculture Organization of the United Nations (FAO), Rome, Italy. http://www.fao.org/DOCREP/004/Y3557E/ Y3557E00.HTM

FAO (2004). FAOSTAT database. http://faostat.fao.org, Rome.

Foereid, B. and Høgh-Jensen, H. (2004). Carbon sequestration potential of organic agriculture in northern Europe – a modelling approach. Nutrient Cycling in Agroecosystems 68: 13-24.

Gilbert, R.A., Sakala, W.D. and Benson, T.D. (2002). Gender analysis of a nationwide cropping system trial survey in Malawi. African Studies Quarterly 6. http://web.africa.ufl.edu/asq/ v6/v6i1.htm

Giovannucci, D. (2005). Organic Agriculture for Poverty Reduction in Asia: China and India focus. Thematic Evaluation. Document of theInternational Fund for Agricultural Development. Final Draft, March 2005. 80 pp. Available online (August 2005) at: http://64.233.183.104/custom?q=cache:4YYRgNz0xKkJ:www.ifad.org/evaluation/public_html/eksyst/doc/thematic/organic/organic.pdf

Gladwin, C.H. (2002). Gender and soil fertility in Africa: Introduction. African Studies Quarterly 6, http://web.africa.ufl.edu/asq/v6/v6i1.htm.

Halberg, N. and Kristensen, I.S. (1997). Expected crop yield loss when converting to organic dairy farming in Denmark. Biological Agriculture and Horticulture 14: 25-41.

Halberg, N. and Larsen, C.E.S. (2003). Participatory development research: Enhancing capacity within applied research – case: Livestock. Paper for Tune workshop: Management of research, communication and change within agricultural sector support programmes. Available online: http://www.husdyr.kvl.dk/htm/php/Tune03/Halb.doc. 14 pp.

Halberg, N., Kristensen, E.S. and Kristensen, I.S. (1995). Nitrogen Turnover on Organic and Conventional Mixed Farms. Journal of Agricultural and Environmental Ethics 8: 30-51.

Hamilton, S. and Fischer, E. (2003). Non-traditional agricultural exports in highland Guatemala: Understandings of risk and perceptions of change. Latin American Research Review 38: 82-110.

Hendy, C.R.C., Kleih, U., Crawshaw, R. and Phillips, M. (1995). Interactions between Livestock Production Systems and the Environment: Impact Domain: Concentrate Feed Demand. Working Document Livestock and the Environment: Finding a Balance. FAO/World Bank/USAID, Rome.

Hermansen, J.E., Strudsholm, K. and Horsted, K. (2004). Integration of organic animal production into land use with special reference to swine and poultry. Livestock Production Science 90: 11-26.

Hildebrand, P.E. and Russell, J.T. (1996). Adaptability Analysis: a method for the design, analysis, and interpretation of on-farm research-extension. Iowa State University Press, 189 pp.

Hossain, M.Z., Choudhury, M.H.K., Hossain, M.F. and Alam, Q.K. (2002). Effects of ecological agriculture on soil properties and arthropod diversity in rice-based cropping systems in floodplain areas in Bangladesh. Biological Agriculture and Horticulture 20, 215-227.

Howard, J., Crawford, E., Kelly, V., Demeke, M. and Jeje, J.J.U. (2003). Promoting high-input maize technologies in Africa: the Sasakawa-Global 2000 experience in Ethiopia and Mozambique. Food Policy 28: 335-348.

Huang, B.K. (1994). Computer simulation analysis of biological and agricultural systems. CRC Press, Boca Raton, 862 pp.

IFAD (2003). The Adoption of Organic Agriculture Among Small Farmers in Latin America and The Caribbean. IFAD Report 1337.

Jiggins, J., Reijntjes, J.C. and Lightfoot, C. (1996). Mobilising science and technology to get agriculture moving in Africa: A response to Borlaug and Dowswell. Development Policy Review 14, 1 March: 89-103.

Knight, C.G. (1974). Ecology and change. Rural modernization in an African community. Academic Press, Inc., New York, 300 pp.

Kristensen, T. and Mogensen, L. (1999). Organic dairy cattle production systems – feeding and feed efficiency. In: Ecological animal Husbandry in the Nordic Countries. NJF-seminar No. 303, Horsens, Denmark 16–17 September, p. 24.

Lockeretz, W., Shearer, G. and Kohnl, D.H. (1981). Organic farming in the corn belt. Science (211), 540-547.

Makowski, D., Wallach, D. and Meynard, J.-M. (2001). Statistical methods for predicting responses to applied nitrogen and calculating optimal nitrogen rates. Agronomy Journal 93: 531-539.

Martawijaya, S. (2004). Bureucrats as entrepreneurs: a case study of organic rice production in east Java. Bulletin of Indonesian Economic Studies 40: 243-252.

McHarry, J., Scott, F. and Green, J. (2002). Towards global food security: Fighting against hunger. Social Briefing no. 4. Towards EarthSummit 2002. UNEP, UN foundation, Stakeholder Forum, 19 pp.

Nou, K. (2004). Personal communication. School of Agriculture, Prek Leap, Cambodia.

Odgaard, R., Høgh-Jensen, H., Myaka, F.A., Sakala, W.D., Adu-Gyamfi, J.J., Vesterager, J.M. and Nielsen, N.E. (2003). Integrating pigeonpea in low-input maize-based farming systems: A step towards increased food production and poverty alleviation in semi-arid Africa. 2: 163-185. Proceedings from conference on 'Local Land Use Strategies in a Globalizing World: Shaping Sustainable Social and Natural Environments'. 21-8-2003.

Onduru, D.D., Diop, J.M., Van der Werf, E. and De Jager, A. (2002). Participatory on-farm comparative assessment of organic and conventional farmers' practices in Kenya. Biological Agriculture and Horticulture 19: 295-314.

Parry, M.L., Rosenzweig, C., Iglesias, A., Livermore, M. and Fischer, G. (2004). Effects of climate change on global food production under SRES emissions and socio-economic scenarios. Global Environmental Change 14: 53-67.

Penning de Vries, F.W.T., Van Keulen, H. and Rabbinge, R. (1995). Natural resources and limits of food production in 2040. In: Bouma, J. *et al.* (eds) Eco-regional approaches for sustainable land use and food production, pp. 65-87.

Pieri, C.J.M.G. (1992). Fertility of Soils. A future for Farming in the West African Savannah. Springer-Verlag, Berlin.
Pingali, P.L. and Rosegrant, M.W. (2000). Intensive food systems in Asia: Can the degradation problems be reversed? In: Lee, D.R. and Barrett, C.B. (eds) Tradeoffs or synergies? Agricultural intensification, economic development and the environment, CABI Publishing, New York.
Ponti, T., de and Pinstrup-Andersen, P. (2005). The feasibility of feeding the world with organic agriculture, now and in the future. Unpublished Manuscript.
Pretty, J. and Hine, R. (2001). Reducing Food Poverty with Sustainable Agriculture: A Summary of New Evidence. The Potential of Sustainable Agriculture to Feed the World (SAFE-World) Research Project, University of Essex.
Pretty, J.N., Morison, J.I.L. and Hine, R.E. (2003). Reducing food poverty by increasing sustainability in developing countries. Agriculture, Ecosystems and Environment 95: 217-234.
Pretty, N. (1995). Participatory learning for sustainable agriculture. World Development 23(8): 1247-1263.
Rasul, G. and Thapa, G.B. (2004). Sustainability of ecological and conventional agricultural systems in Bangladesh: an assessment based on environmental, economic and social perspectives. Agricultural Systems 79: 327-351.
Reason Foundation (2000). Billions Served, Interview with Norman Borlaug by Ronald Bailey, ReasonOnline, available at the Internet (Nov. 2004): http://reason.com/0004/fe.rb.billions.shtml
Rockström, J. and Falkenmark, M. (2000). Semiarid crop production from a hydrological perspective: gap between potential and actual yields. Critical Review in Plant Science 19: 319-346.
Rodale Institute (1999). 100-Year Drought Is No Match for Organic Soybeans, (http://www.rodaleinstitute.org/global/arch_home.html)
Rosegrant, M.W., Paisner, M., Meijer, S. and Witcover, J. (2001). Global Food Projections to 2020: Emerging Trends and Alternative Futures. International Food Policy Research Institute, Washington DC.
Rosegrant, M.W., Cai, X. and Cline, S.A. (2002). World Water and Food to 2025. International Food Policy Research Institute, Washington DC.
Rosegrant, M.W., Ringler, C., Msangi, S., Cline, S.A. and Sulser, T.B. (2005). International Model for Policy Analysis of Agricultural Commodities and Trade (IMPACT-WATER): Model Description. International Food Policy Research Institute, Washington, DC. Available at http://www.ifpri.org/themes/impact/impactwater.pdf
Rounsevell, M.D.A., Ewert, F., Reginster, I., Leemans, R. and Carter, T.R. (2005). Future scenarios of European agricultural land use. II. Projecting changes in cropland and grassland. Agriculture, Ecosystems and Environment 107: 117-135.
Runge, C. Ford, Senauer, B., Pardey, P.G. and Rosegrant, M.W. (2003). Ending hunger in our lifetime. Food Security and Globalisation. Johns Hopkins University Press, 247 pp.
Scialabba, N.E.-H. and Hattam, C. (eds) (2002). Organic agriculture, environment and food security. Environment and Natural Resources Service Sustainable Development Department, FAO, Rome.
Smith, L., El Obeid, A. and Jensen, H. (2000). The geography and causes of food insecurity in developing countries. Agricultural Economics 22: 199-215.

Snapp, S. and Silim, S.N. (2002). Farmer preferences and legume intensification for low-nutrient environments. Plant and Soil 245: 181-192.
Snapp, S.S., Rohrbach, D.D., Simtove, F. and Freeman, H.A. (2002). Sustainable soil management options for Malawi: can smallholder farmers grow more legumes? Agriculture, Ecosystems and Environment 92: 159-174.
Staal, S.J., Baltenweck, I., Waithaka, M.M., deWolff, T. and Njoroge, L. (2002). Location and uptake: integrated household and GIS analysis of technology adoption and land use, with application to smallholder dairy farms in Kenya. Agricultural Economics 27: 295-315.
Stanhill, G. (1990). The comparative productivity of organic agriculture. Agriculture, Ecosystems and Environment 30: 1-26.
Stockdale, E.A., Lampkin, N.H., Hovi, M., Keatinge, R., Lennartsson, E.K.M., MacDonald, D.W., Padel, S., Tattersall, F.H., Wolfe, M.S. and Watson, C.A. (2001). Can organic agriculture feed the world? Agronomics and environmental implications of organic farming systems, vol. 2001.
Surridge, C. (2004). Feast or famine? Nature 428, 360-361.
Tiffen, M. and Bunch, R. (2002). Can a More Agro-ecological Agriculture Feed a Growing World Population? In: Uphoff, N. (ed.) Agro-ecological Innovations – Increasing Food Production with Participatory Development. Earthscan Publications, London, pp. 71-91
Thompson, P.B. (1996). Sustainability as a norm. Techné: Journal of the Society for Philosophy and Technology 2(2): 75-94. Online at http://scholar.lib.vt.edu/ejournals/SPT/v2n2/pdf/thompson.pdf
United States Department of Agriculture (USDA) (2004). USDA Agricultural Baseline Projections to 2013. Office of the Chief Economist, US Department of Agriculture. Prepared by the Interagency Agricultural Projections Committee. Staff Report WAOB-2004-1, 124 pp.
Uphoff, N. (2002). Agro-ecological Innovations – Increasing Food Production with Participatory Development. Earthscan Publications, London.
Vanlauwe, B., Wendt, J. and Diels, J. (2001). Combined application of organic matter and fertilizer. In: Tian, G., Ishida, F. and Keatinge, J.D.H. (eds) Sustaining Soil Fertility in West Africa. SSSA Special Publication No. 58, Madison, USA, pp. 247-280.
Vasilikiotis, C. (2000). Can organic farming 'feed the world'? Available online at www.cnr.berkeley.edu/ ~christos/articles/cv_organic_farming.html
Von Braun, J., Bos, M.S., Brown, M.A., Cline, S.A., Cohen, M.J., Pandya-Lorch, R. and Rosegrant, M.W. (2003). Overview of the world food situation. Food Security: New risks and new opportunities. IFPRI, brief. Available online (November 2004 at: www.ifpri.org).
Weinschenck., G. (1986). Economic versus ecological farm policies (German with English summery). Agrarwirtschaft 321-327.
Yamoah, C.F., Bationo, A., Shapiro, B. and Koala, S. (2002). Trend and stability analyses of millet yields treated with fertilizer and crop residues in the Sahel. Field Crops Research 75, 53-62.
Zeller, M. and Meyer, R.L. (eds) (2002). The triangle of microfinance: financial sustainability, outreach, and impact.: Johns Hopkins University Press; International Food Policy Research Institute (IFPRI), Washington DC, 399 pp.
Zethner, O., Meldgaard, M. and Høgh-Jensen, H. (2003). Organic Farming in Developing Countries. Danish Ministry of Foreign Affairs, Danida and Scanagri. Working paper, 56 pp.

Zollitsch, W., Kristensen, T., Krutzinna, C., MacNaeihde, F. and Younie, D. (2003). Feeding for health and welfare: the challenge of formulating well-balanced ration in organic livestock production. In: Vaarst, M., Roderick, S., Lund, V. and Lockeretz, W. (eds) Animal health and welfare in organic agriculture. CABI Publishing, CAB International, Wallingford, UK. ISBN 0-85199-668-X, Chapter 15, pp. 329-356.

11
Towards a global research programme for organic food and farming

Henrik Egelyng* and Henning Høgh-Jensen

Introduction 324
Examining the basis for a global organic programme: institutional analysis. .. 325
The role of – selected – international organizations 329
Research as a support tool for developing organic farming and food systems . 331
    Institutional history of organic farming: from centres to networks 333
    Human capacity in organic farming and food systems research 335
Towards a development-oriented research programme for organic agriculture 336
Conclusions 339

Summary

The first half of this chapter theoretically explores, from a development policy perspective, the nature of institutional environments for certified organic agriculture. The aims are to understand the conditions required, and the prospects, for organic agriculture to thrive, to present a view of global initiatives for research on organic production, and the current degree of institutionalization of organic farming and organic research at the global level. Through institutional analysis of social incentive structures, or game rules, the chapter analyses how certified organics in the North has been operationalized into a single policy instrument through which multiple development benefits are pursued, i.e. institutionalized in a way that enables and facilitates a governance regime that promotes the use of intrinsically sustainable technologies and methods. Focusing on the needs and prospects for complementing the European situation, where organic farming policies, practices and institutions are now thoroughly studied, and where the institutional landscape of organic research has attained a critical mass, the authors highlight organic institutions in the South as severely under-

---

* Corresponding author: Danish Institute for International Studies, Department of Development Research, Strandgade 56, DK-1401 Copenhagen K, Denmark. E-mail: heg@diis.dk

researched. The chapter outlines the contours of a framework for investigating the social conditions under which organic agriculture is evolving in the South and to explore the extent to which policy instruments and regulations influencing such development exist. The chapter stresses that current trajectories of organic agriculture in tropical countries are often driven by Northern institutions, and as a result these trajectories may not address Southern realities. The chapter includes a brief review of international organizations promoting organic agriculture as a developmental instrument and identifies the potential for advanced international organic research networks contributing to this process. Given the fragmentation of knowledge on organic farming conditions and methods in the South, the analysis includes specifications of the options for, and roles of, research in supporting the development of organic farming and food systems.

Introduction

This chapter lays out the conditions and prospects for global research initiatives for organic production. It does so against a background of the increasing globalization (Knudsen *et al.*, Chapter 1) of organic agriculture (Alrøe *et al.*, Chapter 3) with a potential to contribute to sustainable food security (Halberg *et al.*, Chapter 10) within a global trade framework (Bach, Chapter 5). Considering the distinction between certified and non-certified organic farming (OF) systems and with a view to identifying future research needs, our initial focus is on the nature and process of globalization of institutional environments for organic agriculture, as a prerequisite, or basis, for establishing any global research programme for organic production.

Sustainable methods, such as those preferred by organic farmers, depend on social carriers, institutional foundations and enabling environments or, in short, ecological institutions, in order to be developed and used (Egelyng, 2000). Institutional analysis therefore may help explain why markets for organic agriculture products are mature in some countries whilst only just emerging in others. In European countries subject to the same Common Agricultural Policy (CAP), other factors evidently play an influential role, and the institutional environment for organic agriculture is seemingly less enabling in Greece, for instance, than it is in Austria, Italy and Scandinavia.

Internationally, the picture is diverse enough, as judged by simple indicators such as area under organic production and value of organic sales. Add institutional analysis to the equation and we find that Nicaragua for instance, although having a significant amount of organic land, has no national programme on organic agriculture (Garibay and Zamora, 2003). By contrast, in Costa Rica organic farming is rather well developed and has had specific legislation on organic agriculture since 1995 (Soto, 2003). In some African countries, such as

Tanzania, certified OF is restricted to an export phenomenon and faces internal production, marketing and institutional constraints (Mwasha, 2002). Generally, there is little knowledge available on the nature and quality of OF research in developing countries and, with the possible exception of export cash crops, the body of 'knowledge' on OF methods in tropical conditions is still largely dominated by practitioner's trial and error.

In Europe, organic farming has, in recent years, developed into a single policy instrument through which multiple development and policy goals are pursued. In other words, it has become an instrument through which the positive environmental effects of organic agriculture become public goods produced and paid for by the public through their taxes. In discussions on certified and non-certified organics, it is of paramount importance to understand that this distinction is a prerequisite for society to be able to achieve the multiple benefits of organic agriculture by way of a single public policy instrument. As pointed out in Chapter 6, sustainable agriculture is a 'broad church'. In the absence of organic agriculture certification institutions, policy-makers wishing to promote and support sustainable development and sustainable agriculture will face a very hard time trying to devise efficient and operational policy tools to target the very diverse range of agricultural activities which claim to help improve sustainability.

Thus we would emphasize that, from a public policy and development perspective, certified organic farming has very significant advantages, simply because it has been institutionalized and operationalized in a way that enables and facilitates a governance option. This is a major strength compared to 'LEISA' and similar types of alternative farming regimes (see Parrott et al., Chapter 6). In the absence of public policies, as in many developing countries, the primary benefit of certification may be in yielding market value. But, at the same time, a major segment of non-certified or de-facto organic producers may be uncompensated, despite eco-efficiently supplying the very kind of public goods and social benefits that would justify public policy support of the kind that organic agriculture is given in the North.

Examining the basis for a global organic programme: institutional analysis

Methodologically, this chapter is based on development research including institutional analysis. In the language and tradition of development studies, development research may be defined as the study of social reproduction and transformation processes of developing countries, in conjunction with international factors that influence these processes (Martinussen, 1997). Institutions may be defined as the 'rules of the game in a society' that 'structure incentives in human exchange' (North, 1991; 3-4). Egelyng's work (2002, 2002) provides development studies drawing on ecological economics and applying institutional analysis to understand how policy conditions translate into institutional environments that are more or less conducive to productive application of intrinsically sustainable agricultural methods.

Box 11.1 illustrates how institutional analysis can use elements of classical agricultural policy analysis to identify how policy conditions can translate into an institutional environment that is more or less conducive to organic agriculture.

Institutional analysis can thus help investigate the social conditions under which organic agriculture evolves and explore the extent to which policy instruments and regulations influencing developments in organic agriculture exist in given countries. As countries differ greatly with regard to their agricultural institutional environments, such analysis is needed as a basis for, and within, any global research initiative for organic farming.

Box 11.1. Elements for analysis of institutional environments for agriculture.

Classical agricultural policy analysis focuses on economic objectives and the effects of market interventions. The classical instruments of intervention may include taxes, credits, tariffs, quotas, commodity programmes, research and extension programmes, food grading, cosmetic standards and public infrastructure delivery. The aims may be to influence production, supply, prices, distribution, and/or efficiency (World Bank, 1991; Ellis, 1992; Egelyng, 2000).

Environmental objectives, however, are seldom if ever mentioned in the classical perspective. Our interest here is to identify exactly what the classical analysis omits; how agricultural price policies affect agro-technological trajectories, by its incentive pattern-mediated impact on the generation, promotion and adoption of agricultural technologies.

Inputs such as seeds, irrigation water, fertilizers and pesticides may be subsidized or taxed, legal or illegal, and delivery systems may, or may not, be provided. In effect, all these policy interventions either penalize or encourage adoption of intrinsically sustainable technologies such as crop rotation, soil and water conservation practices and the use of biological and cultural means of pest control (Egelyng, 2000). Even output price policies indirectly determine the agricultural resource use pattern. Consequently, agricultural policies also influence the degree of incentives for private industry to produce agricultural technologies designed to reduce input use and make different farming practices more-or-less feasible and profitable.

Developments in environmental and ecological economics, have led to an increased documentation of positive as well as negative externalities associated with the agricultural sector (Häring et al., 2001). The developmental benefits of organic farming include environmental protection, biodiversity enhancement (providing conservation biological control), reduced energy use and higher quality landscapes (Häring *et al.*, 2001; Scialabba and Hattam, 2002; Dabbert *et al.*, 2004). In the UK, for instance, the social costs of water quality reductions caused by pesticides alone are about € 190 million a year. In the Former Federal Republic of Germany the social cost of pesticides on human health, water quality and species loss has been estimated at € 125 million annually. (Dabbert *et al.*, 2004: 71). In a non-industrialized context, valid evidence on benefits and costs is fragmented. Further research and documentation is needed – involving additional cases and geographic regions and farming systems – for such findings to be generalized beyond Europe.

The recognition of the multiple developmental benefits of OF has led to a realization that opportunities existed for harvesting double, or even multiple, 'dividends' of public policy. More than one dividend is harvested when a public policy is revised so that € 1 spent from the public purse can contribute to two or more policy goals. Originating in the literature on taxes, the concept of double/multiple dividends refers to (theoretical) benefits from replacing conventional taxes with green taxes, producing a positive environmental effect (first dividend) and reducing the overall excess burden of the tax system (second dividend) and producing positive employment effects (third dividend). Since organic farming may contribute to the public policy goals of reducing surpluses and environmental costs and generating positive environmental benefits at the same time, the European Union can be seen as having gradually begun harvesting such double or triple dividends through reform of the Common Agricultural Policy (CAP).

According to a study based on the OECD system of environmental indicators for agriculture, investments in organic agriculture yield more environmental benefits in terms of floral and faunal diversity, soil organic matter and less pesticide pollution than conventional farming. Using OECD indicators for agriculture, Stolze *et al.* (2000) concluded OF is an effective and economically efficient way of achieving environmental goals; OF results in improvements in most environmental indicators and supplies environmental services at lower costs. From a development and public policy perspective, such a positive impact on a wide array of environmental indicators makes certified organic agriculture attractive to policy makers, who wish to improve the state of the environment, while creating jobs and income for the producers (Goulding *et al.*, 1999). However, areas still exist where organic farming faces challenges or where insufficient data exists. This is the case for instance with regard to water use efficiency and animal welfare where organic agriculture may not perform any better than conventional agriculture.

The above analysis is based on European institutions or rules, for the simple reason that Europe has the most advanced policy regime for organic agriculture. In 2002, the USA completed a process leading to an institutional landscape hat resembles that of Europe (see Box 11.2). In China conditions for organic agriculture are significantly improving as well and are co-evolving with the discipline of ecological economics (Ye *et al.*, 2002; Wang *et al.*, 2004).

Institutional analysis can vary in its focus: it may be a matter of exploring whether, and to what extent, the small print of a crop insurance corporation discriminates against organic farming. But it may also be used to design public policy programmes to promote organic agriculture – and to show whether further policy steps can be justified. The institutional environment of a reformed CAP can provide incentives to reflect the greater societal benefits of organic farming, such as sustaining and increasing cultural and ecological values and improving landscapes (Clemetsen and van Laar, 1999). Yet, as of the year 2000, organic farms in Europe received significantly fewer direct (price support) payments per hectare from the first pillar of the CAP than comparable conventional farms did. In contrast, more agri-environmental and Less Favoured Area payments went to organic farmers, while Rural Development Programmes contain an only partially realized potential to support organic farming (Häring *et al.*, 2004).

Box 11.2. The institutional environment for organic farming in the USA.

The US Congress passed the Organic Foods Production Act (OFPA) in 1990, requiring the USDA to develop organic standards, creating a National Organic Standards Board and National Organic Program (NOP) – which did not, however, come into effect until October 2002. Web-facilities include www.organicaginfo.org developed by Organic Agriculture Consortium (OAC)/Scientific Congress on Organic Agricultural Research (SCOAR), through a grant from the Initiative for Future Agriculture and Food Systems (IFAFS). US institutes and agencies with organic farming related activities include the University of California Organic Farming Research Workgroup (www.sarep.ucdavis.ed.u/organic), the Organic Agriculture Research and Extension Initiative (OREI) and the Organic Transitions Program (ORG). The Organic Trade Organisation (www.ota.com) is a North American membership-based business association for the organic industry. The Organic Center has a singular mission: to provide credible scientific information about the organic benefit (organic-center.org). Organic-research at www.organic-research.com has a comprehensive list of addresses and links to organic certification and other organisations worldwide. In the USA, certified organic cropland for maize and soybean doubled between 1992 and 1997 and doubled again by 2001, when organic cropland and pasture reached 2.3 million acres. Organic livestock – poultry and dairy – grew even faster, according the USDA Economic Research Service. With just 0.3% of the farming area, however, the overall adoption level is still very low, compared to Denmark's 5.5% (Statistics Denmark, 2003). For fiscal year 2004, the USDA allocated an unprecedented USD 4.7 million for a new integrated Organic Program.

A recent study by Menrad (2004) suggests that public policy instruments, such as the use of standards and mandatory labelling and control procedures, reduces market transaction costs on the organic market side. Such instruments are not, however, sufficient, for a take-off on the supply side. Technical and market-related risks continue to discourage conventional farmers from converting their farms to organic agriculture. To kick-start any next wave of –

European – conversion, further incentives are needed (Menrad, 2004). Therefore, public policies are of paramount importance for the continued development of organic agriculture – consumers may prove rather powerless in terms of taking organic agriculture beyond its current level.

The implication of all this is that the rules of the game and the conditions under which organic production thrives are, and must be, essential components in any analysis of the history, challenges and prospects for research on organic farming. Discussions about ecolabelling by Bougherara (2004) and Carambas and Grote (2004) alert us to the limitations of market oriented instruments and of the need for politicians to accept responsibility through public policies. The new EU action plan on organic farming emphasizes the role that organic farming can play in rural development. The role of organic farming in a development context is however not adequately understood and, in the context of development in the South, it is important to realize that, while organic agriculture in the North is instrumental for policies pursuing reduction of surpluses and reconcilement of agricultural and environmental policies, organic agriculture in the South is, and must be, driven by a different set of objectives. These may include export promotion, economic self-reliance and rural development through transformation of low-input and low yielding traditional agricultural systems to low-input and higher yielding systems (Scialabba, 2000). However, the current trajectories of organic agriculture in tropical countries are often driven by consumers and institutions in the North and the conventions involved with certification often do not address tropical agro-ecological and socio-economic realities (Barrett *et al.*, 2002; Raynolds, 2004).

The role of – selected – international organizations in promoting organic farming

The FAO – or some FAO departments perhaps – stands out as the most proactive international organization with regard to organic agriculture, seeing organic agriculture as a developmental instrument to address sustainable agricultural and rural development. While the FAO may seem to have commissioned little primary research yet, the FAO has collected and accumulated evidence and information and a number of review studies on organic farming. One source of information on the level and nature of international collaboration on organic farming is the 2000-04 meeting lists, hosted by FAO (available at www.fao.org/organicag). This lists not only hearings on the EU action plan for Organic Farming, but also Asian roundtables on organic production, organic trade fairs meetings of the organic agriculture harmonisation Task Force, and regional efforts (Organic Agriculture in Central America) etc. The International Fund for Agriculture (IFAD) has also explored the potential of OF as a strategy for diversification of production among small farmers (IFAD, 2003). The United

Nations Conference on Trade and Development has published analyses identifying the potential of OF to contribute to economic and ecological diversification, poverty reduction and environmental protection. ITC and CTA have also produced some studies (market analyses) on OF.

In contrast, the World Food Programme (WFP) has no direct activities related to OF, although it did sponsor a minor study in 2002 on Harmonization of Organic Production Systems and the question of whether the Caribbean can meet the worlds' organic production standards. The UNDP also has only a limited involvement in supporting organic farming activities, but is involved in a global programme for 'Ecoagriculture' Research and Development in selected biodiversity hotspots.

Given increasing awareness of the rationale for developing countries to adopt organic agriculture, the World Bank would appear to have a unique role to play in promoting organic agriculture globally, although this has not yet been widely exploited. The reality in Africa is that agricultural products are already produced with practically no fertilizers and often without pesticides and in a context where governments cannot afford 'perverse' subsidies (except for energy). It would seem logical, therefore, for a global development bank to assist African nations to benefit from this position: supporting and promoting, for instance, certification schemes allowing Africans to harvest more price premiums on agricultural products that are de-facto organic, and only needing certification to prove it in the world market. A quick search of the World Bank database, however, produced no such projects, nor, for that matter, any other organic farming projects. In contrast, analyses by a number of civil society groups were found claiming that the bank has avoided investments with a focus on organic and fair trade certification.

Similarly, the International Agricultural Research System such as the CGIAR could be expected to be actively involved in promoting organic agriculture in developing countries. However, the CGIAR does not seem to have any major programme on certified organic farming and few, if any, CG research activities are designed to strengthen certified organic farming in its client countries. Little, if any CGIAR related research on certified organic farming exists. It is not clear what has prevented the CGIAR being more active in this regard. The CGIAR has already for long been under pressure from donors to adapt to the sustainability agenda and some of the centres have significantly changed their focus – from that of pursuing technological solutions to more livelihood-based approaches (see e.g. www.cimmyt.org and www.ilri.org). Given that many countries in the South have weak national research communities, the capacity and resources of the CG centres could play an influential and pivotal role in identifying and developing research topics for certified organic farming.

At the same time, research on organic farming is in a process of institutionalizing itself at the global level; one indication of this has been the recent (2003) establishment of the International Society of Organic Agriculture Research (ISOFAR). ISOFAR has a major potential to become a truly global organization with activities and major impact worldwide, providing it can encompass the new development-oriented researchers based in the South. Its first publication series aims to provide an 'Overview of long-term experiments dedicated to Organic Agriculture worldwide'. The Danish Research Centre for Organic Farming (DARCOF) has organized a series of workshops (upon which this book is based) exploring the possibilities for globalizing its activities and creating partnerships of relevance also for donors wishing to actively support sustainable development through organic agriculture.

Recognizing that several areas of public policy action are needed for organic agriculture in developing countries to flourish, it is intended that this initiative will strengthen existing moves in this direction. For example, DANIDA's international environmental support funds include a component with potential to promote research and development of organic farming as part of a broader strategy of promoting sustainable development. Our review of the DANIDA programme and projects showed that in recent years it supported at least six projects in South Asia with an organic agriculture component. Other donor organizations, such as GTZ (Germany) and HIVOS (the Netherlands), to mention but two, also give a high priority to supporting organic agriculture from an explicit sustainable development perspective.

International development agencies can contribute to the development of organic farming in the South through a range of activities, including for instance the following:

- Assist farmers overcome the 'entry barrier' of the initial high costs involved with certification.
- Assist farmers in overcoming the transition period from converting until premium prices can be obtained on their products.
- Assist governments in enhancing their capacity to harvest the double development dividends from organic farming.
- Contribute to increasing understanding of and helping embed organic ideas, principles and practices at relevant levels of agency, i.e. among public and private agencies as well as farmers' organizations.
- Help reform policies and programmes to become conducive to the development of organic farming.

Research as a support tool for developing organic farming and food systems

The development of organic farming and food systems has been supported, at different times, by both public and private research. Both the purposes and the methodological approaches have changed over time. Table 11.1 shows a number of phases through which the development of knowledge about organic farming passed. Each knowledge phase can be associated with specific methodologies. The earlier phases were dominated by trials for demonstration purposes, or for simple screening purposes. Such trials are still conducted and reported at scientific gatherings but increasingly systems comparisons are taking their place. Comparative studies for demonstration purposes are considered less important and efforts during the last decade have increasingly been on increasing the understanding of underlying processes and incorporating these into model tools. Further below, we will elaborate on the question of methodologies in the South.

As outlined in Table 11.1, the earlier phases of knowledge development were strongly driven by practitioners. This legacy is still reflected in the current weak documentation of organic farming systems. Much knowledge was documented in local farmers' magazines, but only a limited amount in international journals. In 1982 the first volume of *Biological Agriculture and Horticulture* was published and shortly after (in 1986), the *American Journal of Alternative Agriculture* was established. The emergence of these two journals illustrates the growing academic interest in organic farming at around this time, as well as a growth in available public and private research funding. However, further expansion of dedicated organic journals has not occurred, although other journals have adapted to the changing agenda and have focused more on sustainable practices. At the same time some organic research does reach journals with a single disciplinary focus.

Table 11.1. Knowledge phases over time. Inspired by Conford (1988, 1992).

Period	Knowledge emergence in North	Knowledge emergence in South
1920–1960s	A weak alternative practice but philosophy developed (e.g. Howard, Balfour, Steiner)	There was no clearly visible alternative practice although some of the experiences developed in the South were transferred to the North (e.g. Howard)
1970–1985	Organic farming developed by practitioners	No clear development
1985–1997	Organic farming developed by practitioners supported by state measures, including support for research	Organic farming developed by practitioners
1997–today	Organic farming developed by practitioners supported by state and EU measures, including support for research	Organic farming developed by practitioners and the state is now starting to take an interest in generating a framework to support market developments

Today, much 'know-how' on organic farming exists in the public domain and on the Internet and is theoretically, at least, available for exchange and transfer within and between developing countries. However, in reality knowledge transfer is hampered by a number of constraints:

- Much research in organic farming is of an applied type that is difficult to publish in academic journals.
- Many small organic organizations have still not fully entered the information age.
- Much knowledge on the Internet contains no quality assurance.
- Fragmentation of existing scientific knowledge.
- Insufficient quality in research and educational activities.
- Lack of tradition in the existing organic movement for prioritising international communications.
- Much knowledge is published in the grey literature of project reports, evaluations etc., and is therefore neither widely available, nor peer reviewed.

Some of these characteristics have their roots in the historic development of the organic movement, which has often found itself marginalized from, and in opposition to the 'mainstream', often including the academic world. Others are due to limitations in resources.

Institutional history of organic farming: from centres to networks

During the late 20th century, a series of private organic research institutes has emerged (Table 11.2). Most of these institutes were established by outstanding professionals, some of whom gave their name to the institutes. These dedicated individuals took the scientific development of organic farming to a new level of professionalism, which led to the start of a series of international organic scientific conferences under the auspices of IFOAM – with the first held in Switzerland in 1977. The 15th such conference will be held in Adelaide this year. These conference proceedings already constitute a valuable resource.

Several of the preceding chapters stressed that knowledge about the potential of organic farming in the South is still fragmented and that this hampers the transfer or exchange of knowledge on organic farming between countries and regions. To the best of our knowledge, there is as yet no major study outside of the industrialized world that analyses the ideas, practices and institutions which comprise the increasingly diverse global organic agri-food network (Raynolds, 2004: 729). The developmental – livelihood and sustainability – implications of organic globalization are therefore severely under-researched. The little research that has been done in the Southern hemisphere raises questions on all scales (Bougherara, 2004; Dorward *et al.*, 2004: 83; Raynolds, 2004). This contrasts with the European situation, where organic farming policies, practices and institutions have been, and are being, thoroughly studied (Dabbert *et al.*, 2004; Häring *et al.*, 2001) and where the institutional landscape of organic research is now firmly established (Niggli and Willer, 2002; Willer and Yussefi, 2004).

The historical development of the research in organic farming in centres (Table 11.2) has enhanced this tendency towards fragmentation (above), partly because the mainstream research and education funding bodies have, to some extent, ignored these centres and partly due to a weak publishing culture. It is important that these weaknesses are not repeated in any organic research frameworks that emerge in the South. Research and advocacy of organic farming in the South has not followed the same path as that in the North (described in Table 11.2). However, some networks have recently emerged.

Table 11.2. Some important organic farming research institutions.

Year of establishment	Name and location
1950	Institute for Bio-dynamic Research, Darmstadt, Germany
1974	Research Institute for Biological Agriculture, Switzerland (FiBL)
1976	Lois Bolk Institute, The Netherlands
1980	Ludwig Boltzman Institute for Biological Agriculture and Applied Ecology, Austria
1982	Elm Farm Research Centre, UK
1986	Biodynamic Research Institute, Sweden
1987	Norwegian Centre for Ecological Agriculture (NORSOK), Appelsvold. This centre is being merged with the conventional research system in Norway

PROSHIKA in Bangladesh is one such example. It is one of the largest NGOs in Bangladesh, working with the poor and landless, and actively promotes organic farming practices as an integrated element of their development and educational efforts. KIOF (Kenyan Institute of Organic Farming) was established in 1986 and promotes OF through a series of educational activities to reach its aim of sustainable agriculture. ECOAS (The Egyptian Centre of Organic Agriculture Society) was established in 1998 and is the dominant NGO for the organic sector in Egypt, working on inspection and consultations. ECOAS attracts several academic staff members from higher research institutes like Ain Shams University in Cairo. ECOAS frequently facilitates, research on topics of relevance to the organic sector by MSc students from the universities. OAASA (Organic Agricultural Association of South Africa) established in 1994 actively promote organic practices. Similar organizations and networks are emerging in many, if not most developing countries. They share in common an insistence on development issues and prioritizing the livelihoods of poor smallholders. Research on the extent to which certified organic farming is rooted in society at large and in these groups does therefore seem highly relevant.

Human capacity in organic farming and food systems research

The relative absence of coordinated, high quality, research and educational programmes is a major limitation to the development of organic farming in the South. The authors want to stress the importance of filling this gap, the need for new initiatives and new forms of collaboration.

A few universities, like Makerere University, Uganda, and Ain Shams University, Egypt, have identified organic farming as an important area for research but lack the funding to do this on the desired scale. By contrast, DARCOF II (2000–2004) employs about 150 research scientists, working at 20 institutes on more than 40 research projects. DARCOF is one of three founding partners in the Danish Research School for Organic Agriculture and Food Systems (SOAR; www.soar.dk). Such networks function well and, with appropriate collaborative structure with partners in the South, could be a model for developing research and educational programmes that operate in a more global context.

Global inventories and maps of the 'organic' policy and research activities and practices in the South are beyond the scope of this chapter. However, the authors wish to stress the urgent need for such an overview which could provide the same sort of strategic vision that was achieved in the agricultural sciences in Europe in the early 1990s (see e.g. Kølster and Sigsgaard, 1990) and again later for the social sciences (see e.g. Waibel, 2001; EU-AgriNet, 2002; Dabbert *et al.*, 2004).

DANIDA has traditionally supported research education through its Junior Programme Officer (JPO) programme. While this has not been a PhD programme, it has given some JPOs the possibility to do research and/or for combining their work with a PhD or research assignment. The potential exists for DANIDA to develop their programmes along this line and fund positions in CG-centres, allowing Danish as well as Southern agricultural researchers to undertake research in tropical organic agriculture.

The international research agenda is currently undergoing a period of profound change. Increasingly, livelihood issues are being coupled to market and policy developments. A new initiative, the Sub-Saharan Africa Challenge Programme (SSA CP) started in April 2005, provides an example of an interdisciplinary and inter-institutional research initiative that aims to couple smallholders to markets and thereby induce new patterns of development in many rural areas. This approach closely parallels that of the organic farming sector, which we will discuss in more detail further below.

The SSA CP builds on a new research paradigm called Integrated Agricultural Research for Development (IAR4D; see International Centre for Development Oriented Research in Agriculture; www.icra-edu.org), a research paradigm integrating natural resource management, intensification of production, new partnerships including the producers and policy makers, and takes its point of departure in market conditions. It is based on the underlying assumption that the livelihoods of smallholders can be improved by better integrating them into appropriate markets.

The IAR4D recognizes the challenges of knowledge fragmentation, created by the framework within which scientific knowledge currently is accumulated – through universities, graduate schools, research centres and NGOs, scientific journals, etc. It attempts to address this problem by insisting on inter-disciplinarity and inter-institutionality. More generally, intra- and inter-organizational networking is attracting increasing interest with knowledge centres, companies, policy-makers, and local people building local, regional and global networks and working partnerships. We wish to emphasize the importance of utilizing existing research and educational structures to develop coordinated networks that can steer regional and global initiatives that focus on human capacity building. Too often young researchers work in isolation without being part of broader supportive educational networks. Such changes are integral and necessary elements of a development-oriented research programme for organic agriculture.

Towards a development-oriented research programme for organic agriculture

Organic agriculture is increasingly being perceived from a worldwide development perspective, with a major role to play in meeting multiple societal goals associated with sustainable development. According to EU policy, organic farming is an important tool in the strategy of environmental integration and sustainable development, which are key principles of the first and second pillars of the CAP and of the Rural Development Policy (EU, 2004a). It is not only European policymakers however who have discovered the developmental dimension of organic farming and who pursue it as a societal goal. Some international organizations with responsibilities for promoting global development now promote organic agriculture in developing countries. They recognize its local and national development benefits in supporting livelihoods and enhancing environmental sustainability. Thus, there are some indications that a globalization of organic agriculture and research on organic farming is gathering critical momentum.

From a global development perspective, promoting organic agriculture worldwide appears an attractive strategy for pursuing double dividends for development and reaping the global, national and local benefits of more sustainable agricultural systems. However, tailor-made public policy programmes – including research programmes – are needed to realize these potentials. As universally defined by CODEX, EU (2092/91/EEC) and IFOAM, and regulated under national laws and certification programmes, the essence of certified organic agriculture is that most synthetic inputs are prohibited and that crop rotations are the main source of soil fertility. Consequently, any agricultural science component of a global research programme for organic agriculture in developing countries therefore needs to focus very strongly on increasing biological options for pest and weed management and soil fertility.

The EU experience has provided valuable lessons on the creation of enabling environments needed for organic agriculture to thrive. In the absence of regulatory frameworks and support measures, it is unrealistic to expect developing countries to pursue an organic agricultural development trajectory. A leapfrogging option theoretically allows developing countries to avoid the costly and ecologically inefficient historical path of EU and US agriculture. Options not available to or not chosen by EU countries and the USA, could well prove to serve other regions or countries in terms of development values or social and environmental returns.

Given that the European Commission has begun harvesting the double dividends that organic farming offers and given that similar global dividends and win-win opportunities exist, does our analysis suggest that we are currently moving into a global system and maturation stage of development of organic agriculture? Our quick answer is that winds are favourable, but major challenges remain ahead.

The 2004 EU action plan on organic farming aims its 21 actions at the European level, but acknowledges that the two current systems under which organic products are imported need improvement. Action 20 therefore stresses the need to 'Support capacity-building in developing countries under the development policy of the EU by facilitating information on the possibilities offered by more general support instruments to be used in favour of organic agriculture' (EU, 2004a).

The plan also emphasizes the role that organic farming plays in rural development (EU, 2004a). A similar paradigm may be applied to the tropics. Different paradigms of networking have become increasingly visible over the last couple of decades, in particular in the literature of 'regional development'. Two competing networking paradigms (Light House versus Dynamo) offer different explanations of the prospects for a global research programme for organic farming. While these initiatives differ in detail from each other, and from the research paradigm of the IAR4D, they share a common focus on development and in involving a multiplicity of actors in the research and development process.

As argued throughout this chapter, an enabling institutional environment is a prerequisite for any organic agricultural research programme reaching critical mass. Social sciences and development research will therefore need to play a central role in any global organic agriculture research programme, to help overcome, to paraphrase Dorward *et al.*'s (2004) terminology, 'agricultural investment dilemmas'. For instance: the second generation Agricultural Sector Programme Support for Africa may spend hundreds of million euros by 2009 to promote agricultural development across the continent. Institutions and individuals in countries with a solid experience in developing enabling environments for sustainable agriculture, including certification systems and labels, a significant organic industry, and including use of science-intense production related problems, could be utilized to consider possible ways of supporting a pro-poor and sustainable agricultural trajectory for Africa. However, the existing levels of support for development research on institutional and policy environments for certified organic agriculture and on the supply and use of intrinsically sustainable inputs has been, and remains, limited.

What may be needed is a global learning alliance for organic research. Such an alliance would need to include analyses of socio-economics and sociological aspects of organic farming, initiatives to improve the transfer of knowledge between countries and regions, concrete scientific research improving organic techniques and organic seeds, human capacity building, improvement of organic farming systems through participatory action research and research on the marketing of certified and non-certified organic products.

As organic farming in its certified form is market oriented (Knudsen *et al.*, Chapter 1), we suggest that a global research initiative on organic farming could build on the following pillars:

- Participation of all actors in the food chain, including farmers, traders, NGOs and policy makers.
- Inter-institutionalizing the programme in order to mobilize the expertise needed, including inputs from the social sciences, socio-economics and human sciences
- Cross- and inter-disciplinary research (development studies) to build bridges between categories of researchers, between disciplines, between science and society, across traditions and institutional locations (agencies and sectors).
- The long demonstrated capacity for technical innovation within CG centres coupled with an active development of local, regional and global organic markets.
- Integration of the enhancement of extension and education as part of such a programme, including research education.

In agreement with the principles of organic farming, the programme should aim to improve the livelihoods of smallholders and pastoralists, who produce much of the developing world's food. Although the majority of the world's population will live in urban areas by 2030, farming populations will remain at approximately the same size as today. For the foreseeable future, dealing with food supplies in the developing world means confronting the problems that small farmers and their families face in their daily lives (Dixon *et al.*, 2001; Pinstrup-Andersen and Pandya-Lorch, 2001). As DARCOF III takes off (see www.foejo.dk), its scope may broaden beyond Europe and become global. One important stepping stone for a global initiative could be a global meta-synthesis, collection and systematization of relevant organic agriculture scientific research results (i.e. establishing a database facility). Such an endeavour would imply the need for collaboration with one (or more) major international organization(s). Such a programme would need to take into account that the frontiers in agricultural research have undergone significant changes in the last decade, partly due to the globalization of our food systems. This includes changes in intellectual property rights as one aspect of the institutional environment for agricultural research (Egelyng, 2002). It also includes increased product differentiation, greater demands for ecosystem services from agriculture, and large changes in farm and market structures. Such changes have transformed agriculture into a sector that, in addition to efficiently producing food and fibre, must also deliver public health, social well-being, (rural) development and a sound environment.

Conclusions

In concordance with the other chapters in this volume, our analysis indicates that the current globalization of agricultural production, policy regimes and agricultural research has reached sufficient critical mass for a global initiative on organic agricultural development and research to be highly relevant. While orchestration of such a global research programme may not be possible for any single global agency, and while currently few international organizations actively support the ongoing globalization of organic agriculture, our review highlights the growing need for such an exercise. Our analysis showed how development studies and institutional analysis may provide a stronger foundation for understanding the conditions of, and prospects for, organic agriculture in the South. Given the essence of certified organic agriculture, that most synthetic inputs are prohibited, we also stressed how the scientific component of such a programme would need to focus on increasing the biological options for pest and weed management and soil fertility management.

Within Europe a coordination effort started in 2004 to enhance the quality, relevance and efficiency of research into organic food and farming (CORE

Organic). We envisage a future activity inspired by the CORE Organic and the SSA CP that will aim at gathering the globally dispersed expertise on OF to act as the basis to enhance the prospects for the development of organic farming and supporting research. The approach to supporting such a global initiative would need to be inter-institutional and interdisciplinary and focus on both public policy and market incentives, including promotion of organic certification systems. Such an effort arguably has the potential to attract the attention of many development stakeholders to benefit millions of poor smallholders in rural areas all across the world.

References

Barrett, H.R., Browne, A.W., Harris, P.J.C. and Cadoret, K. (2002). Organic certification and the UK market: organic imports from developing countries. Food Policy 27: 301-318.

Bougherara, D. (2004). Ecolabeling: private provision of a public good? Presentation given at the 8th Biennial Scientific Conference, International Society for Ecological Economics. Montreal. 11–14 July 2004.

Carambas, M.C. and Grote, U. (2004). Economic and ecological impacts of eco-labeling in the agriculture sector of asian developing countries: Thailand and the Philippines. Presentation given at the 8th Biennial Scientific Conference, International Society for Ecological Economics. Montreal. 11–14th July 2004.

Clemetsen, M. and van Laar, J. (1999). The contribution of organic agriculture to landscape quality in the Sogn og Fjordane region of Western Norway. Agriculture, Ecosystems and Environment 77: 125-141.

Conford, P. (1988). The Organic Tradition. An Anthology of Writings on Organic Farming, 1900-1950. Green Books, Bideford, UK.

Conford, P. (eds) (1992). A future for the Land. Organic practice from a global perspective. Green Books, Bideford, UK.

Dabbert, S., Häring, A.M. and Zanoli, R. (2004). Organic Farming: Policies and Prospects. Zed Books, London and New York.

Dixon, J., Gulliver, A. and Gibbon, D. (2001). Farming Systems and Poverty. Improving farmers' livelihoods in a changing world. FAO and World Bank, Rome and Washington DC.

Dorward, A., Kydd, J., Morrison, J. and Urey, I. (2004). A Policy Agenda for Pro-Poor Agricultural Growth. World Development 32(1): 73-89.

Egelyng, H. (2000). Managing agricultural biotechnology for sustainable development: the case of semi-arid India. International Journal of Biotechnology 2(4): 342-354.

Egelyng, H. (2002). Sui generis Protection of Plant Varieties in Asian Agriculture: a Regional Regime in the Making? In: Evenson, R.E, Santaniello, V. and Zilberman, D. (eds) Economic and Social Issues in Agricultural Biotechnology. CABI Publishing, pp. 31-42.

Ellis, F. (1992). Agricultural Policies in Developing Countries. Cambridge University Press, UK.

EU-AgriNet (2002). Organic farming – Supporting the new demand. Available online (August 2005) at europa.eu.int/comm/research/agriculture/research_themes/organic_ farming.html

EU (2004a). European Action Plan for Organic Food and Farming. Commission Staff Working Document. Brussels. SEC (2004) 739. Annex to the Communication from the Commission COM (2004) 415 final.

EU (2004b). Impact of CAP measures on Environmentally Friendly Farming Systems: Status quo, analysis and recommendations. The case of organic farming. ENV.B.1/ETU/2002/0448r.

Garibay, S.V. and Zamora, E. (2003). Produccion Organica en Nicaragua: limitaciones y potencialidades. www.orgprints.org

Goulding, K., Jarvis, S. and Macdonald, D. (1999). Integrating the environmental and economic consequences of converting to organic agriculture: evidence from a case study. Land Use Policy 16: 207-221.

Häring, A.M., Dabbert, S., Offermann, F. and Nieberg, H. (2001). Benefits of Organic Farming for Society. Paper presented at the European Conference – Organic Food and Farming, 10–11 May 2001, Copenhagen, Denmark.

Häring, A.M., Dabbert, S., Aurbacher, J., Bichler, B., Eichert, C., Gambelli, D., Lampkin, N., Offermann, F., Olmos, S., Tuson, J. and Zanoli, R. (2004). Impact of CAP measures on Environmentally Friendly Farming Systems: Status qou, analysis and recommendations. The case of Organic Farming. Institute of Farm Economics, University of Hohenheim.

IFAD (2003). The Adoption of Organic Agriculture Among Small Farmers in Latin America and The Caribbean. IFAD Report 1337.

Kølster, P. and Sigsgaard, L. (1990). Danish Research in Ecological Agriculture 1990. In: Besson, J.M. (ed.) Biological Farming in Europe: Challenges and Opportunities. Proceedings of Expert Consultation, FAO regional Office for Europe (REUR) and Swiss Federal Research Station for Agricultural Chemistry and Hygiene of Environment, Liebefeld, Bern, 28–31 May 1990, pp. 229-232.

Martinussen, J. (1997). Society, State and Market: A Guide to Competing Theories of Development. Zed Books.

Menrad, K. (2004). The impact of regulation on the development of new products in the food industry. Paper presented to the 8th ICABR International Conference on Agricultural Biotechnology: International Trade and Domestic Production Ravello (Italy), 8–11 July 2004.

Mwasha, A. (2002). Organic agriculture development in Tanzania. Ministry of Agriculture and Food Security. Paper presented to the workshop on the development of organic agriculture held in Bangkok, Thailand on 9/7/2002 to 20/7.

Niggli, U. and Willer, H. (2002). Organic Agricultural Research in Europe. Current Status and Future Prospects. (http://www.ewindows.eu.org/Agriculture/organic/Europe/ Report).

North, D.C. (1991). Institutions, Institutional Change and Economic Performance. Cambridge University Press, Cambridge.

Pinstrup-Andersen, P. and Pandya-Lorch, R. (2001). Meeting food needs in the 21st century. In: Wiebe, K., Ballenger, N. and Pinstrup-Andersen, P. (eds) Who will be fed in the 21st century. Challenges for science and policy. IFPRI, Washington, USA.

Raynolds, L.T. (2004). The Globalization of Organic Agro-Food Networks. World Development 32(5).

Scialabba, N. (2000). Factors Influencing Organic Agriculture Policies with a Focus on Developing Countries. Paper presented to the IFOAM Scientific Confence on 28–31 August, Basel, Switzerland.

Scialabba, N.E. and Hattam, C. (eds) (2002). Organic Agriculture, environment and food security. FAO, Rome.

Soto, G. (2003). La situación de la agricultura orgánica en América Central. In: Soto, G. (ed.) 2003. Agricultura Orgánica: una herremienta para el desarollo rural sostenible y la reducción de la pobreza. Memoria del Taller. FIDA, RUTA, CATIE, FAO, Turrialba, Costa Rica.

Statistics Denmark (2003). www.dst.dk

Stolze, M., Piorr, A., Häring, A. and Dabberts, S. (2000). The environmental impact of organic farming in Europe. Organic Farming in Europe: Economics and Policy, Vol. 6. University of Hohenheim, Germany.

Waibel, H. (2001). Organic farming in Europe: Re-orienting agricultural policy and implications for developing countries. Introduction to the issues involved and review of recent literature on the topic. Quarterly Journal of International Agriculture 39(3): 269-277.

Wang, S., Li, M. and Wang, D. (2004). The Emergence and Evolution of Ecological Economics in China over the Past Two Decades. Presentation given at the 8th Biennial Scientific Conference, International Society for Ecological Economics. Montreal, 11–14 July 2004.

Willer, H. and Yussefi, M. (eds) (2004). The World of Organic Agriculture: Statistics and Emerging Trends. 6th revised edition. BioFach, Bonn.

World Bank (1991). The Political Economy of Agricultural Pricing Policy, Krueger, A.O., Schiff, M. and Valdes, A. (eds) Volume II, Asia, Washington-/Baltimore.

Ye, X.J., Wang, Z.Q. and Li, Q.S. (2002). The ecological agriculture movement in modern China. Agriculture, Ecosystems and Environment 92(2-3): 261-281.

12 Synthesis: prospects for organic agriculture in a global context

Niels Halberg, Hugo F. Alrøe and Erik Steen Kristensen*

Introduction	344
Three perspectives on the challenges and prospects of organic agriculture	345
How may certified organic farming meet the challenges of the increased globalization of organic food chains?	347
The dilemma: the globalization of certified food chains may compromise organic principles	348
The concepts and the tools to revitalize the OA sector	349
What solutions do certified and non-certified organic agriculture offer to sustainability problems in the global food system?	354
Large-scale conversion to organic farming: impacts on food security and sustainability	355
Non-certified organic agriculture in developing countries	358
Improving soil fertility	360
Organic farming as showcase for recycling nutrients from households?	361
Reducing medicine use and residuals in livestock products through organic practices	362
Conclusions	363
The challenges for organic agriculture	363
The prospects of organic agriculture	364

---

* Corresponding author: Danish Institute of Agricultural Sciences, P.O. Box 50, Blichers Allé 20, DK-8830 Tjele, Denmark. E-mail: Niels.Halberg@agrsci.dk

Introduction

As discussed by Knudsen *et al.* (Chapter 1) there is a rapid development in the global food chains towards increased trade and competition over long distances and very large corporate retail chains taking control over large parts of food trade. This global competition leads to a downward price pressure and demands for large volumes, standardization, specialization and high production efficiency and productivity for agricultural systems all over the world.

This industrialization of the agricultural sector leads to increased externalities in terms of emissions of nutrients and pesticides, loss of biodiversity and reduced animal welfare. Moreover, even though food production has increased significantly over the last decades and most countries engage in the global food trade, there are still ¾ billion food-insecure people globally many of who live in countries with net food exports. Add to this the fact that some regions like Europe have a surplus of food production based on high level of subsidies. Some of this surplus is being dumped at foreign markets at prices with which the local producers cannot compete, leading to unfair pressure on local farming systems.

Organic agriculture (OA) is an alternative, which builds on a non-industrialized understanding of the relationship between food production and nature. Organic farming therefore has a potential for a more sustainable development. However, challenges from global trade with organic products may also threaten the sustainability in organic agriculture and threaten to dilute some of its basic principles and ideas and the benefits that it holds. Trade in organic food across continents is increasing and organic products from developing countries like Brazil, Egypt and Uganda are being exported to e.g. Europe. Increasingly governments in developing countries are creating conditions in support of organic export (Scialabba and Hattam, 2002). Moreover, agricultural development organizations such as IFAD (IFAD, 2002; Giovannucci, 2005), FAO and NGOs are becoming open to the idea that OA should be considered as one beneficial development pathway for smallholder farmers.

The increased globalization is a consequence of the general political and economical development in the world, which has both positive and negative socio-economic effects and there are no simple solutions to avoid the negative effects in the agricultural sector. However, through increased knowledge and understanding of these complex relationships we hope that it may be possible to point out more sustainable pathways and – in this context – to better define the potential role of organic agriculture.

As described in the preface, the work that led to this book was initiated by five questions. We can now, on the basis of the work, synthesize these to two key questions, namely:

1. How may certified organic farming meet the challenges of the increased globalization of organic food chains in order to offer a significant alternative to mainstream food production in the future?

2. What prospects and solutions do certified and non-certified organic agriculture offer to general sustainability problems in the global food system and to the improvement of smallholder farmers' livelihood in developing countries?

The two strands of organic agriculture, certified and non-certified, face very different challenges and offer different opportunities. 'Non-certified organic agriculture' is characterized by the same agro-ecological principles as certified organic agriculture, and therefore results in the same benefits for soil fertility etc. But the production is consumed locally and not based on market premiums; the costs of certification do not apply; and the practice is governed by other means than organic certifiers. The present chapter gives an overview and synthesis of the previous chapters, treating the two key questions in turn in the two main sections of this chapter. In accordance with the book title we first discuss the challenges and then the prospects, since the future prospects of organic agriculture will depend very much on how the present challenges are met. The book presents a rich picture of different perspectives on the questions and different ways to address them. Wanting to synthesize this rich picture, it seems clear that no homogeneous message can be found. In order to provide a fair treatment of the above questions, we therefore need to work consciously with the range of perspectives in the book. This is the subject of the following section.

Three perspectives on the challenges and prospects of organic agriculture

In Chapter 2, Byrne *et al.* describe three positions with different perspectives on globalization and sustainable development: growth without borders, growth within limits, and growth and ecological injustice. This threefold distinction has played an important role in the creation of the present book and provides a useful structure to retain the range of perspectives in this synthesis. Table 12.1 shows the three perspectives and examples of how the key questions above are answered from each perspective.

Table 12.1. Three basic perspectives on globalization and sustainable development.

	1. Growth without borders	2. Growth within limits	3. Growth and ecological injustice
Focus	Market solutions	Ecological system limits	Individuals and local communities
Relevant discipline	Neo-classical and environmental economics	Ecological economics	Political ecology
Characteristic concepts	Free trade, internalizing external costs	Sustainable scale, finite ecosphere, functional integrity	Ecological justice, fairness with regard to the common environment
How may certified organic agriculture meet the challenges of globalization?	Develop globally recognized principles and regionally adapted standards; create a space for organic agriculture in free trade institutions, e.g. the 'green box' in WTO	Enforce principles of ecology and sustainability in the organic certification standards to resist ill effects of market pressures	Include ecological justice in the organic certification standards to resist ill effects of e.g. distant trade, corporate involvement and large-scale cash-cropping
How can certified organic agriculture offer a solution?	Provide alternative products in the market and increase consumer choices	Provide means to promote sustainability in non-localized food systems with global trade	Provide means to promote ecological justice in non-localized food systems; create alliance with fair trade
How can non-certified organic agriculture offer a solution?	Through institutional protection of vital local primary production systems and markets	Provide a more sustainable strategy to development of local agriculture in low-income countries	Provide local food systems that promote ecological justice; institutional support for their further development

The first perspective, growth without borders, is the dominating perspective of modern, Western culture, whereas the second and, in particular, the third are generally less prevalent. Within the organic movement the two last are of key importance, even though organic agriculture is also very much seen through the glasses of the first and influenced by it. They both focus on problematic aspects of the present development that are not visible in the first: the ecological limits to economic growth and differences in how environmental deterioration impacts on different individuals and local communities. From the first perspective, certified organic agriculture simply provides yet another alternative in the market to the benefit of consumer choice and the standards are an alternative to internalizing the external costs of production. The main requirement for free trade is to distinguish organic products in the market by way of internationally recognized non-discriminatory standards. The challenge to certified organic agriculture is then to work for preferential treatment in free trade institutions on this basis, and to argue that 'non-discriminatory' in the context of organic agriculture implies regional differences in the standards due to importantly different conditions.

From the second and third perspective, the challenge for certified organic agriculture is to fully address the problems of sustainability and ecological justice in its principles and standards. Certified organic agriculture suggests a way to solve these problems in a globalized market by way of certifying the process behind the products and providing the consumers with a choice of more sustainable and just products. But this can only work in so far as the proper goals and principles are actually implemented successfully in the organic practices and in so far as the consumers are aware their choices and willing to take on the responsibility of buying the environmentally and socially friendly, but more expensive, products. Non-certified organic agriculture have a prospect of solving the same problems in less economically developed areas by building on local resources as an alternative to high-input agricultural development strategies such as the 'green revolution'. The first perspective on the other hand, with its focus on market solutions, seems blind to the prospects of non-certified organic agriculture. It seems clear, however, that these prospects will depend on the willingness to create institutional protection of vital local primary production systems, on the basis of their special status in contrast to industrial production, and in parallel to e.g. the WTO 'green box'.

In the following sections we will discuss the two questions building on the framework in Table 12.1.

How may certified organic farming meet the challenges of the increased globalization of organic food chains?

There is a tendency for OA – like conventional food systems – to be influenced by globalization trends such as global competition and increasing free trade linked with demands for harmonization and supply-on-demand leading to long distance trade, specialization, economics of scale, commodification of common goods and lack of transparency. Entering into global trade on these terms threatens to dilute the special characteristics of OA. In this section we discuss these challenges and suggest ways to address them by strengthening the principles of OA.

The dilemma: the globalization of certified food chains may compromise organic principles

The organic food system has over the past two decades been transformed from a loosely coordinated local network of producers and consumers to a globalized system of formally regulated trade, which links socially and spatially distant sites of production and consumption (Dabbert *et al.*, 2004; Raynolds, 2004). Organic farming is also included in 'Codex Alimentarius' (Chapter 5). Though preferences for local organic food persist, Northern countries are increasing their reliance on organic imports, particularly from the global South, including products competing with locally produced conventional products (Rigby and Brown, 2003; Raynolds, 2004; see also Chapter 1). At the same time, supermarket sales of organic products have been increasing, dominating sales in the UK, Switzerland and Denmark (Dabbert *et al.*, 2004). This will again put pressure on organic farms towards increased specialization and large-scale production, which will then be at the cost of diversity both in terms of farms and in terms of numbers of enterprises per farm.

This development could result in the same basic social, technical and economic characteristics: specialization and enlargement of farms (Milestad and Darnhofer, 2003), decreasing prices, increasing debt loads with increasing capital intensification, increased use of external inputs and marketing becoming export-oriented rather than local (Hall and Mogyorody, 2001; Milestad and Hadatsch, 2003). The 'conventionalization' of organic farming thus includes different issues that may be linked but each can be considered on its own: lack of transparency and trust among producers and consumers, increasing food miles and dilution of the 'nearness' principle, specialization and concentration of production at the cost of smallholders and reduction in diversity in crops and farm types. Finally, to the extent to which the organic products are produced as a response to demands expressed in the Developed countries through intermediaries there might not always be a strong ownership to the organic certification schemes or embedment of the organic principles among the local farmers. As discussed by Hauser (Box 3.1), some types of organic production do not build on a thorough understanding of organic principles embedded in the local farmer organization.

The problems of conventionalization has been recognized and discussed within IFOAM for a number of years (Schwartz, 2002; Woodward *et al.*, 2002; Rundgren, 2003) and efforts are made to find solutions that facilitate the strengthening of small-scale farmers' role in the certification process and their ownership to organic principles. But it seems difficult for the movement to agree on specific policies against the conventionalization pressure arising from supermarket sales and large-scale production entering the global organic market. The situation raises questions concerning the future sustainability of OA such as:

- What is the risk that fundamental organic principles on nearness and ecology are compromised? and
- How may this be avoided by adopting new principles such as ecological justice or fair trade?
- What are the different possibilities for OA to secure a specific and sustainable role in the global trade system as seen from different perspectives such as free trade rules (WTO), fair trade and principles of ecological justice and the commons regime?

In the following we will discuss the potential tools and measures that could support a positive development of organic farming from the viewpoints of different perceptions of global trade and sustainability as presented in Table 12.1 and based on the previous chapters. The aim is to synthesise how a strengthened organic sector and movement may contribute to solving some of the problems in the conventional food chain as presented in Chapter 1, based on the concepts of fair trade, resource sufficiency, functional integrity, asset building, resource maintenance and ownership. We will argue that very different views and measures are needed vis-à-vis the export oriented certified OA in developing countries vs. the non-certified OA in developing countries.

The concepts and the tools to revitalize the OA sector

Certified organic farming as a 'green box' in WTO

It is not easy as a consumer to get an overview of the farm structure in the organic sector – especially related to overseas trade – nor to which degree this matters for the social and environmental benefits that consumers may expect to support when paying the price premium. Therefore, in order to secure the trustworthiness of the OA sector in the long run, more clear principles and rules should be developed. And more studies of the actual positive effects on e.g. biodiversity and socio-economics of the price premium on certified organic products from developing countries are needed. There is a need for a strategic renewal of organic principles and rules in light of the global development and for a debate of the feasibility of encompassing all forms of OA in one concept.

One strategy could be to strengthen and clarify the organic principles and seek to establish a special niche within the global trade in food products. Some proponents argue for OA to find its place in the world of increasingly free trade, e.g. by working to get a special treatment under the WTO. Bach (Chapter 5) explains why the free-trade rules under the WTO agreements may be an obstacle for the goals to secure organic products a special treatment in the global trade: due to the non-discrimination principle which is the core of the international trade law it is generally not allowed to discriminate between products that are physically alike. This means that it is difficult to get acceptance of measures, which gives special treatments of products based on the way they are produced if this is not expressed in the appearance of the product itself. Obviously, this is a potential barrier for future initiatives of eco-labelling and special trade and tax policies favouring organic products (e.g. lower taxes or import toll on organic products). On the other hand, recently there has been some positive developments in the definition and use of standardized voluntary eco-labelling schemes that are either life-cycle based or based on rules for the production methods, where OA would fit under the second type. Under the overall term Multilateral Environmental Agreements it may be possible to establish rules with priority over the general WTO rules thus allowing for special treatment of environmentally friendly products. But this requires as a minimum that 'Environmental standards and policies must be developed in the appropriate environmental organizations but recognized by the WTO as being necessary and consistent with the WTO rules' (Bach, Chapter 5). Therefore, from this perspective the organic movement should seek to strengthen international agreements and definitions of organic farming including clear objectives and criteria. A starting point may be the organic food guidelines under the 'Codex Alimentarius', which acknowledge that organic farming standards are a legitimate means of recognizing product quality rather than a technical barrier to trade.

In order to get a general acceptance of agri-environmental support schemes targeted at organic farming in e.g. the EU by WTO members, stronger efforts are needed to convince also developing countries, that this is not a new form of protectionism against their products. This is more likely to succeed if efforts are made to help developing countries to establish their own organic certification schemes, which are affordable and manageable for smallholder farmers. Therefore, it is a challenge for the organic movement to strengthen the international dialogue and supply assistance and capacity building to developing countries in order to secure an acceptance of the organic niche within the global free trade rules. IFOAM has supported the development of this for a while but more efforts are needed to help building national and regional institutions and regulations that may support the development and certification of organic farming in developing countries.

This obviously presupposes that increased global trade is considered a positive scenario for the development of organic farming, which is an opinion not necessarily shared by all stakeholders. Contrary to the attempts to secure OA a special niche in the global trade, that there are important parts of the OA movement find that organic food systems should focus on local markets, both in form of non-certified OA in developing countries and in developed countries as expressed by the 'foodshed' movement (Boxes 1.4 and 1.5).

The idea of ecological justice and the tools of ecological economics

There is, as explained, no consensus regarding the future direction of OA development. This points to the need for a conceptual discussion and for tools to analyse the advantages of OA in the global context. Ecological justice (EJ) may serve as an overall (normative) framework for such a conceptual discussion. EJ covers many of the issues at stake when discussing organic farming and its potential to offer alternative solutions to the problems associated with the globalization of agriculture and food systems (Alrøe *et al.*, Chapter 3). In combination with the idea of managing important local resources under a commons regime (Byrne *et al.*, Chapter 2), EJ may prove a step in the direction of revitalizing the specific features of organic farming. This is based on a view of OA as an alternative to mainstream agriculture with a potential to solve a broad range of environmental, animal welfare, food security and livelihood problems in agriculture. Ecological Justice has been proposed as one of four basic principles in the ongoing work to redefine IFOAM's principles (Alrøe *et al.*, Chapter 3).

Having accepted this idea the next question is then, to what extent does OA in its many different forms conform with the idea of EJ, how is this best assessed and involving which stakeholders and how is this ensured in the future development of OA in all its different forms. In other words, how is EJ to be specified? We see two distinct and complementary ways of implementing EJ within the OA movement. The first is through a more detailed set of principles and rules guiding both the development of organic farming systems and food chains (column 3 in Table 12.1). The second is through analyses of different choices and strategies for organic food chains based on the tools of ecological economics (column 2 in Table 12.1).

One main aim of OA is to find another balance of the conflict between economical pressures to reduce nature's time and labour time using external inputs on the one side and the resulting impact on biological and environmental systems on the other as described by Kledal *et al.* (Chapter 4). This raises two relevant questions: (i) How may the methods used in ecological economics provide a better framework for describing the benefits of organic farming? (ii) How may such positive attributes of OA be acknowledged within international trade rules allowing for a special treatment?

Ecological economics offers itself as the scientific tool (or discipline) for this purpose, encompassing both quantitative questions related to the scale of production and related externalities and qualitative questions related to the definition of multiple objectives and involvement of stakeholders and institutions. Ecological economics includes tools for combined assessment of resource use and production from an economic and environmental point of view using a range of measures or indicators to compare the outcome of a production system and chain with multiple objectives. Examples of analyses are the comparisons of energy use efficiency in organic and conventional farming (Refsgaard *et al.*, 1998) and other assessments of the relation between intensity of production, resource use efficiency and externalities (Halberg *et al.*, 1995, 2005; Rasul and Thapa, 2004; see also Chapter 1).

The discussion of the acceptability of long distance transport may benefit from life cycle based assessments of the importance of the environmental impact of the transport itself in comparison with other environmental impacts from food production (Halberg *et al.*, 2005). Life cycle assessment will possibly show that intercontinental sea transport – in contradiction to air transport – has only marginal significance compared with the total environmental impacts from a food chain. And such analyses may again be used to determine rules for transport forms and distances within certification schemes. This way a more precise definition of 'nearness' may help to distinguish between organic products where long distance trade is less a problem than others. In this respect one should distinguish between products and the different choice situations consumers face. A large group of the organic products traded long distance in fact substitute comparable, imported products, which are exotic at the place of consumption (e.g. organic vs. conventional bananas or coffee purchased in Europe). In this respect OA may have a positive environmental and socio-economic effect in the country of origin given the right certification scheme without affecting total global trade or increasing food miles. Another group of imported organic products compete with or replace locally produced organic or conventional products of the same time (e.g. apples are grown widely in Europe but still organic apples are imported from outside Europe). These are the products that may compromise OA principles and increase the total food miles. As described by Knudsen *et al.* (Chapter 1), large amounts of conventional substitutable food products are cross-traded, that is, products such as beef or cereals are imported AND exported to and from countries in different continents.

There are however a number of conceptual and methodological obstacles for such a 'scientific' approach to the characterization and regulation of organic food chains.

First of all, certification of OA was never meant to be based on assessments of the outcome but on the production methods used. These are supposed to build on the intentions to enhance specific values and objectives by adhering to certain principles such as recycling, improving soil fertility as a mean to improve health of plants etc. Therefore, according to some proponents, future certification of OA should not focus on the analysis of specific product chains and their relative environmental benefits using e.g. LCA but rather on an extension of the generic rules and a local adaptation of rules to cover the overall organic food system.

Second, a number of social and environmental aspects are not easily measured with current methodologies and would therefore not allow for a product-based assessment of OA. This includes the goal of securing the farm family's livelihood and social justice, enhancement of biodiversity (Noe *et al.*, 2005) and soil fertility (Schjønning *et al.*, 2004). Moreover, this scientific – or technocratic – approach to defining 'nearness' may not satisfy all critics of global trade.

Third, there are some potential impacts related to the implementation of certain techniques, which are very uncertain or hitherto unforeseen. This is why OA adheres to the precautionary principle (Alrøe et al., Chapter 3) and thus avoids techniques such as GMO without engaging into a specific risk analysis or cost benefit assessment. Therefore, the possible window for OA in WTO regulations and the credibility of OA among consumers cannot rest on documentation of environmental and social benefits for each specific product. There is a need to acknowledge the specific principles and rules within OA as the basis for a 'green box' or trade rules as discussed in Chapter 5.

Having said this, there is still from our point of view a need to develop the future principles and rules for organic farming in a continuous assessment of the potential outcomes in relation to social and environmental goals. This may be controversial within the organic farming movement and so far there has been little will to strengthen the rules regarding trade etc. (Rundgren, 2003). As discussed by Woodward and Vogtman (2004), there has been a tendency to dilute the organic principles for both recycling, demands for use of local resources and for animal welfare. As an example of stronger principles, some supermarket chains in e.g. UK have formulated a policy favouring local products over imported organic products, where such a choice in fact exists, e.g. procuring fruits and vegetables from conventional UK growers rather than importing similar organic products (Wright, 2004). It would be wise for the organic movement to formulate its own policy on this question, which could give guidance to the retail sector.

Some inspiration for a policy, which includes social and ecological justice more explicitly, may be gained from the fair trade movement as it has been debated for some time within IFOAM (Cierpka, 2000). Certified fair trade is an alternative form of trade that, like certified organic, has the potential to work across globalized food networks in distant trade relations, and which goes some way towards meeting the principle of ecological justice (see Alrøe et al., Chapter 3; Byrne et al., Chapter 2). But both organic and fair trade fall short of the target in some respects. Fair trade goes further in specifying the social conditions and costs of production, but is lacking in ecological considerations. Organic trade, on the other hand, goes further in detailing the ecological conditions and costs of production, but is lacking in social considerations. However, the organic and fair trade movements cannot simply combine forces to meet their ecological and social ideals and fulfil the prospects of ecological justice. Both standards omit, for instance, considerations on distant transport. Major challenges in the development of a future organic fair trade policy are to secure ecological justice to those outside the trade network as well as those within, and to resolve the potential conflict between the benefits of fair global trade to low-income areas and the inherent disadvantages of distant trading.

Assuming that the OA movement finds ways to counteract the conventionalization of organic farming systems and food chains, there is a potential also in the future to demonstrate an alternative agricultural development, which is more in harmony with societal goals for environment, resource use, animal welfare etc. One such development could be to link with attempts to recycle nutrients from urban settlements and using alternative household waste management (see below).

What solutions do certified and non-certified organic agriculture offer to sustainability problems in the global food system?

As discussed above and in Chapter 1, the development of mainstream food systems are strongly influenced by the globalization process, which creates both socio-economic improvements and negative socio-economic and ecological externalities. One example of these complex relations is the recent developments in Brazil and Argentina where huge areas have been converted from either traditional mixed farming systems or natural vegetation into soybean mono-cropping over a period of 5–8 years. More than 14 million ha of land in Argentina alone are now grown with a few Glyphosate-tolerant GMO soybean varieties in a pesticide-dependent system, which seems to be an unplanned ecological experiment of overwhelming dimension (Pengue, Box 1.1). Besides the very low degree of diversity in this agro-ecological system and its dependence on a specific pesticide there is also evidence of a rapid concentration of land on fewer and larger farms (due to the quickly adopted and very efficient no-till technology linked with the Glyphosate tolerance), rapid loss of a traditional farming system and of P-mining of the soils. The harvest and mining of P with soybeans mainly for export is linked with the opposite problem of P-surplus and eutrophication in the soy-cake importing countries in Europe (Knudsen *et al.*, Chapter 1). This is thus a good example of the short sighted, economically powered globalization of food systems without recognition of the externalities, which is where ecological economics offers another assessment of the costs and benefits (Kledal *et al.*, Chapter 4). From the political ecology point of view this global one-way transport of P is another example of ecological injustice (Byrne *et al.*, Chapter 2). Organic farming – presuming a closer adherence to ecological justice principles – may offer an alternative development pathway in relation to this example.

Even though the origins of OA go back long before environmental problems came on the agenda (Woodward and Vogtman, 2004), it is fair to say that the broad public support and spread of OA – especially in Europe – has been linked with the increased awareness of environmental and animal welfare problems connected to intensive agricultural production in combination with the food safety problems in the 1990s (Dabbert *et al.*, 2004). Supporting OA has been proposed as a means to internalizing externalities in the food chain and reducing food surplus in high input agricultural areas (Dabbert *et al.*, 2004; see also

Chapter 1), and giving back some control over the food chain to smaller agents (Box 1.4 and 1.5). The motives for supporting OA differs from Northern to Southern Europe and probably also between political decision members and the individual consumer. At the policy level the problems of overproduction and land abandonment have been motives for supporting conversion to organic farming and the EU commission now uses the inventories of 'agricultural area with OA' as an environmental indicator (EEA, 2005). At the level of the individual consumer the motives for buying organic products include animal welfare aspirations, personal health and food safety objectives and expectations of better taste (Dabbert *et al.*, 2004). In developing countries with low input agriculture OA may improve local food production (Chapters 6 and 10) and asset building (Box 3.1) in smallholder families (Box 1.4) and soil fertility (Chapter 8). It is a challenge for OA to deliver on all these prospects and this will need a strengthening of the adherence to the principles as well as increased efforts in research, innovation and development (Egelyng *et al.*, Chapter 11).

Documented benefits of organic farming include lower resource use, environmental protection, reduced energy use, preservation of biodiversity and landscape values (Refsgaard *et al.*, 1998; Hansen *et al.*, 2000; Scialabba, 2000; Stolze *et al.*, 2000; Scialabba and Hattam, 2002; Dabbert *et al.*, 2004). In a non-industrialized context, valid evidence on these benefits is fragmented (Pretty, 2002; Scialabba and Hattam, 2002; Uphoff, 2002; Rasul and Thapa, 2004; Parrot *et al.*, Chapter 6). While lower environmental impact of OA as compared with intensive conventional farming systems has been documented when evaluated per farm or hectare, this is not always the case when comparing the environmental impact per kg product (De Boer, 2003; Halberg *et al.*, 2005). The general question of the environmental benefits of OA has been discussed in Chapter 1. In the following sections we will focus on selected topics related to how OA may play a significant role in defining viable alternatives to the current mainstream development. First, the consequences for global food security of large-scale conversion to OA are considered followed by a discussion of the potential for non-certified OA in low input regions. Then follows sections on the soil fertility aspects of introducing organic farming and the long-term possibilities for recycling nutrients in household wastes based on the respective chapters. Finally, the potential of reducing medicine use and residuals in livestock products through organic practices is summarized.

Large-scale conversion to organic farming: impacts on food security and sustainability

The organic movement's own vision is to eventually convert all agriculture into organic farming; this is not surprising. However, from other sources outside this movement it is also considered to support a development towards a large-scale conversion to organic farming in both Europe and North America and developing countries motivated by other objectives as explained above. Organic farming would – from one perspective – fit into a European policy of reducing agricultural surpluses, maintaining important semi-cultural landscapes and biodiversity and reducing environmental pressure from intensive farming (Chapters 1 and 10). But conversion of large parts of high input/high yielding agricultural areas would result in decreased yields in the years immediately after conversion and could therefore potentially have a negative impact on poor people's food security – in areas with net import of food – due to increased world prices. The long-term effect would depend on the proportion of land converted to OA and the future growth rates to be realized in OA crop yields. There may be a potential to increase organic yields per hectare faster compared with conventional over the next decades by increased research into soil fertility and plant nutrition, plant breeding and improved weed management in organic systems.

If large areas in Europe and North America were converted, corresponding to a mixed crop rotation with sufficient grasslands to act as both a main fodder input to most of the dairy production and soil fertility building for cash crops in the rotation, the overall impact on production levels would probably be acceptable. Such a partial but large scale conversion of for example 40% of agricultural land in Europe and North America would be possible without significant impact on world market prices and global food security in the year 2020 compared with a baseline development as projected with the global food policy model IMPACT (Halberg *et al.*, Chapter 10). This, however, presupposes that yields in OA increase 25–50% more than in conventional farming over the next 1–2 decades, which would be a challenge to research and development of OA. If yields in OA continue to be only 65% of conventional yields then world market prices will be app. 5–10% higher than in the baseline scenario for cash crops such as maize and wheat in 2020. But even this would have very little impact on food security on a global scale according to IMPACT because other factors than global food availability are determining the number of food insecure families.

As described by Knudsen *et al.* (Chapter 1), many countries with large populations of food-insecure people depend on food import (and some of these actually export food). It is too simplistic to add up a nation's food production and divide by the population in order to calculate food security (Conway, 2001; Halberg *et al.*, Chapter 10). Food security is not only a matter of producing sufficient food. The majority of food-insecure people are poor rural inhabitants living in South Asia and sub-Saharan Africa. These families are lacking food mainly due to poverty and lack of knowledge and assets to produce either food or gain sufficient income. These large groups of people will benefit from increased local food production (and/or employment opportunities) and this is where non-certified organic farming – as well as other types of locally adapted low-input farming systems – has a major role to play.

Scenarios for large-scale conversion to OA in low input regions such as sub-Saharan Africa (SSA) shows that it is possible to improve the overall food supply and reduce dependence on food imports towards the year 2020 if the many positive experiences from case studies are possible to scale up. This builds on the assumptions of moderately higher yields in OA after conversion of low-input farming systems and maintaining identical or higher yield growth rates compared to the yield improvements in conventional crops. A conversion of 50% of agricultural land under these assumptions would lead to a combination of 2–12% lower world prices on stable food and increased local consumption according to the IMPACT model. However, in its present form the model cannot take into account distributional effects between income groups within the regions or the positive effects on the sustainability of farming systems. From the *functional integrity* point of view on sustainability (Chapters 3 and 10) OA has an advantage in so far that soil fertility and farm-level biodiversity is being enhanced. From this perspective it seems less important if the yields for some time mostly will suffice for home consumption (self-sufficiency farming).

Viewed from a *food sufficiency* or *resource productivity sustainability* perspective the focus should be on the need for high yields per hectare in order to secure a surplus for the growing urban populations in developing countries. The discussion on whether to build development strategies in e.g. SSA on increasing the use of external input (HEIA) or on low external input sustainable agriculture (LEISA) and OA strategies has both a pragmatic and a quasi-ideological dimension (Halberg *et al.*, Chapter 10; Pender and Mertz, Chapter 8). The different views build on different perceptions of sustainability and (possibly) on different interpretations of the potential to overcome structural and economic problems to make fertilizer etc. accessible and affordable for smallholder farmers. But the fact is that the fertilizer use and the food production per capita in SSA have not increased significantly over the last 3 decades (Knudsen *et al.*, Chapter 1; Pender and Mertz, Chapter 8; Halberg *et al.*, Chapter 10). Therefore, as long as this is not the case for huge numbers of smallholder farmers with poor market access, there are good reasons for focusing development efforts on the improved use of local resources as in organic farming. This presupposes that methods are developed to secure and restore the soil nutrient availability, including methods to improve the low phosphorus status on weathered soils in e.g. SSA.

Moreover, the agro-ecological methods and the capacity building towards improvement of agro-ecological knowledge are not in themselves a barrier to a switch to more input intensive agricultural practices if this should be attractive for the smallholder farmers at a later stage, when they can afford the risk and if market conditions are in favour. Therefore, supporting OA seems as one development pathway, which should be considered along with other LEISA options. There are, however, important constraints for a large uptake of OA as discussed below.

If organic farming should represent a sustainable solution for a large scale conversion worldwide it will have to develop and adjust both its principles and its technical performance including productivity in order to produce sufficient food. A key factor will be the involvement of local stakeholders in the adaptation of OA methods to the large variation in agro-ecological conditions existing especially across the African continent using experiences from participatory methods for development of natural resource management (Sutherland et al., 1999; Onduru et al., 2002). A key question not addressed here is the need to create low-cost certification procedures for – groups of – smallholder farmers as part of securing local ownership and control over the marketing channels. Official interest in organic agriculture is emerging in many countries and 57 countries have a home-based certification organization (Willer and Yussefi, 2004), but producers often have to comply with foreign standards not necessarily adapted to their country conditions (Scialabba, 2000). The development of certified OA in some tropical countries has been driven by demands from companies and organizations with the aim of supplying consumers in the North and conventions involved with certification often do not address tropical agro-ecological and socio-economic realities (Barrett et al. 2002; Raynolds, 2004).

Non-certified organic agriculture in developing countries

As argued above, organic agriculture has a high potential in particular in developing countries. It is, however, important to be aware that the societal benefits of organic farming practices are different in developing countries and the domestic market for certified organic products is very small. Because the organic standards are mostly developed for the Northern markets there is also a need for adaptation of the standards for tropical conditions. It may therefore be relevant to introduce the term 'Non-certified organic agriculture' (NC-OA) as a concept specifically suited for promoting and protection of organic agricultural production, which is marketed locally without premium prices.

NC-OA is characterized by the same agro-ecological principles as certified organic agriculture, and therefore results in the same benefits for soil fertility etc., but the production is consumed locally and not based on market premiums; the costs of certification do not apply; and the practice is governed by other means than organic certifiers. As discussed by Hauser (Box 3.1) non-certified OA may comply more with IFOAM principles than certified organic cash crop schemes. This is especially true in relation to the different degrees of nearness, where the non-certified OA obviously are *localized food systems* where the proximity of consumers and producers allow a relatively deep insight in the production forms (Figure 3.2). This is by far always the case for the newly globalized certified organic farming systems as discussed above.

For smallholder farmers using no or very little fertilizer and pesticides, NC-OA have a potential for improving yields given the right training and local adaptation of methods. In areas where the focus is on improvement of the farming systems, diversifying and increasing yields of crops for home consumption or local markets, the role of NC-OA is to organize capacity and asset building around the principles of agro-ecology. This has the potential for improvement of the food security and stability of many smallholders in large parts of SSA, Asia and Latin America and thus can lead to an improvement of their health and livelihood. FAO and others have reported how the introduction of input intensive and capital intensive methods among poor smallholders without appropriate micro-finance schemes may worsen their economic problems and lead to a focus on short term returns as opposed to asset building (increased soil fertility, diversity and food security) on the smallholder farms (Knudsen et al., Chapter 1). These are important reasons for some development agents to focus on improved agricultural methodologies that depend less on capital and external inputs such as NC-OA.

Empirical evidence shows that it has been possible to raise yields considerably under very different agro-ecological conditions and in different farming systems with relatively low cost techniques such as the 'Zaï' system where Sorghum is planted in pits supplied with manure and water, rather than broad sowing (cit. f. Pretty, 2002, p. 91). Non-chemical pest management has also been developed, such as the push and pull method to avoiding maize bugs and parasitic weeds in smallholder low input farms (Khan et al., 2000). But more research is needed to verify to what extent these case stories may be feasible solutions for larger regions, even after adaptation to agro-ecological differences.

The organic farming movement and proponents of LEISA share the same vision to 'maintain and encourage agricultural and natural biodiversity on the farm and surrounds' (IFOAM, 2002) as a method for stabilizing yields and promote natural regulation of pests. For the organic movement this is also part of the responsibility for recognizing the wider ecological impact of farming on the natural cycles and living systems, i.e. in compliance with principles of ecological justice. It is, thus, very important to state that our suggestion for NC-OA goes beyond the farm and includes local consumption of the products in order to ensure local development and food security. Maintaining biodiversity contributes to resilience in yields and can be considered an insurance against catastrophically poor yields due to climatic fluctuations or pest epidemics (Perrings, 2001). However, promoting biodiversity systematically and adopting soil improvement technologies and agroforestry methods is a challenge for poor smallholder farmers and uptake of these methods is often slow and small if they do not give benefits to the farm family in the short run (Pender and Mertz, Chapter 8). Many poor families cannot afford to take risks or to manage their land with a long-term fertility perspective if it does not also yield an immediate food output. Other constraints to adoption of LEISA and non-certified OA technologies are labour constraints, too low return to labour use or other costs and lack of seeds or seedlings of key: e.g. nitrogen fixing) species and a well functioning extension service.

Summing up, the concept of NC-OA is targeted towards developing local sustainable food systems that also protect the environment and avoid overexploitation of local resources. To make the concept viable there is a need for support to capacity building, improving local marketing and increased participatory research in order to develop locally adapted agro-ecological methods.

Improving soil fertility

In large areas of Africa south of the Sahara (SSA) soil fertility is being depleted due to declining use of fallow in combination with insufficient application of nutrients and organic matter among others (Pender and Mertz, Chapter 8). Organic farming can play a positive role in reversing negative soil fertility developments and thus solve productivity constraints, but only as part of the solution and care should be taken to assess the socio-economic conditions as described above. Pender and Mertz (Chapter 8) state 'in order to be environmentally sustainable [soil improvement and farming] technologies must not undermine future productivity by degrading the resource base or the supporting ecosystem'. In areas with low input agriculture NC-OA may improve soil fertility, increase and stabilize yields and improve poor farmers' asset building as discussed in Chapters 5, 8 and 11. Pender and Mertz (Chapter 8) suggest that non-certified OA can be considered as one form of LEISA, but with more specific requirements for using soil and water conservation and recycling to improve soil fertility. LEISA and OA methods according to Pender and Mertz (ibid.) offer solutions to the needs for intensification under difficult agro-ecological conditions in Africa, Asia and Latin America and is especially relevant in areas with low to medium yield potential and limited access to markets.

The development of OA may have significant positive impacts on e.g. soil fertility, but in the long run this may not be sufficient to sustain high yields on poor soils due to lack of nitrogen and possibly phosphorus. Specifically, large areas in sub-Saharan Africa are phosphorus (P) deficient from a plant nutrition point of view and OA limits the options for P inputs. As discussed by Pender and Mertz (Chapter 8), P fertilizer in the form of rock phosphate may be profitable to use in areas near sedimentary deposits but deposits of suitable quality are scarce in large regions of the developing world. Transporting rock phosphate with low concentrations of P over long distances on poor roads in e.g. Africa is neither economically realistic nor wise from a resource use point of view. It may be a solution to use agro-forestry principles (e.g. trees with deep roots that can recycle P from below the root dept of annual crops) in some areas. But this will probably not be a possibility in all areas with P-deficient soils and therefore it should be considered to use more concentrated (by chemical treatment) P fertilizer in situations where P is severely limiting yields – and given that this is economically feasible for farmers. Such a non-dogmatic approach to the adjustment of OA principles and rules is necessary and recommendable also in light of the history of OA as developed and formulated in the North under different agro-ecological conditions from the older and more weathered soils in many tropical countries. This should be less problematic in projects promoting non-certified OA but in projects with certified production for marketing under a label it would be necessary to grant exceptions and allow for (chemically) treated P fertilizer if this appears to be the only solution to sustained OA.

Organic farming as showcase for recycling nutrients from households?

While OA builds on ideas of improving soil fertility and recycling nutrients this has almost only been considered at the farm level or as collaboration between farms. Few, if any, OA projects have focused on recycling of nutrients leaving the farms with products, i.e. to get the nutrients in household waste back to the soils. There is an increasing focus on the development of techniques for the treatment of human waste (so-called black water) and household waste (grey water) with the aims of recycling nutrients and scarce water resources to crop growth and reducing costs in waste water treatment. Examples of this given by Refsgaard *et al.* (Chapter 7) demonstrate the feasibility of such an approach in a wide diversity of countries based on a combination of traditional waste collection systems and new technologies. The larger part of N and P excreted by humans is found in urine, while faeces together with the grey water could add more organic material to soils in order to enhance long-term fertility. This is interesting in developing countries because of the urgent needs for low-cost solutions to handle black and grey water in the rapidly increasing urban settlements. Hygienic considerations should be taken seriously and methods for this exist. Managed correctly, the development of such technologies could be a win-win situation for urban settlements and for farming systems and the OA movement should consider to become part of this development. Likewise, as several examples show, there is a potential for saving large costs to establish traditional household waste water facilities in developing countries if alternative handling systems were to be used more widely in new settlements and when old structures needs total restoration.

The topic of recycling nutrients between cities and the soils could be an interesting area for OA to demonstrate alternative solutions. But in areas with intensive livestock production and surplus of N and P in manure the incentive for recycling human waste will be low. Moreover, there are health aspects to consider seriously together with risks of odour when treating and spreading the waste products. But the most important barrier for the use of recycled human waste in OA is probably the risk for contamination of soils with other, toxic substances. Presently, rules for OA in e.g. Denmark prohibit the reuse of sewage sludge from urban settlements even though the national law favours the spreading of sludge on farmed land within a maximum limit of P-supply and the National Environmental Agency declares that risks of e.g. heavy metal contamination are very small.

Moreover, the question of whether to consider nutrients in human waste an organic fertilizer or a non-organic import (parallel to imported pig slurry from conventional farms) has to be solved before such an idea could be implemented.

Reducing medicine use and residuals in livestock products through organic practices

The widespread use of antibiotics as growth promoters and preventive medicine in intensive livestock production has been questioned because of the negative side effects such as increased antibiotic resistance in pathogens. This topic is less debated in most developing countries even though high levels of preventive medicine use such as antibiotics and acaricides may create similar disadvantages for farmers and consumers in the long run. Examples of high levels of antibiotic residues in livestock products and of increased prevalence of resistant bacteria in developing countries are given by Vaarst *et al.* (Chapter 9). Organic livestock production can help reduce these problems, because the organic systems emphasize disease prevention and maintain explicit standards designed to reduce the use of medicine through breeding, feeding, housing, appropriate flock and herd sizes and active health care (chapter 9). Vaarst *et al.* state:

> We consider it a very important strength and potential of organic animal farming that there is an explicit and strong focus on the health of the whole animal production system, the animals and their interactions with humans, other animals and the wider farm system. This is viewed as the primary way of reducing the risk of disease outbreaks and medicine use.

There is an important difference between the health risks to livestock in Europe and North America and in smallholder farms in tropical countries. Many diseases such as mastitis in intensive systems are related to the management and environment of livestock and the organic approach has been developed in response to this. Thus, it has been demonstrated by several studies that comparatively low levels of medicine use is in fact possible in livestock production without drastic reductions in productivity and without negative effects on the animals health or welfare as discussed in Chapter 9. However, livestock in tropical smallholder systems face severe health threats from vector borne and epidemic diseases, which are harder to control by individual management efforts except if livestock is kept in zero-grazing systems. But free range and scavenging livestock keeping is widespread in tropical countries together with pastoral and transhumance systems. In both cases medicine use is limited due to limited accessibility and financial constraints and much of the efforts to control these diseases already build partly on ecological principles. Community approaches are an important part of these efforts in disease prevention, due to e.g. communal grazing systems. Choice of appropriate, often traditional, animal species and breeds can reduce problems of disease, medicine use and, thereby, antibiotic resistance.

On this basis one major challenge for the development of organic livestock production systems is to think the organic principles into a wide range of diverse livestock systems with very different conditions. Vaarst *et al.* (Chapter 9) propose that this should build on criteria such as the risks for different types of diseases in different environments and the potential for reducing these threats using organic methods while still maintaining the animals' opportunity to fulfil their natural behaviour.

Conclusions

In order to illuminate the challenges and prospects of organic agriculture in a global context, we need to consider different perspectives with different views of the role of globalization, growth, trade and sustainability.

The challenges for organic agriculture

- Certified OA faces a pressure from the globalization of food systems, which threatens to dilute the specific characteristics of organic food by increased specialization and reduction in diversity, standardization, long distance trade and lack of transparency.
- For certified OA to represent a sustainable alternative to mainstream food production and food chains new principles based on both social and ecological justice should be adopted, which can guide the increasing global trade in organic food.
- Attempts should be made to distinguish certified OA to such a degree that specific trade regulations may be adopted under the WTO.
- The concept of ecological justice should be better implemented in the organic standards, incorporating for instance means to avoid the commodification and unjust appropriation of land and other local natural resources, externalities connected to distant trade by different ways of transport, and securing the functional integrity of exporting production systems.
- The realization of a fair organic trade should build on experiences from the fair trade movement and include considerations regarding the livelihood of smallholder farmers and workers and the involvement of local stakeholders in the certification process. However, a simple combination of organic and fair trade standards will not be adequate to meet the aims of ecological justice.

- More analyses of the potential effects of organic farming for environmental and socio-economical sustainability should be carried out using methods from ecological economics and based on analyses of whole food chains.
- Within the overall principles of OA the specific regulation and certification should be further developed and adapted to local agro-ecological and socio-economical conditions in order to secure and promote the local embedment of the organic ideas and certification processes.
- The major challenge in organic livestock production systems is to think the organic principles into a wide range of diverse systems with very different conditions. There is a need for local adaptation of principles and rules and for integrating community-level approaches to organic livestock production and disease prevention into certification schemes or non-certified organic projects.

The prospects of organic agriculture

- The two strands of OA, certified and non-certified, offer different opportunities and prospects, which should be dealt with consciously by the organic farming movement.
- Given that the challenges are met by implementing the necessary new principles, then certified organic agriculture constitutes a way to promote ecological justice through the global market by providing alternative products to consumers.
- Certified OA may provide alternative products in the market and increase consumer choices thus demonstrating in practice alternative development ways for agriculture.
- There is a potential for reducing the use of veterinary medicine and preventive use of antibiotics by promoting organic animal husbandry, especially when this is integrated with land use and food production.
- Organic agriculture has the potential to integrate the nutrients from household wastes in the nutrient cycling as a long-term goal.
- Non-certified OA is a potential development tool in areas with low input traditional farming and food insecurity and may improve the resilience and yield stability of smallholder farms as well as local communities.
- When capacity building of farmers is an integrated part of such programmes, it may improve the long-term asset building on smallholder farms and communities.
- Non-certified OA methods should be developed to accommodate the specific agro-ecological conditions including soil types. In some cases, especially with problematic soil types, LEISA types of smallholder farming will be more suitable than organic farming according to current standards.

- Large-scale conversion to OA in regions with low input agriculture has the potential to improve the food security among resource-poor people.
- Sufficient food production from OA to improve urban populations' food needs in 2020 is possible given that relatively high yield growth rates in organic crops compared with conventional can be achieved.
- Large-scale conversion to OA in high input regions will increase world food prices only slightly if organic yield per hectare improves faster than conventional yields are expected to grow. Large-scale conversion to OA in Europe/North America will have very little effect on food availability and food security among resource poor in SSA.

References

Barrett, C.B., Place, F., Aboud, A. and Brown, D.R. (2002). The challenge of stimulating adoption of improved natural resource management practices in Africa agriculture. In: Barrett, C.B., Place, F. and Aboud, A.A. (eds) Natural Resources Management in African Agriculture. CAB International, pp. 1-21.

Cierpka, T. (2000). Organic Agriculture and Fair Trade two concepts based on the same holistic principal. Online at http://www.ifoam.or, see 'Social Justice' and next 'Organic Agriculture and Fair Trade'.

Conway, G. (2001). The Doubly Green Revolution: Balancing Food, Poverty and Environmental Needs in the 21st Century. In: Lee, D.R. and Barrett, C.B. (eds) Tradesoffs or Synergies? Agricultural internsification, Economic development and the Environment. CAB International.

Dabbert, S., Häring, A.M. and Zanoli, R. (2004). Organic Farming: Policies and Prospects. Zed Books, London and New York.

de Boer, I.J.M. (2003). Environmental impact assessment of conventional and organic milk production. Livestock Production Science 80: 69-77.

EEA (2005). Agriculture and environment in EU-15 – the IRENA indicator report. European Environmental Agency, Copenhagen, 2005, Forthcoming.

Giovannucci, D. (2005). Organic Agriculture for Poverty Reduction in Asia: China and India focus. Thematic Evaluation. Document of the International Fund for Agricultural Development. Final Draft, March 2005. 80 pp. Available online (August 2005) at: http://64.233.183.104/custom?q=cache:4YYRgNz0xKkJ:www.ifad.org/evaluation/public_html/eksyst/doc/thematic/organic/organic.pdf

Halberg, N., Kristensen, E.S. and Kristensen, I.S. (1995). Nitrogen turnover on organic and conventional mixed farms. Journal of Agricultural and Environmental Ethics 8: 30-51.

Halberg, N., van der Werf, H.M.G., Basset-Mens, C., Dalgaard, R. and de Boer, I.J.M. (2005). Environmental assessment tools for the evaluation and improvement of European livestock production systems. Livestock Production Science (in press).

Hall, A. and Mogyorody, V. (2001). Organic farmers in Ontario: an examination of the conventionalization argument. Sociologia Ruralis 41(4): 399-422.

Hansen, B., Kristensen, E.S., Grant, R., Høgh-Jensen, H., Simmelsgaard, S.E. and Olesen, J.E. (2000). Nitrogen leaching from conventional versus organic farming systems – a systems modelling approach. European Journal of Agronomy 13: 65-82.

IFAD (2002). Thematic Evaluation of Organic Agriculture in Latin America and the Caribbean. Evaluation Committee Document. Rome. Document 290404.

IFOAM (2002). IFOAM – Norms for Organic Production and Processing. International Federation of Organic Agriculture Movements, Bonn (www.ifoam.org).

Khan, Z.R., Pickett, J.A., van den Berg, J., Wadhams, L.J. and Woodcock, C.M. (2000). Exploiting chemical ecology and species diversity: stem borer and striga control for maize and sorghum in Africa. Pest Management Science. 56(11): 957-962.

Milestad, R. and Darnhofer, I. (2003). Building farm resilience: The prospects and challenges of organic farming. Journal of Sustainable Agriculture 22(3): 81-97.

Milestad, R. and Hadatsch, S. (2003). Growing out of the niche – can organic agriculture keep its promises? A study of two Austrian cases. American Journal of Alternative Agriculture 18 (3): 155-163.

Noe, E., Halberg, N. and Reddersen, J. (2005). Indicators of biodiversity and conservational wildlife quality on Danish organic farms for use in farm management. A multidisciplinary approach to indicator development and testing. Journal of Agricultural and Environmental Ethics 18: 383-414.

Onduru, D.D., Diop, J.M., van der Werf, E. and de Jager, A. (2002). Participatory on-farm comparative assessment of organic and conventional farmers' practices in Kenya. Biological Agriculture and Horticulture 19: 295-314.

Perrings, C. (2001). The Economics of Biodiversity Loss and Agricultural Development in Low-income Countries. In: Lee, D.R. and Barrett, C.B. (eds) Tradesoffs or Synergies? Agricultural internsification, Ecnomic development and the Environment. CAB International.

Pretty, J. (2002). Agri-Culture: Reconnecting People, Land and Nature. Earthscan, London, 261 pp.

Rasul, G. and Thapa, G.B. (2004). Sustainability of ecological and conventional agricultural systems in Bangladesh: an assessment based on environmental, economic and social perspectives. Agricultural Systems 79: 327-351.

Raynolds, L.T. (2004). The Globalization of Organic Agro-Food Networks. World Development 32(5): 725-743.

Refsgaard, K., Halberg, N. and Kristensen, E.S. (1998). Energy utilization in crop and livestock production in organic and conventional livestock production systems. Agricultural Systems 57: 599-630.

Rigby, D. and Brown, S. (2003). Organic food and global trade: Is the market delivering agricultural sustainability? Discussion Paper, ESEE Frontiers II Conference. 21 pp.

Rundgren, G. (2003). Trade and Marketing, No simple solution. Ecology and Farming 34: 6-8.

Schjønning, P., Elmholt, S. and Christensen, B.T. (2004). Soil Quality Management - Synthesis. In: Schjønning, P., Elmholt, S. and Christensen, B.T. (eds) Managing Soil Quality. Challenges in Modern Agriculture. CAB International, Wallingford, UK.

Schwartz, A. (2002). Redesigning our Food System: Certified Organic Isn't Enough. In: Thompson, R. (ed.) Cultivating Communities. Proceedings of the 14th IFOAM Organic World Congress 21–24 August 2002, Canada, pp. 162-163.

Scialabba, N.E. (2000). FAO perspectives on future challenges for the organic agriculture movement. Proceedings from the 13th IFOAM Scientific Conference, Basel, Switzerland, 28–31 August 2000.

Scialabba, N.E. and Hattam, C. (eds) (2002). Organic agriculture, environment and food security. Environment and Natural Resources Service Sustainable Development Department, FAO, Rome.

Stolze, M., Piorr, A., Häring, A. and Dabbert, S. (2000). The environmental impact of organic farming in Europe. Organic Farming in Europe: Economics and Policy, vol. 6. University of Hohenheim, Germany.

Sutherland, A.J., Irungu, J.W., Kang'ara, J., Muthamina, J. and Ouma, J. (1999). Household food security in semi-arid Africa – the contribution of participatory adaptive research and development to rural livelihoods in Eastern Kenya. Food Policy 24: 363-390.

Uphoff, N. (2002). Agroecological Innovations – Increasing Food Production with Participatory Development. Earthscan, London, 306pp.

Willer, H. and Yussefi, M. (2004). The world of organic agriculture – statistics and emerging trends 2004. International Federation of Organic Agriculture Movements (IFOAM), Bonn, Germany. Online at http://www.soel.de/oekolandbau/weltweit_grafiken.html

Woodward, L. and Vogtman, H. (2004). IFOAM's organic principles. IFOAM. Ecology and Farming, May–August 2004, 24-26.

Woodward, L., Pearce, B.D., Hird, V. and Jones, A. (2002). Food and Fuel: The Environmental Impact of Increasing 'food Miles' of Organic Food. In: Thompson, R. (ed.) Cultivating Communities. Proceedings of the 14th IFOAM Organic World Congress 21-24 August 2002, Canada, p. 200.

Wright, S. (2004). Organic profile in British retailing. O and F Consulting. Sainsbury and Planet Organic at Økologi-Kongres 2004. Odense Congress Center.

Index

Acaricides 246, 255-8, 264
Africa *(see also agricultural productivity, disease, Ethiopia, food security, Kenya, Uganda)*
    livestock farming in 245-6, 249, 255-69
    organic farming in 155-7, 159-60, 165, 228-31, 246, 287-9
    soil fertility depletion (in) 215-40, 360
    sub-Saharan African challenge Programme 336
Agriculture (al) *(see also green revolution, high external input, industrialization, intensification, low external impact sustainable)*
    development 5
        expenditure on 338
        technological factors 3, 5, 64
    future scenarios 282-92, 302-23
    production 5, 216-7
    productivity *(see also yields)* 5-6, 121-6
        in sub-Saharan Africa 216-21
Agri-environmental schemes 350
Agrochemicals *(see also fertilizers, herbicides, pesticides)*
    levels of use of 5-6,
Agro-ecology 160-4, 287-9, 314-5, 359-60
Agro-ecological conditions
    significance for organic yields 302
Agroforestry 217, 224, 315, 359
Aid *(see donor agencies)*
Animal(s) *(see also livestock)*
    behaviour, natural 242-3
    multifunctionality of 243
    production (trends in) 8, 242
    welfare 86, 123-5, 189, 243, 270-2
        in organic farming 247-8, 317, 354-62
Antimicrobial drugs 4, 242-3, 248, 255
    problems with 189
    levels of use of 8, 252

    reductions (in use of) 243, 249, 264-7
        through organic approaches 251-5, 362-4
    residues 249
    resistance (to) 250-1
Argentina
    soya production 20-4, 354
    organic farming in 155-8, 160-1, 285
Asia
    organic farming in 155-7, 159-60

Beef
    global market in 24
    production in Brazil 24-6
    sustainable production of 26
Bhutan
    livestock production 245, 247, 266
Biodiversity
    (the) Convention on (Biological Diversity) 145-6
    conventional farming
        and the reduction of 4, 10-1, 15-6
    (and) organic farming 31-2, 169, 359
Biodynamic approaches 161
Biosafety 146
Biotechnology 6
(see also Genetically Modified Organisms)
Breeding
    livestock for disease resistance 243, 247, 264-6

CAP 291, 324, 327
Capital
    man-made, 118
    natural, 118
Care
    (of) livestock 272
    (the) principle of 78
Carbon sequestration
    (from) organic farming 31-2
Cartagena Protocol *(see biosafety)*
Cattle *(see livestock)*

Cash crops
    (and) food security 281
    (and) smallholder systems 18-9
Cereals
    demand for *(see also food security)* 9, 284, 292-3, 304
        from livestock systems 284, 288, 307-10
        production 6, 19, 125, 223, 285-6, 292
        under organic systems 291-2, 305-6
Certification *(see also labelling)* 33, 93-4 141-4, 158, 351-2
    (as) barrier to markets 33, 142
    costs 142-3, 229
    (and) fair trade 66, 352
    local ownership of 348, 350, 358, 364
    (and) social justice 33
CGIAR Institutes 162, 330
Codex Alimentarius 148, 348, 350
Common Agricultural Policy *(see CAP)*
Commons (the) 55-63, 90-1, 260
    characteristics of 56-9
    governance of 59-64, 261-4
    (as) social institutions 57-9
Composting 165, 225-6
Commodification 63-8, 90-2, 362
Commodity prices
    (in) Africa 222
Communal *(see grazing systems)*
Communication technologies
    impacts on trade and agriculture 3, 17
Community participation 244, 261-4, 312
Community Supported Agriculture (CSA) 39-41
Conventionalization of organic agriculture 77, 348
Cooperatives 36-7
Credit *(see also smallholder farmers- constraints)*
    influence on agricultural practices 12, 18, 33, 221-5, 233, 235, 298, 301
Crops *(see cash, rotations)*

DARCOF 331, 335, 339
Dairy production 36-9, 251-4, 290
DANIDA 331, 335
Deforestation 4, 11, 23, 24-6

Denmark
    100% organic scenario 290
    organic dairy sector 36-9, 252-4
Desertification 162
Dietary changes 9, 282-5
    (see also meat: growing demand for)
Development policy 323-5, 329-31
Disease(s)
    cross species transfer 250
    (in) livestock *(see also animal welfare, antimicrobial drugs, Foot and Mouth Disease, Rinderpest, veterinary practices)*
        management of 243, 255, 264-7
        organic approaches 248, 251-4, 264-76
        prevention of 243, 254-8
        resistance 259-60
        vector borne 246, 255-61
            control of 243, 255-8, 261-8
Doha Round *(see WTO)*
Donor agencies
    (and) organic farming 229, 246, 328-30
Drugs *(see antimicrobial)*

Ecolabelling *(see labelling)*
Ecological economics *(see also time, scale)* 55, 113-31, 326, 346, 352-5
    (and the) laws of thermodynamics 116-7
    (and) neo-classical theory 114, 128
    (and) sustainability 117
Ecological engineering 184, 191-209
Ecological injustice(s)
    examples of 53
Ecological justice *(see also political ecology)* 50-71, 76-108, 119
    (and) organic farming 69-71, 75, 97-109, 346-8, 352-4, 362
    (in) practice 63-70, 81, 351-4
    (vs.) sustainable development 52-5
    theoretical basis of 53-63, 84-90
Ecological modernization 52, 64, 79, 153
Ecology, principle of 78
Economic(s) *(see also ecological economics, environmental economics, market economics)* 184
    growth 54-5, 285, 345

Education (as factor in agricultural productivity) 265, 271, 298-300, 335-8, 339
Efficiency 351
  (vs.) equity 64
Environmental benefits (of organic farming) *(see also biodiversity, carbon sequestration, land reclamation)* 155, 169, 327-9, 354-5, 359
  assessments of 355
Environmental economics 54, 82, 127, 326, 351, 355
Environmental indicators 327
Environmental problems
  (connected with) agriculture 4, 10-16, 51
  (see also biodiversity– reduction of; desertification; deforestation; eutrophication; globalization – environmental impacts; global warming; resource use; soil degradation; water pollution)
Epidemiology 255-62
Equity
  (and) efficiency 64
Erosion (see also land and soils) 10-11
Ethics 85-7, 90
Ethiopia 224-6
Ethnoveterinary practices 268
European Fair Trade Association 66
Eutrophication 13, 15
Extension services 164, 265, 272, 279,
  gender bias 297
Externalities 63-4, 77, 90, 92-4, 118, 352, 354

Fairness, principle of 78
Fair Trade *(see also certification, European Fair Trade Association)* 2, 33, 65-8, 71-3, 100-1, 348, 353
  (and the) commons
  Labelling Organization (FTLO) 66
  (and) organic production 100, 108, 168, 353, 362
FAO 5-9, 218-9, 265,6, 281, 286 329, 344
Farmer Field Schools 252, 272
Fertilizers *(see also Nitrogen, Nutrients Phosphorus, soil fertility)*
  factors affecting use of 222-3, 297

  levels of use of 4, 14, 23, 164, 219-21, 293-4
  problems associated with 6, 10-11, 221-3
Fish farming 166, 198, 202
Food *(see also processing, retailing, storage and distribution, supply chains trade)*
  imports
    dependency on 4, 19, 356
  insecurity *(see food–security)*
  miles 139, 352
  safety 316, 354
  security *(see also population growth)* 4, 10, 16, 19-20, 278-322, 344
    agro-ecological influences 292-6
    entitlements 167, 281
    and exports 281, 300
    long term factors 285
    modelling 278, 282-4, 290-2, 301-43
      (and) organic systems 32, 34, 102, 153, 160, 165-7, 176, 278-317, 355-60, 364
    projections of *(see modelling)*
    regional aspects 286
    socio-economic influences *(see also food security–entitlements)* 278, 280-91, 296-301
    staple foods 282-6
    (in) urban areas 315, 365
Foot and Mouth Disease 242
Fossil fuels
  contribution to global warming 10
  levels of use in agriculture 10
Free Trade 56, 64-7, 346-9
Functional integrity 314, 349

GATT *(see Tariffs, WTO)*
Gender
  adoption organic farming 297
  labour and roles 297
Genetic diversity *(see also biodiversity)*
  reduction in 9-10
Genetic ecology 250
Genetically Modified Organisms (GMOs) 172
  soya in Argentina 20-4
  (and) organic farming 298

Globalization *(see also free trade, socio-economic impacts)*
  concentrating effects 4, 65, 344
  definition of
  environmental impacts 3, 10-17, 50, 59-63, 344
  implications for / effects on agriculture 3-25, 65, 135-8, 354
  (and) organic production 2-3, 26-49, 51, 64-7, 78-84, 344-50
  strategies of resistance (to) 63-71, 350-4
  (and) sustainable development 52, 79-84
  (and) trade 64-5, 95-7, 344
Global warming *(see also carbon, sequestration, fossil fuels, greenhouse gases)* 4
  agriculture's contribution to 10
  and food security 285-6
  and organic farming 31-2
Glyphosate 20-3, 354
GMOs *(see Genetically Modified Organisms)*
Governance
  (of the) commons 58-63
  of organic farming 324-9, 347-63
Grazing Systems *(see also Zero-grazing)*
  communal 242, 273, 362
Green Revolution 5, 18, 221-4, 279, 287-92
Greywater *(see waste)*

Health *(see also disease)*
  (and) organic farming 78, 169
  (and) sanitation 198-9, 201, 361
Herbicides *(see Glyphosate)*
High external input farming
  (in) Africa 218, 221-4
  conversion to organic agriculture
  food security effects 303
  relative yields 290
High yielding varieties 5-6
HIV/AIDS 171, 229, 296
Homeopathy 251, 254
Husbandry *(see livestock)*

IFOAM 51, 76, 78, 155, 159-60, 228, 268, 348, 350, 353

IFPRI 282-6
IMPACT model for food projections 282
Improved varieties *(see Green Revolution: high yielding varieties)*
Indigenous
  breeds 247
    (and) disease resistance 264-8, 362
  farming practices 217, 226
    (and) organic approaches 67, 162-4
  knowledge 255, 266-8
  medical practices 248, 266-8
Institution(s)/(al) *(see also, organic farming – institutions)*
  environments 325-6, 336-40
  in developing world 334-9
  in Europe 325-6, 356
  in USA 328
  role of in organic farming 323-4, 329-31, 333-5, 346
Integrated Soil Fertility Management 231-3
Intensification (of agriculture) 6, 20-6, 119-24, 344
Intercropping 297
Internationalization *(see globalization, and trade)*
International Trade Law 140
Industrialization (of agriculture) *(see intensification)*
Intellectual Property Rights *(see TRIPS)*
Interdisciplinarity 169, 336
  in organic research 176, 340
Irrigation *(see water)*
ISOFAR 330

Kenya
  livestock farming 260-1
  organic farming in 161-2, 334
Knowledge *(see extension, indigenous organic, research)*

Labelling *(see also certification)* 135, 141-4, 150, 329, 349-50
Labour *(see also organic-labour)* 220, 296, 359
Land *(see also soils)*
  degradation 10-11, 283
  loss (of agricultural land) 4, 10-11, 283

pressure on 4, 23-6, 260, 283
reclamation 162, 217
tenure 170, 227, 291
use
    (and) disease management 244, 255, 261-4
    future scenarios 291
Landscape(s) 327-8, 355
Large scale conversion 290
    (and) food security 301-12
Latin America *(see also Argentina)*
    organic farming in 155-7, 159-60, 165, 298, 329
LEISA *(see Low External Input Sustainable Agriculture)*
Liberalization *(see also globalization; Internationalization; trade)* 54-5, 63-5, 149
Life Cycle Assessment 349, 351
Limits to growth 55, 104-5
Livelihoods *(see also smallholders, socio-economic impacts)* 170-1, 297
Livestock 244-76, 32-3
    husbandry
        (in) Northern systems 251-5
        (in) Southern systems 251, 255-64
        intensification of 285
    impact on food security 286
    organic
        disease management 251-76, 362
        potential of 244-7
        transition to organic 271-2, 290-2
Local *(see also breeds; varieties)*
    knowledge *(see indigenous)*
    markets 38
    production and trade 82, 94-5
    (and) organic systems 69-70, 101-102, 158, 168, 350, 358
Low external input areas
    conversion to organic farming 305
    effect of organic agriculture 287
Low External Input Sustainable Agriculture 102, 224-6, 268, 299, 314-5, 365
    adoption problems 227
    experiences with 224-8
    (comparison with) organic farming 161, 172, 228-9, 231-3, 359-60
    principles of 224

yields 225

Marginalization
    (of) land *(see also land)* 4
    (of) smallholders *(see also smallholders)* 4
Manure(s)
    green 169-70, 226-7
    use of 5-6, 231
Market(s) *(see also globalization, liberalization, local; organic (farming); trade)*
    access (to) 137-8, 140, 142-3, 221, 229, 288
    distortions of 4, 138-9
    economics 90, 346
    international *(see also trade; international, globalization)* 98, 345-6
    local
        (for) organic produce 28, 32, 359
        (in)stability of 64-5, 98-9, 136
Material Flow Accounting 113, 128-9
Meat
    growing demand for 4, 9, 282, 285
Mechanization 5-6, 8
Mixed farming 245
Monocropping 8-9, 354
Multicriteria analysis 113, 129
Mulches 169-70
Multifunctionality 58
    (of) organic systems 168-70, 325-7
Multilateral Environmental Agreements (MEAs) 145-7, 350

Nature
    intrinsic value of 83-90
    services from 91, 94-7, 327
Nearness principle (the) 69, 108, 162, 348, 352
Neo-classical *(see market – economics)*
Neo-liberalism *(see liberalization, Free trade)*
Networks 171
NGOs
    role in organic farming 159, 171, 229, 246, 334-5, 344
Nitrogen
    biological fixation of 14, 169-70, 288
    efficiency 188-90

environmental impacts of 12-3
nitrates in groundwater 15
global nitrogen cycle 12-5
transformation of 12, 14
Non governmental organizations *(see NGOs)*
Nutrient(s) *(see also Nitrogen, Phosphorus)*
cycles, 31
global 12-5, 31, 354
urban–rural 32, 181-209, 361-4
deficits 216-40
depletion 23, 228
global transfers of 24, 354
(in) organic farming 31, 182-6, 230
recycling 184-209, 361-4
and livestock 243
surpluses 183-8, 354

OECD 65,327
Organic (farming/agriculture)
asset building 105
certified *(see also certification)* 26-9, 154-9, 337, 345-6, 352-8, 362-5
in sub-Saharan Africa 229
challenges to 51, 97-102, 136, 348, 363
'like products' 140-2, 349
consumption 26, 30, 32, 355
conventionalisation of 77, 348-53
development benefits of 326, 329, 336
differences between North and South 154-5, 164-5, 171, 287-92, 328-9, 333-5, 355-65
diversification (in) 299, 348
economic returns to 15, 164, 167-8, 229, 287-92, 294-301
environmental benefits of 31-2, 155, 169, 327-9, 354-5, 359
evolution of 28-9, 94, 155-6
export oriented 2, 26, 28-30, 32-8, 102, 106, 108, 145, 288, 299, 329-30, 344-6, 348-9, 362
(and) globalization 26-30, 51, 64-7,78-84, 325, 344-7, 353, 362
heterogeneity in 156-7, 280, 287, 324
institutional support for 28, 171, 174-6, 229, 329-31, 337, 346-7, 356-7
knowledge base 154, 163, 175, 298, 325
labour requirements 168, 171, 254, 296
legal issues 140-8
(and) Low External Input Sustainable Agriculture 228, 299, 325-40, 359-60
markets 27-9, 158-9, 168
access to 135
opportunities 27-9, 103-6, 138-9, 365
pressures 29, 105-6
non-certified 70, 97, 102-6, 159-64, 228, 254, 299, 316, 325, 345-6, 356-60, 364-5
perceptions of 153, 164
policy evaluations *(see also institutional environment, State–role of)* 325-7
premia 155, 159-60, 166, 169-70, 173-6, 178, 228-9, 325, 347, 351, 360
principles of *(see also organic: values of)* 29-31, 78-9, 101, 347-9, 351-3, 362-5
applied to livestock disease management 244, 362
productivity of 125, 165-7, 176, 287-96, 299-301, 303-13, 356, 359
profitability *(see organic- economic returns)*
(as) provider of public goods 301, 325
(and) smallholder livelihoods *(see also smallholders, socio-economic impacts)* 31, 103-106, 155, 158, 170-2, 228-31, 245-6, 287-90, 352, 356, 364
(and) social capacity 32, 69, 106, 296
(and) soil fertility *(see also soil fertility organic)* 31, 228-31, 292-6, 356, 359-62
spin-offs to conventional agriculture 107, 175
standards 76-9, 171, 228, 349-51
(and) sustainability 51-3, 79-84, 102-6, 155, 325, 336, 346, 355-66
concerns about 230, 280
transitions to
farm level 157, 164-7, 174, 245-7, 271-2, 278, 287, 293, 299, 328, 334-5
global level 302-17, 355-60, 365

374

trends in 26-30
values of *(see organic: principles of)*
veterinary approaches 247
yields *(see organic farming - productivity)*

Participatory approaches 130, 279, 298, 359
Pastoralism 245
Pathogens 198, 250
Permaculture 161
Pest control 292
   biological means 340, 359
Pesticides
   levels of use of 4, 6-7, 354
   problems associated with 4, 15-6, 279, 327
Phosphorus 226-8, 230, 360-1
   depletion of 23, 189-90, 354
Policy *(see institutional environment)*
Political ecology 49-74, 82, 346, 354
Population growth
   and agricultural production 5, 83
   and food security 219, 286
Precautionary principle 83-4, 146, 353
Preferential treatment *(in trade - see tariffs)*
Preventative treatment (of disease) 272
Processing 4, 34-5
Procurement (institutional) 41
Production cycle
   economic view of 121-23
Property rights 61
Public goods 325

Recycling *(see nutrients)*
Regional development 337-8
Research 102
   (and) fragmentation of knowledge 177, 336
   Integrated Agricultural Research for Development (IAR4D) 336
   in organic farming 161-4, 328, 330-3, 355-6
      an agenda for 173-6, 324-40
      documentation of 332-3
      prospects for 324, 337-9
      shortcomings of 333-5
   partnerships for 336

Resource consumption *(see also energy use)* 3, 60, 182
Retailing 348
Rinderpest 242, 261-3
Rice Intensification Systems 225, 297
Risk 88, 146, 220
   in organic farming 292, 296
   use of external inputs 294
Rotation (of crops) 225, 337
Rural development 337-8

Sanitation *(see ecological engineering, waste water)*
Sasakawa Global 2000 223-4
Scale *(see also intensification)* 119-26, 352
Seasonality and disease management 244
Set aside schemes 292
Smallholders
   (and) cash crops 17, 228-9, 298
      constraints on *(see also credit)* 5, 215, 220, 226-7, 287, 296-301, 359
   indebtedness 224, 279
   (and) organic approaches 155, 173
      potential of 20, 67, 103-6, 158, 170-2, 287-90, 306-13, 359
   resilience of 217, 287, 299, 364
   strategies of 163, 232, 293-4, 290-9
   use of inputs 6, 163-5, 222-4, 227, 261-4, 359
Socio-economic impacts
   (of) conventional agriculture in the North 4, 16-26
      (of) globalization 50-1
      (of) green revolution 17-8, 279
   (of) organic agriculture
      (in) the North 26, 303-8, 325
      (in) the South 67, 280, 287-9, 293-4, 308-13, 325, 359
   (of) waste treatment 186-8, 196-7, 200-8
Soil(s) *(see also Africa)*
   degradation of 10-12
   erosion of 10-12
   fertility 218-20
      in sub-Saharan Africa 215-40
      restoring 220-32
   (and) organic farming 31, 290-6,

    360-2
    (as) global commons 62-3
Soya production 20-4, 285, 354
Specialization
    agricultural 8-9, 347
Standards *(see certification, labelling)* 142-3
Staples *(see food security)*
State (the role of) *(see also institutional environment, subsidies)* 186, 219-21, 247
Stocking rates 262
Storage and distribution 19
Subsidies 64, 135-8, 328-30, 344
    domestic support 136-9
    distorting effects of 136-9, 148-9, 296
    environmental 138, 148-9, 328
    export 136-7, 313, 344
    (to) organic farming 328
Supply chains 4, 346-55, 362
    concentration of 17-8, 345
Surpluses (dumping of) 4, 137
Sustainable development 3, 50-5, 114-7
    definitions of 118-9
    (vs.) ecological justice 51-6,75
    and globalization 52-71, 79-84
    (and) organic farming 31-2, 51-3, 79-84, 102-6, 155, 325, 336, 346, 355-66
    targets 50, 70
    World Summit for 142

Tariffs *(see trade: barriers to)*
Technologies
    factors affecting uptake 165, 196-8, 220-1, 256
Thise Dairy Cooperative 36-9
Tick-borne *(see diseases)*
Time 120-1, 351
Trade *(see also fair trade, free trade, globalization, local, nearness principle, transport)*
    alternative networks *(see fair trade)*
    barriers to 64, 77
        non-tariff barriers 140-7, 350
        tariffs 137
            preferential treatment 144-5
    (and the) environment 22-4, 59, 139-149

globalization (of) 4, 19, 20-6, 28, 57, 64-7, 135-6, 344-7, 352
    impact on agriculture 3, 18
informal 263
internationalization *(see trade: globalization of)*
local *(see local: markets, production and trade)*
long distance 4, 9, 90, 94, 347, 362
negotiations 135-7, 362
(in) organics 2-3, 24, 28, 31, 34, 42, 54-5, 66, 72, 76-108
(and) spread of disease 263
(and) supply chains 4, 9, 33, 54-5, 77
(and) sustainability 2, 3, 59
Traditional: breeds / knowledge / varieties *(see indigenous)*
Transhumance 246
Transport *(see also food miles, fossil fuels, local production, globalization, trade - international)*
    (effects on) agriculture 6, 9, 224, 229, 232, 288, 362
    and certification 351-2
    costs of 103, 110, 224, 229, 232
    environmental effects 3, 17, 32-3, 60, 76, 354
    (and) globalization 3, 9, 19, 66-7, 70, 98-100, 103, 108
    international *(see long distance)*
    (and) land-use 60-1
    long-distance 3, 66, 70, 77, 354
    (and) markets 2, 17, 66, 96, 103, 108
    (and) natural resources 61, 101
    (and) organic food 32-3, 70, 77, 98-100, 102, 108, 354-5
TRIPS 145-6
Trypanosomiasis 246, 259-60
Tsetse 246, 259-60

Uganda
    livestock farming 245, 263
    organic production and processing 34-5, 103-6, 155-8, 226, 230, 292
Urban agriculture 209
Urban recycling systems 188-209
USA
    Local foodsheds (in Iowa) 39-41
    Organic livestock systems 251-4

Variety *(see also high-yielding, improved, traditional and indigenous varieties)*
   (of) crops
   (of) diet
Vaccination 268
Vector-borne diseases *(see diseases)*
Veterinary practices 243-76
   (and) antimicrobial drugs 242-54

Waste *(see also socio-economic impacts)*
   greywater 194-6, 200-9, 361
   perceptions of 184-6
   (constraints on) recycling 196-200, 205-9
   treatment 186-208
   urban 183, 190-1
Water
   harvesting 168, 170, 281, 300
   irrigation 168, 221, 224, 227, 296
   pollution 11,
   quality (of drinking) *(see also eutrophication)* 285
   resources 282-3
Women
   and food security 280, 285, 299
   and land rights 227
   in organic farming 33, 297
World Bank 330
WTO 65, 98, 135-141, 144-8
   and organic production 136-50, 346, 350, 362

Yields *(see also agricultural productivity, cereals, organic productivity)*
   changes over time 5-6, 283, 293
   comparison organic vs. conventional 291-3
   (and) food security 164-7, 283-96, 356-8
   regional differences 5-6, 215-7, 293-4

Zero grazing 263-72, 362

377